Completely Morgan

Four-wheelers 1936 to 1968

First published in 1994 by Veloce Publishing Plc., Godmanstone, Dorset DT2 7AE, England. Fax 0300 341065

ISBN 1 874105 33 2

Readers with ideas for automotive books, or books on other transport or related hobby subjects are invited to write to the editorial director of Veloce Publishing at the above address.

British Library Cataloguing in Publication Data -
A catalogue record for this book is available from the British Library.

Typesetting (Bookman, 9pt), design and page make-up all by Veloce on Apple Mac.

Printed and bound in England.

IMPORTANT NOTE. While the author has prepared the practical information in this book with diligence and care, it is possible that it contains errors or omissions. It is not possible to foresee every possible area of personal danger or possibility of mechanical damage and warn readers accordingly. The author, publisher and retailers therefore accept no responsibility for personal injury or mechanical damage which is in any way related to the use of this publication. If you use the practical information contained within this book, you accept full personal responsibility for the consequences.

Completely Morgan

Four-wheelers 1936 to 1968

Ken Hill

VELOCE PUBLISHING PLC
PUBLISHERS OF FINE AUTOMOTIVE BOOKS

Morgan

ACKNOWLEDGEMENTS

No book of this type can be written without the unstinted co-operation of many people.

Firstly, I must again record my sincere thanks to Peter Morgan, Charles Morgan and the staff of the Morgan Motor Company for their unfailing assistance with the provision of the works production records, photographs and a mass of material from their archives.

I am also indebted to Michael Ware and Linda Springate of the National Motor Museum for placing their expertise, library and records at my disposal and to Sarah Joselin LBIPP, LMPA for her expert photographic advice and assistance.

Thanks to Motor Racing Publications, Motorbooks International and Shire Publications Ltd who kindly allowed me to use previously published material.

Sincere thanks to The Morgan Plus 4 Club, The Morgan Sports Car Club and John H Sheally II for invaluable assistance and permission to use original copyright material. Also to T W Barney, Chief Technical Editor of Glass's Guide Service Ltd, for supplying and granting permission to reproduce the copyright Technical Data Sheets, and Lucas after-sales service for assistance in compiling the electrical data and wiring diagrams.

In addition, a very special thank you to Mike Duncan, Jerry Willburn and John Worrall for all their advice and assistance, and for sharing their technical expertise which helped so much in compiling the technical chapter.

Many of the photographs used in this book originated from members of the Saga Travel Club, who responded so wonderfully to my request for information and photographs. In all, over 320 people have contacted me and lent over 200 photographs from family albums. It is impossible to list them all individually, but I wish to place on record my grateful thanks to them all.

I have obtained information from notes left by the late Dick Pritchard. The material had been collected by him over forty or more years and the exact origin of much of it is unknown. I know that he assembled material from most motoring magazines, including *Light Car & Cycle Car*, *Motor Cycle*, *Motor Cycling*, *Autosport*, *Motor Sport*, *Motor* and *Autocar*, all of which he read regularly. In addition I have collected information over the past twenty years from magazines, Morgan clubs and Morgan owners from all over the world. In many cases it is now impossible to establish the exact source of certain material, photographs and information, and therefore I offer my heartfelt thanks to all such writers, magazines and photographers.

Finally, as usual, I'd like to thank my wife, Janet, who has once again helped me so much in the research and compilation of this book.

Ken Hill

Morgan

CONTENTS

INTRODUCTION

Over the past eighty years, hundreds of thousands - if not millions - of words have been written about the Morgan sports car and the Morgan Motor Company Limited, a few thousand of which, over the last thirteen years, were contributed by me.

Throughout this period I have continued with my research of Morgan history. This research, along with information gathered from Peter Morgan and my numerous Morgan friends worldwide, has created a necessity to add to, or correct, some previously-published material. The majority of my previous work remains accurate, though much of it is now out of print. Some parts of this book are therefore direct reprints of this previously-published information.

The chapters covering maintenance and practical tips have been greatly enlarged, although it is not intended they be regarded as workshop manual information but rather to act as a basic guide to practical Morgan ownership.

It is impossible to cover every practical subject and fact relating to Morgans; in writing this book I have endeavoured to incorporate as much information as possible within the limitations of size. Much of its content has never before been available to the Morgan owner.

To own a Morgan, no matter what age or model, is to become a member of the worldwide 'Morgan brotherhood'. This fraternity has no rules or regulations, no committees, no annual membership dues, no divisions based on the number of wheels or type of radiator, no geographical or political boundaries, and its members are united by a common love. In fact, the only qualification needed is the exposure to - and contracting of - the incurable disease 'Morganitis'.

If you have read this far, I am sorry to inform you that you have just shown the first symptoms. Welcome to the wonderful world of Morgans!

Ken Hill
Brockenhurst, England

Publisher's note on photo reproduction

As Ken Hill points out in his acknowledgements, many of the photographs used to illustrate this work have come from private photo albums. Of course, this means that the originals are a little fuzzy and of poor tonal balance; nevertheless, because of the unique and rare nature of this material, we have reproduced such pictures in this publication. We hope you will agree that the inclusion of these otherwise delightful photographs is worthwhile.

I

MULTUM IN PARVO

The history of a small company with a huge reputation

Henry Frederick Stanley Morgan, the man who was to guide the destiny of the Morgan Motor Company Limited for almost fifty years, was born at Morton Jeffries Court - the Manor House of the village of Morton Jeffries in Herefordshire, England - on August 11, 1881 to Florence and the Reverend Prebendary Henry George Morgan (known as George to his family), who had succeeded his father Henry George as Rector of Stoke Lacy in 1871. It was in this tranquil setting that Harry, or 'HFS', as he was to become known throughout the motoring world, spent his early childhood, in company with his younger sisters, Freida, Ethel and Dorothy.

HFS was educated at Stone House, Broadstairs, and then at Marlborough College. However, as with most public schools of this period, conditions were austere and the food frequently almost inedible. Slowly young Harry became undernourished and his health began to suffer, becoming so ill that his parents removed him from the college.

Amongst his many talents was a natural ability for art, which he had obviously inherited from his father, an accomplished amateur artist. Shortly after he was withdrawn from Marlborough, his parents took him to Italy. It is not certain if this was to help speed his recovery or an attempt to encourage his artistic talent. The first reason is the more likely because, in spite of his family's religious background, HFS felt no call to join the church, nor did his parents try to force him to do so: with what was, for the time, very forward thinking, his parents allowed him to choose freely between art and engineering. He chose the latter and continued his education at the Crystal Palace Engineering College in South London. Whilst there, he was allowed to design, develop and build all manner of things. One of the best of these was a bicycle which he designed and built himself, which proved to be almost invincible in the College races and established a new speed record.

HFS's father was devoted to his children and would often fly in the face of public opinion if he felt it right for them. At that time it was very much frowned upon for a clergyman's son to go into trade. He was in later years, after his son had established himself as a car manufacturer, to become a familiar sight in his top hat, watching, with ill-concealed pride, the performance of Morgan cars in competitive events. When his son left engineering college at the age of 18, he secured for him an apprenticeship to William Dean and subsequently C J Churchward, the chief engineers of the Great Western Railway at Swindon, where HFS worked in the drawing office for seven years.

By this time, Britain's railway companies had become a productive training ground for car people. At the same time as HFS was at Swindon, W O Bentley was serving his apprenticeship at the Doncaster works of the Great North Railway, the same company which, a few years earlier, had enabled Henry Royce to develop his engineering skills at its Peterborough works.

For most motorists, memories of their first attempt to drive a car tend to be of a pounding heart and feeling of thankfulness when it was all over and they, their car and everything around them were still intact. But for HFS it was to be rather different. His first drive in 1899 was in a 3 1/2hp Benz which he

Henry Frederick Stanley ('HFS') Morgan 1881-1959. Founder of the Morgan Motor Company Limited.

had hired from an unsuspecting Mr Marriot, the first motor trader in Hereford. All went well until he started to descend the 1-in-6 gradient of a hill between Bromyard and Hereford and the Benz took charge and ran away with him. He emerged intact, but the damage to the Benz was to cost his father £28.

However, Mr Marriott could not have been too put out by the incident as he continued to hire cars to HFS from time to time over the next two years until young Morgan had saved enough money to buy a motor vehicle of his own - an Eagle Tandem. The Eagle, a three-wheeler, was driven by a water-cooled 8hp de Dion engine. It had a motorcycle-style rear wheel, with a car-type front axle. A wickerwork passenger seat was sited between the front wheels.

This vehicle was to stimulate HFS's interest in the design and development of the three-wheeler. The Eagle was also to give him his first conviction for a motoring offence when he exceeded the 12mph speed limit. In 1902 the Eagle was replaced by a four-wheeler, a 7hp two-cylinder car called The Little Star, which was purchased with part of the £200 that his godfather had left him on becoming of age.

The dream becomes reality

At the end of his apprenticeship, HFS was offered permanent employment in the Swindon works but decided instead to leave the GWR and open a garage in the area of his home. After some thought and research, he decided on Malvern Link as the right location and opened the garage in partnership with his close friend Leslie Bacon. Initially, It was far from plain sailing. Motor cars were still regarded as noisy, smelly nuisances, and for a clergyman's son to be selling them was beyond belief. He was to receive many rude letters; HFS silenced the writers by displaying their correspondence in the garage window!

The business soon began to flourish and they were able to acquire agencies for Wolseley and Darracq cars. In addition, he started a bus service with a 15-seater 10hp Wolseley from Malvern Link to the Wells and from Malvern to Gloucester. This, however, was not too successful and eventually the bus was sold and they started a car-hire business, which did well.

Before long HFS was able to devote more time to his experiments with design. He started work on a new type of car with another very good friend, W Stephenson-Peach, the engineering master of Malvern College and a direct descendant of George Stephenson who, in 1814, designed the famous 'Rocket' locomotive. About this time HFS bought a 7hp twin-cylinder Peugeot engine with the intention of installing it in a motorcycle, but he was becoming increasingly attracted to the idea of three-wheeled transport. This led him to construct a machine which was neither a motorcycle nor a car, but something halfway between the two - a cyclecar.

There can be little doubt that with-

Fascinating, but a very simple design. A 1910 cartoon depicts the great interest shown in Morgan's single-cylinder prototype at the Olympia Motor Cycle Show.

out the facilities of Malvern College's engineering machinery, HFS would not have had time to build the experimental car and run his business. In all it took a little under a year for the car to be built, with completion early in 1909. The prototype was a three-wheeled vehicle of very simple design. The main chassis member was a single large diameter tube, which had the clutch mounted at the front end, the bevel-box at the rear and the propshaft running through the centre. Further tubing was used for bracing from the bevel box to the centre pins of the front wheels. These front wheels were mounted on a unique design of independent front suspension. Two other tubes, mounted parallel and running longitudinally, completed the chassis and also acted as exhaust pipes, with silencers mounted on the front of each. The drive to the rear wheel was by chain, via two speeds, high and low, which were engaged through a simple dog clutch arrangement, and a lever next to the bevel-box. The bevel-box also served as the seat. Braking was obtained by foot and/or hand pressure to levers which operated constricting bands mounted on each side of the rear wheel. Steering was by means of a tiller. The bodywork was minimal, with a bucket seat of sorts, a front-mounted fuel tank and a mudguard over the rear wheel only. Mounted at the front was the previously mentioned 7hp Peugeot engine.

The prototype's good power-to-weight ratio, combined with the rigid frame and ingenious independent front suspension, have remained part of the Morgan hallmark ever since. These features ensured a remarkable performance for such a small vehicle when it was run for the first time; its first outing on public roads was to Shelsley Walsh, not to compete in the hill-climb but simply to raise interest.

Although built primarily for his own use, HFS soon found he was being approached by others with requests for him to make similar cars: so great was this pressure that he began to think in terms of the possibility of manufacturing cars as a business. Yet again, Reverend George Morgan was completely supportive of his son's endeavour and provided the necessary capital to equip the garage with the few machine tools necessary for HFS to start production on a limited basis.

With the information gained from driving the prototype around Malvern, HFS was able to incorporate one or two improvements to these production cars, which included the fitting of tie bars to the crosshead at the front of the vehicle, and silencers modified to the 'coffee pot' design which was to become so well known.

Having patented his design (the

The twin-cylinder 1910 Olympia Show car had an 8hp engine and tiller steering; Morgan offered a steering wheel for an extra charge.

As a result of comments received at the Motor Cycle Show debut, Morgan constructed a two-seater prototype, considerable improvements to which were made before the 1911 show.

patent drawings were produced by a bright young man called John Black, later Sir John Black, who was to contribute to Morgan history again 28 years later in his role as head of the Standard Motor Company) and spurred on by his local success, HFS and his partner formed the Morgan Company and decided to exhibit at the First International Motor Cycle Show at Olympia, London in November 1910. They displayed two versions of the car - one fitted with a 4hp JAP 85.5mm x 85mm single- cylinder engine and the other with a twin-cylinder 8hp version of the same engine and a bare chassis. Both cars were single-seaters and were fitted with tiller steering. Although the cars and their unique design attracted a lot of attention and good press coverage, the number of orders received - 30 in all - was disappointing, and Leslie Bacon decided to withdraw from the partnership. However, he remained friends with HFS for the whole of his life and Peter Morgan remembers going to stay with him on holiday.

This could well have been the end of the Morgan company if help had not come from an entirely unexpected quarter. The managing director of Harrods, the famous London store, was one Richard Burbridge, who was far-seeing enough to recognize that, in spite of its primitive design, the Morgan Runabout had great potential. In return for being granted the first dealership, the Knightsbridge store injected much-needed capital into the business and guaranteed orders. Encouraged by these developments HFS decided to carry on. It became obvious that he must do two things if he was to succeed. First and foremost, he must prove his product in competition (a principle still practised by the family today) and, secondly, he had to build a two-seater version of his design. This he did in ample time for the Motor Cycle Show at Olympia in November 1911. There, on Stand 127, he exhibited:

Two-seater, fitted with hood and screen, price £96.00
8hp two-seater, price 85 guineas
8hp Torpedo single-seater, price 85 guineas
8hp single-seater, tiller steering, price 85 guineas
8hp two-seater chassis, price 80 guineas

The steering on all but the one mentioned was by wheel. The combination of early competition successes and the two-seater version of the car ensured an amazing number of orders were taken at the show. A caption associated with a contemporary drawing of the two-seater read, *'The Morgan Magnet attracts all young couples by its double seating accommodation'*. After the show, *Motor Cycling* commented *'The interest displayed in the Morgan Runabouts was phenomenal. As all our readers know, these machines have competed with unqualified success in all the big trials of the last 18 months, gaining a reputation of which the makers may justly be proud'.*

So great was the demand that it far outstripped the production capacity of his little works and HFS was obliged to seek a tie-up with a major manufacturer. Fortuitously, as it turned out, every company he approached turned him down, and so, with the help of deposits accepted from potential buyers, he had to buy more machine tools and make further extensions to the additional premises which his father had financed the previous year in order that the initial production of Morgan Runabouts could begin.

Behind every great man ...

1912 was to be a momentous year for HFS with three major events taking place. This was the year in which the Morgan Motor Company Limited was formed and it made a small but significant profit - £1314. It was also the year when, in June, he married Ruth Day, the daughter of the Reverend Archibald Day, vicar of St Mattias, Malvern Link. He had met Ruth at a dance at the Grand Hotel in Malvern, which later became the Malvern Girls' College. It has often been said that behind every great or successful man, there is a strong woman. By all accounts Ruth was certainly this. Besides being very attractive, she was also capable, giving practical advice and encouragement whenever necessary.

Ruth Morgan was HFS's passenger in trials. Here, at Brooklands during the 1921 Auto Cycle Union Six Day Trial, she stands behind her husband. On the left is EB Ware, chief experimental engineer of JAP Engines.

In the early years, Ruth accompanied HFS on many trials and competitions, acting as a very good 'bouncer' on the steep hills. Somehow she still managed to fit into this busy schedule time to have five children; four girls, Sylvia, Stella, Brenda and Barbara (known as Bobbie), and one son, Peter. In the case of Peter it would be true to say that he was born to competition; press reports of the period tell of Ruth acting as trials passenger to HFS less than three months before his birth. As far as is known, she never drove in competitions herself but did own a variety of different cars as her own personal transport, including a Fiat, Alvis 12/50, Alvis Speed 20 and a Hillman Minx.

Ruth made a great impact on the motor sporting scene and, following her death in 1956, a magazine editor wrote: *'With sorrow I heard of the death recently of Mrs HFS Morgan. In the early days of the Morgan three-wheeler, HFS, piloting one of his own cars, was seldom absent from any big trial, and his wife was always at his side. In a personal letter I received from Mr Morgan he said, "You will remember how keen she was and how much she helped me in trials work".*

'Only those who took part in them can remember the severe conditions under which those trials were often run, and to sit quietly and stoically in an open three-wheeler, come rain or snow, beside a skilful driver was often in the nature of a major ordeal. Mrs Morgan took it all in her stride, and always there was a welcoming smile and a helping hand for those with whom she came in contact. HFS achieved many successes; the glory was always with his passenger to whom he owed so much.'

Keeping the pot boiling

The chairman of the Morgan Motor Company Limited was Preb H George Morgan. He was to become a familiar figure at Brooklands, in his long coat and top hat, and many other events where his son was competing or attempting to break records. Completely dedicated to the company (though he had made a small personal fortune by investing in gold shares on the stock market), Morgan Senior was not unknown to write, under a pseudonym, controversial letters to the motoring press, to stimulate public interest in the marque. Should this bring little or no response, he would answer himself, in a manner best likely to stimulate debate, using another pseudonym. What is not generally known is that he also wrote several articles, in particular for *Light Car* or *Light Car & Cyclecar* magazines. With an excellent head for figures, he kept the company's books for several years and still found time to indulge his hobby as an artist. (An excellent self-portrait hangs in the entrance hall of the Morgan factory). George Morgan died on November 10, 1936, aged 84. Many of the motoring magazines printed tributes to him, with the *Light Car & Cyclecar* magazine heading its obituary, *'A Pioneer Passes Away'*.

On the formation of the company HFS had become managing director, but on the death of his father he succeeded him as chairman. George Goodall, who had become works manager in January 1927, was joined by his son Jim in 1936, although it was not until 1938 that George was appointed managing director. At about this time HFS established a new home at Bray in Berkshire, and handed over the day-to-day running of the factory to George Goodall, although he still visited the factory once a week. George Goodall did an excellent job and the move made for a far more relaxed life for HFS.

George Goodall, works manager of the Morgan Motor Company, was known as 'Uncle George'. Here is is having fun on a Brough Superior 11-50 at Brooklands in 1935.

Morgan chassis frames and the people who made them, at the Worcester Road works, 1912.

From little acorns ...
Continued outstanding successes in all forms of competition by HFS and an increasing band of private owners ensured the expansion of the company. At this time motor manufacturers had to make 95% of all the parts for their product as there were very few outside manufacturers of components. Slowly, the signs of increased productivity began to show at the Malvern Link works. The machine shops were extended, albeit only a little, and various storage and painting sheds were erected behind the main works. These were not large affairs at all. For example, the painting was carried out by hand using brushes, a practice which continued up to the outbreak of war in 1939, and did not require a large factory area. For several years after the war, painting was by sprayed coach paint for speed but the rubbing down and build-up of paint was still labour intensive

Profits increased. The published figures were £4797 in 1913 and £10,450 in 1914. The latter figure was somewhat reduced from what it might have been, due to the outbreak of war.

When the First World War began, sales had never been higher, with almost 1000 cars being built in 1913. These included export sales to France, Russia, India and North and South America. As with all engineering works, a major part of the company's production had to be made over to the production of shells and other munitions for the war effort. To avoid any possibility of profiteering from the government contracts, a fixed profit allowance - generous by any standards - was legislated. When one considers that in addition to this the government also supplied, or helped substantially with the cost of, new machinery, there can be little wonder the company continued to make increased profits throughout the war years.

Throughout the war very limited car production was maintained and the repair shop remained open. Experiments and improvements continued which included the production of an experimental four-wheeled model which, for various reasons, was discontinued. As the war drew to an end, HFS realized that the sooner he could get his factory back into full production after peace was declared, the greater number of cars he could sell in the postwar rush. So successful was he with this that the factory was back into full production long before any of his major

Pickersleigh Road in the early 1920s - the erecting and finishing shop.

Coach painters take a lunch break - Pickersleigh Road, circa 1922.

competitors.

Besides careful planning and foresight, the other factors which helped with this success was, without doubt, the simplicity of the cars' design and construction and the fact that the vehicle was not dependent on any single engine manufacturer (a status which the factory still strives to maintain today).

All vee-twin engines have a basic banjo-shaped crankcase. It is therefore only necessary to fabricate a simple baseplate, drilled out to match both sets of mounting holes, engine and chassis front tubes to give the car its power source. Although engines were not plentiful at this time, HFS was able to buy enough to keep production at maximum rate and eventually to build up sufficient stocks to sustain production when the depression arrived and engines became difficult to obtain.

The immediate postwar boom in demand for mechanized transport was caused by the war itself. Before 1914, most people had never driven in or on a motor vehicle of any type; the war was to change all that. Ordinary men and women were trained to drive all manner of vehicles, and when they returned to civilian life they wanted to be able to continue driving. Secondhand motorcycles and cyclecars were almost unobtainable and, as with all cases where demand far outstrips supply, prices shot up. In the three years immediately after the war the cost of a new Morgan was greater pro-rata than at any other time in its history. At the declaration of war, the highest priced car was £98. At the 1920 Motor Cycle Show the cheapest was £206, and at the 1932 Show the lowest price was back down to the prewar figure of £98.

Immediately after the war, production became severely restricted by the small size of the factory in Worcester Road. In April 1919 HFS was walking through the turning shop and, whilst passing a capstan lathe, slipped on a small patch of grease. Instinctively, he threw out his right hand to save himself and caught it in the machinery. As a result of this accident the first two fingers of his right hand had to be amputated. A new site in Pickersleigh Road was purchased, which HFS opened on October 16, 1919, three weeks before Peter Morgan was born on November 3, 1919.

The new premises were used for the finishing work of painting and trimming. The production of the chassis with all its associated machinery continued at the Worcester Road factory. The part-completed cars were then loaded on a Crossley lorry and transported to Pickersleigh Road. This was not an efficient or convenient arrangement, and by 1922 a major building programme was underway. Early in 1923 the whole factory was relocated to the new site, with the exception of the machine shop which was not moved until 1929. To mark the occasion, HFS held a party for over 500 people in one of the new buildings. In his speech he praised his workforce for their loyalty and hard work in spite of all the disruptions and a call for increased productivity.

Just how prosperous the company and HFS had become is reflected in the fact that purchasing the new site and building the factory was achieved without loans. In 1925 his family moved to a much larger house in Malvern Link named Fern Lodge, which is now Seaford Court School. Also in 1921 he purchased a Rolls-Royce 40/50 to replace his Prince Henry Vauxhall. It is not certain whether this Rolls-Royce was bodied at the Morgan works. However, in 1932 HFS arrived at the Motor Cycle Show at Olympia in a Rolls-Royce Phantom II, which was fitted with an all-weather body built in the Morgan factory. A report in *Light Car & Cyclecar* magazine went as follows: *'I happened to call there whilst it was in the course of construction, and very odd it looked in a shop filled with three-wheeler bodies. Mr Morgan designed the body himself, and had no difficulties in getting it accepted for guarantee purposes by the hyper-critical Rolls-Royce engineers'.*

After the Morgan success in the French Cyclecar Grand Prix in July 1913 demand for the car increased enormously, particularly in France. It was reported that the Paris agent sold 150 examples in 1914 and placed an order for 500 in 1915. Postwar, the demand for small, reasonably-priced cars remained high, both here and in France. In 1914 HFS granted a licence to Roger Darmont (later Darmont et Badelogue) to build Morgan cars in Paris. The success of the Darmont Morgan was assured when Paul Houel scored a memorable victory in the Cir-

Pickersleigh Road paint shop, 1929/30. The mottled dashboard of the family model in the foreground was introduced in 1929. On the left, halfway down the workshop, is a delivery van.

cuit de L'Eure on August 16, 1919.

The Depression and mass-production

The postwar boom could not last forever and by 1923 it was over. It was followed by the General Strike and Depression of the mid-to-late 1920s and early 1930s, which were to have a devastating effect on the manufacturers of light cars and cyclecars, both three and four-wheelers, in Britain and on the Continent. The less well-off were the main purchasers of this type of vehicle, and it was these people who were most affected by the new economic climate. Moreover, mass-production was on the increase, with both Austin and Morris joining Ford in being able to produce cars on a continuous line system. The prices of all light cars and cyclecars fell steadily throughout the 1920s as their manufacturers attempted to compete with mass-production. Many went bankrupt. Morgan survived because of numerous factors: the loyalty of the workers, who did not strike; stocks of engines of different makes, which avoided supply problems when strikes affected their production; continued competition and record-breaking successes and, above all, the shrewd business acumen of HFS. Unlike so many of his competitors, he slowly reduced his production and lowered his prices. At the same time he realized that he must also make his product even more attractive to the now limited market. By careful planning and correct timing he was able to introduce several improvements through these lean years, matching them against those introduced by the mass-production companies. Most of the improvements, such as front-wheel brakes, electric lighting, and self-starters, were at first offered as optional extras and later adopted as standard fittings.

Litigation

In November 1921, Charles J Fox, a piano manufacturer, purchased what was left of the British Anzani Engine Company. S F Edge of the AC company had bought a large shareholding in the company and then after a few months withdrew AC's order of Anzani engines and changed the AC's power unit to a

Old-time timepiece: the punch clock that was used for years to 'clock in' employees at the Morgan factory.

Morgan-bodied Rolls-Royce; this 1935 Phantom II was the second of the marque that HFS owned. He designed its body, which was built at the Morgan works. Note the Morgan stork mascot, replacing the Spirit of Ecstasy, also used on his earlier Rolls-Royce 40/50.

Cubitt engine. In this way he effectively killed off the Anzani competition but ensured that AC and Cubitt (he was a major shareholder in both) made a good profit. Fox named his new company the British Vulpine Engine Company Limited and continued to produce exactly the same range of engines as before, but in much reduced numbers.

In June 1926 the Vulpine company went into liquidation. It was claimed that this was due to the fact that Lea Francis had cancelled its orders for the side-valve engine. This was true but Lea Francis was taking so few engines per month that the stoppage would have had little effect on the profits of the company. The real reason was the result of a lawsuit that Fox had taken out against the Morgan Motor Company. What had happened was that because of market uncertainty HFS had refused to place orders a year in advance for any of the engines used in his cars. He ordered his engines 100 at a time from JAP, Blackburne and Anzani. In this way he was also able to play off one company against the other and obtain the best possible terms. Vulpine, in order to remain competitive, had to reduce its prices. One of the ways in which this was achieved was by reducing the quality of the steel used in the manufacture of the exhaust valve. This resulted in a large number of broken exhaust valves on Morgans during the guarantee period. Vulpine only replaced the valves. However, Morgan customers demanded and got the replaced valves fitted free, in accordance with the Morgan Motor Company's guarantee which, of course, meant that HFS was losing money every time. When he contacted Vulpine about this, Fox refused even to discuss the matter, saying *'HFS wanted a cheap engine and he got it.'* So HFS cancelled his contract with Vulpine and Fox took Morgan to court for breach of contract. He lost the

Morgan Three-wheelers replaced motor cycle and sidecar combinations for police patrols. Here, two 1933 Family models are delivered to the Newport Monmouthshire Police by Alex Thom, the Newport motor agent.

case and also had costs awarded against him. This litigation ruined the Vulpine company and Fox had no option but to go into liquidation. Once again HFS's shrewdness as a businessman had saved the company from financial disaster.

Demise of the three-wheeled cyclecar

In 1931 Morris introduced the first *'£100 motorcar'*, closely followed by the two other mass-producers, although it was rumoured that the companies sold below cost to stay in contention. I recently asked Peter Morgan if this had ever happened in the Morgan Motor Company. He replied that it certainly never has whilst he has been in charge and, as far as he knew, had never happened in his father's time, although profit margins had at times been reduced to the minimum to enable the company to stay competitive.

Amongst the other ploys used by the company to maintain its position in the market was the clever use of advertising. The Morgan Motor Company produced a new advertisement on average every two weeks which listed the major competition or record-breaking achievements of the marque for the previous fortnight. Many other companies employed the same basic idea, but their cars were not achieving the same success as the Morgan.

HFS also realized he must continue to exploit that part of the market which had served him so well in the past. He knew there was a huge potential market for his cars with the family man; his advertisement for the family model invariably included the words *'Speed, comfort, economy and reliability combined'*, with the latest competition results. As mass-production cars began to take more of the market, the emphasis of the company's advertising shifted to *'tax, economy, comfort, cost'* and comparisons with motorcycles and cars.

The extent of the postwar boom, the subsequent depression and the effect that it had worldwide can be gauged by these figures: in John Bolster's

A cartoon of HFS which appeared in the Junior Car Club Gazette in the early 1930s.

Over to Four-wheelers. The designation '4-4' referred to four wheels and a four cylinder engine, the latter initially being a Coventry-Climax engine.

French Vintage Cars he states that in the early 1920s there were at least 350 light car and cyclecar manufacturers in the Paris suburbs alone. By 1929 there were just 23. In Britain in the early 1920s, there were at least 25 three-wheeler manufacturers. At the 1929 Motor Cycle Show 11 exhibited; in 1931 six; 1933 just three and by 1936 only one - the Morgan.

The major cost-saving derived by buying a three-wheeler was its low taxation, and this was used to maximum effect in Morgan advertising. Until 1920, taxation of motor vehicles was based roughly on a guinea per 4 horsepower, in addition to a petrol tax of 3d per gallon. On January 21, 1921, the Motor Taxation Act, which taxed private cars at £1 per horsepower, was enacted. This was calculated by using a complicated RAC formula, which in simple terms reckoned 1 horsepower to be equal to 2 square inches (12.9sq cm) of piston area. This was known as the 'Treasury Rating' and is basically why so many long-stroke engines were designed in the 1920s and 1930s. Three-wheelers of the Morgan type were taxed regardless of engine size, provided that they weighed less than 8 hundredweight (406kg) at a flat rate of £4 per annum. This compared to £9 for an air-cooled Rover, £7 for a Jowett and £8 for an Austin 7, all classed as four-wheeler light cars. An independent report showed that, based on an annual mileage of 6000 (9650km), the difference in fuel costs between the three-wheeler and the three light cars was approximately £1 10s to £2. Oil consumption was very similar. Tyre life on the three-wheeler was less than on the four-wheeler, with its more evenly distributed load, but the extra wear was more than balanced by the cost of the additional wheel and tyre. The three-wheeler was also less expensive to garage. The report found that the running costs of the four vehicles, which took into consideration petrol, oil, tyres, insurance and a club subscription (JCC), were as follows: Morgan £54 12s 3d; Rover £68 15s 5d; Jowett £65 18s 0d; Austin £65 14s 0d. A saving of almost £1 per month to the three-wheeler owner was a great deal in those days!

Only three companies came anywhere close to seriously challenging Morgan's supremacy in three-wheelers. In 1933 Raleigh had produced a true three-wheeler with a four-seater open body. Then, in 1935, it offered the Safety Seven Saloon. The first attempt at a 'passenger saloon' was really an extended van fitted with windows and a side opening door in the centre of the nearside. There was no door on either side for the driver. Although Sir Harold Bowden had serious ideas of attacking Morgan sales, the whole enterprise failed completely and he was pleased to sell out his car division to T L Williams, who then made commercial vehicles at Tamworth under the name of Reliant.

In 1929, BSA entered the market with a good-looking three-wheeler powered by an 1100cc overhead valve, vee-twin Hotchkiss engine. It was fitted with an attractive two-seater or family body, electric starter, car-type controls and all-round weather protection equipment; truly a car on three wheels. It performed well on the road and in competition, and nearly 2000 examples were produced. In 1933 the company introduced a water-cooled, four-cylinder 60mm x 95mm, 1100cc-engined car, still with front-wheel-drive, but within a year had added a fourth wheel and named the model the Scout. By the end of 1935 all three-wheeler construction had ceased.

The Morgan's most serious and successful competitor was without doubt the Coventry Victor. This company was very well known as an engine manufacturer which, in 1926, started making its own car. The car followed the 'proper' car layout, with a horizontally-opposed twin-cylinder water-cooled engine mounted in a conventional manner immediately behind the radiator, through which projected the starting-handle shaft. A plate clutch with fabric universal joints at each end conveyed the drive to a bevel-box, bolted at the rear to the channel-section crossmember. It was produced in various forms, including a family model and, eventually, a fully enclosed version, the Avon Bodied Sports Coupé. It continued in production until 1938, albeit only in very limited numbers for the last two years.

Both Raleigh and BSA produced a commercial van version of their three-wheelers and these sold well. HFS decided to join in the battle for this part of the market with his own van. It was not

Immediately after the war sheet steel was at a premium and many Morgans were bodied by other companies. This Woody Drop Head Coupé is one example; it's a 1949 4/4 with a Standard Special engine.

at all successful and only around 40 were produced over five years.

HFS was a very shrewd businessman, as has already been illustrated, and his interpretation of the market trend in the early 1930s was brilliant. At the 1933 Motor Cycle Show (a show was not held in 1932 so HFS staged his own at the Malvern works, where he exhibited eight models to invited guests) he introduced the F-type model which was a complete departure from his original ideas. By producing a three-wheeler model which was more in line with the design of a 'proper' motor car, with a water-cooled side-valve Ford engine mounted in a Z-section chassis behind an inverted vee-shaped radiator, he was within one easy step of another major development.

Demand for the Morgan three-wheeler was dropping fast. In 1929 total production consisted of 444 Family models; 297 Aeros; 140 De Luxe; 108 Super Sports; 11 vans; 1 box carrier and 1 Convertible Family, making a total of 1002. By 1934, this was down to 659 and in 1935 to just 286. The introduction of the F-type did little to stem the decline; it was obvious that the motoring public was turning away from three-wheelers.

Way back in 1915, HFS had experimented with a four-wheeled version of his car, but had abandoned the idea. Now was the time to rethink the whole concept. The design and layout of the F-type lent itself to easy conversion into a four-wheeler. After building three experimental cars, the new four-wheeled Morgan was announced at Christmas 1935. In keeping with his maxim that his cars must prove themselves in competition, HFS entered the final prototype in the London-Exeter Trial and timed the announcement to coincide with this. Although production of the new model did not commence until March 1936, the company produced 130 in the period to the end of December. By then, annual three-wheeler production was down to just 137 units.

Predictions

In October 1926, *Light Car & Cyclecar* magazine asked prominent persons in the motor trade to predict what light cars would be like in 10 years' time. Most of the predictions made by HFS came true, but he did not employ many of them in his own cars. Of all the opinions published, his was the closest to what actually happened:

"Mr H F S Morgan, who has forgotten more about cyclecars than most other people in the world have ever learned, expresses views with which many will agree.

"My convictions are that the next 10 years will not see any startling changes in the design of light cars. Price will still be, as it is now, the most important factor and will prevent the use of eight-cylinder engines and four-wheel-drive which, no doubt, will become popular with those who can afford the luxury.

"The popular car for 1936 will probably have a lower chassis than those used today and larger wheels. A four-seated body with really good all-weather equipment will be in demand and glass side screens will be used instead of celluloid. Steel press work will be very largely used instead of wood, this method bringing the total weight to about 12cwt. Shock absorbers will be discarded; a better springing will have been achieved by improved designs.

"The engine will be about 9hp RAC rating. Overhead valves will be used, operated by an overhead camshaft. The engine will turn over at 3000rpm and the petrol consumption with be 40mpg.

"Superchargers will certainly not be used on the cheap light car. The gearbox will remain much as it is at present, also the back axle, except that it is possible that worm drive will be-

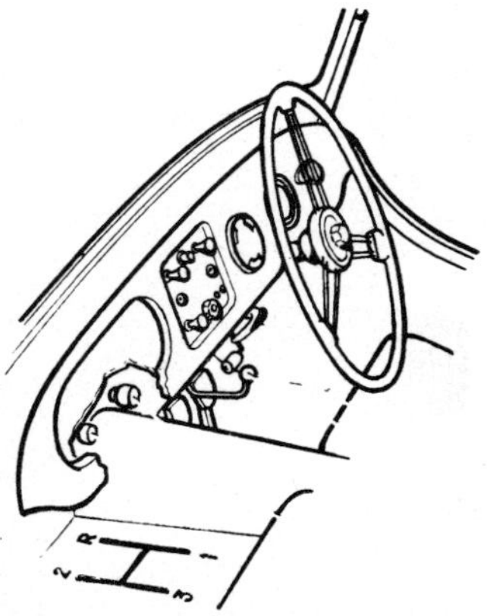

come popular.

"Hydraulic four-wheeler brakes may possibly be used and a single electrical instrument will serve for starting, lighting and ignition. The cost of such a car will be about £150 complete."

The final death knell for light cars and cyclecars, both three and four-wheelers, was sounded in 1936. On April 28, Neville Chamberlain, the Chancellor of the Exchequer, announced that the government intended to abolish the Road Fund Tax on all motor vehicles in 1937. Although this did not actually happen, it was too late to save all but a handful of manufacturers. By 1939, Morgan's output of three-wheelers was down to 29, whereas the production of the 4-4 model was 234 and increasing steadily at the outset of the Second World War.

Once again successes in competition played a big part in the increase of orders, particularly in the 24-hour race at Le Mans and the RAC Rally. Clearly, the timely introduction of the 4-4 saved the company from going the same way as Morgan's three-wheeler competitors. The company was financially buoyant as it entered the war.

Early in 1939 HFS pondered the question whether it would not be better for the company to manufacture its own engines rather than buying them in. To this end he commissioned a small overhead-cam engine to be designed by a Mr Hatch who, at that time, was working in the engine design department of AJS. (The Morgan factory still have the original drawings of this engine.) With the benefit of hindsight, it was to be a good thing for the postwar survival of the company that various factors - such as the offer from the Standard Motor Company to produce a special 1267cc engine just for Morgan - delayed any further development of this engine.

The war years

Almost immediately war was declared, Morgan's machine shops were made over to the manufacture of essential precision parts for the Oerlikon gun and various hydraulic undercarriage components, gauges and compressor parts for aircraft. There was to be no production of cars at all in this war, although the repair shop, offices and stores did operate for some time on a very limited basis. Part of the factory was rented out to HFS's old friend, John Black, of the Standard Motor Company, who built carburettors there. The remainder was rented to Sir Alan Cobham, and it was here he developed his brainchild, in-flight refuelling. For a large part of the war the various buildings housed several complete aircraft, including a Lancaster bomber. During this time the head of security for the factory complex was Charles (Reg) Lisseman. It was he who earned the doubtful distinction of being the man who refused entry to no less a person than Winston Churchill because he did not have the correct pass. The matter was soon sorted out and the great man made his official inspection. It was obvious, however, that all Lisseman's superiors were very upset over the incident and were going to deal severely with him when the visit was over. However, as he was leaving, Churchill made a point of congratulating Lisseman on his efficiency!

At the end of the war, the Morgan company was in good financial shape as a result of the rentals and the government's fixed profit payments for the production of parts for military equipment. However, the return to car production was to be totally different from that achieved after the First World War, for a number of reasons. The removal of the tenants could not take place overnight and was delayed until the middle of 1946. The slow return of the workforce was another important factor. The armed forces were kept at a high level of readiness, due to the continuing war in the Far East and the occupation and policing of Europe. It was therefore to be almost 18 months before the workforce was complete again.

As they returned, the workers were employed in re-organizing the factory. Financially, the machine shop was able to support the wage bill and cost of re-organization by the continued production of aircraft parts. (The company remained licensed to produce aircraft parts until 1960.)

During the war the government installed almost 100 machines which, due to the fastidiousness of HFS, were still in excellent condition when the war ended. Eventually, HFS was able to purchase most of them at a very reasonable price, so helping to reduce the overall cost of re-equipping the factory for car production.

Production re-starts

Throughout this period a few cars were built from spare parts stored during the war and a few chassis were also produced from the limited amount of parts available. These were fitted with whatever body the purchasing garage could invent. Some of these were excellent, whereas others left a lot to be desired ... Alternatively, they were sent away for bodying by another company. Amongst those companies used for this purpose were Motorists of North West London and Peamore Garage, Exeter, the main reason for which was the

Peter Morgan was an accomplished competition driver. Here he is in September 1948 competing in a TT replica.

limited availability of sheet steel. Allocation was based mainly on the respective company's ability to export its products.

Morgan's orders far exceeded anything they could hope to produce. In order to obtain more steel, and then to get back into full production as soon as possible, the company established overseas agencies in America and France. Morgan had exported cars on a fairly regular basis before the war, either directly or through exporting agents. Now the government slogan for British Industry was 'Export or Die'. By 1950 the company had more than doubled its exports and by the mid-1950s they amounted to over 66 per cent.

Overseas markets now had little interest in three-wheelers and sales in the home market were very low. Reluctantly, HFS decided to end production of the three-wheeler. On March 31, 1952 the last F4 model Morgan - no. F1296, Ford engine no. C61153, gearbox no. XC85 - and on July 29, 1952 the last F Super - no. F1301, Ford engine no. C611562 and gearbox no. XC107 - left the factory to go to Morgan agent Bennett's of Southampton.

Standard's Sir John Black, who had the reputation of being a very hard man in business, remained friends with HFS. It was therefore mainly on a personal basis that the Standard Motor Company supplied engines to Morgans. Just before the war, the Triumph Motor Company had gone bankrupt, and in 1946 Sir John Black acquired the remains of the company, with Triumph becoming a subsidiary of Standard. In 1951 Sir John Black made an exceptionally good offer to buy out the Morgan concern: he was searching for a quick means of entry into the lucrative sports car market in America. The Morgan company was financially sound enough to resist this takeover bid and, eventually, Sir John developed Triumph to produce the very successful TR series, the engine of which Morgan adopted for the Plus 4 model with such outstanding success. Again, the supply of the TR engine was a personal deal between Sir John and HFS. Then, early in 1954, Sir John resigned as managing director of Standard. Officially this was due to the effects of a serious motor accident, but it is believed that fellow directors on the Standard-Triumph board demanded his resignation because of his dictatorial attitude. This led to a reduction in the supply of engines to Morgans for a while, until a change of policy at Standard-Triumph reinstated the supply, this time with the more powerful TR3 engine.

Family links maintained

HFS's son Peter Morgan was educated initially at Link School, Malvern, and then at the famous Oundle Public School in Northamptonshire. On leaving Oundle in 1936 he had a choice of continuing his education at engineering college or university. Selecting the former, he went to Chelsea College of Automobile Engineering, where he stayed until the outbreak of war. Initially, he went to Admiralty House and attempted to enlist into the Royal Navy as an artificer but was rejected. He was then interviewed for the Army and was selected for the Royal Army Service Corps, where he became a workshop officer for a motor coach company. Posted overseas, he went from the Royal West African Frontier Force Workshop to Nairobi, where he remained until the end of the war. He was released from the Army in late 1946, having reached the rank of captain, and joined the family company as development engineer, draughtsman

Peter Morgan

Morgan's chief tester, Charlie Curtis, long retired, with the author's Le Mans Replica.

- and a member of the board of directors.

Peter Morgan took over as managing director when George Goodall died in 1958 and, when in the following year his father died, aged 78, he became chairman.

Peter learnt to drive before the war and on his father's insistence graduated to racing, through trials and rallies. The competition results reflect his competence.

After the death of his father, Peter Morgan's additional responsibilities meant he had little spare time available to continue competition driving but he did return to trials for a short while, driving the prototype Plus 4 Plus.

Over the years Peter Morgan has enjoyed various hobbies and interests; he played golf, had a miniature railway layout in his garden, enjoyed ski-ing and, until his faithful Red Setter, Kerry, died in 1986, did a lot of walking. He still enjoys gardening, which he finds relaxing, but his lifelong hobby is stamp collecting, specializing in the Seychelles in particular, and postal history.

HFS's farewell

The content of HFS's will reflected his love for his family and his sense of fairness, both of which had been very obvious during his life. Handwritten on two sheets of paper it read: 'I have made this will as simple as possible and hope it will give my dear family the least possible trouble ... 'I bequeath to my son, Peter, rather more than his sisters, as he has the trouble and worry of running my old business (now under the control of my family), which may have to be sold owing to the increasing difficulty of running a private business. I have had a long and happy life, and words cannot express my thanks to my children, in-laws, and grandchildren for their love and kindness to me. May you all live long in happiness together. God bless you all.'

In all, he left £250,287 (duty £146,349). This meant that the business was left in trust to his four daughters and his son, with Peter Morgan and Major Kendall appointed trustees and Peter Morgan in charge of the company. It was obvious that if the Morgan Company was to survive, there had to be an all-out effort to increase production and sales and reduce costs. To this end, Peter Morgan called a meeting of the entire staff and explained the position; with typical loyalty they responded to meet the challenge.

Saved by the chequered flag

In 1961 Morgan's main market, the United States of America, endured depression and the factory was forced to reduce production whilst exploring other export markets. The company did amazingly well to survive at all during this period. Its saviours were really two new models plus the outstanding success of Chris Lawrence in the Le Mans 24-Hour race. This led to the production of the Plus 4 Super Sports, tuned by Lawrence, and the introduction of an entirely new concept in Morgan design, the Plus 4 Plus. Whereas the former was an unqualified success , the Plus 4 Plus did not sell well, although its revolutionary design helped to draw the attention of the motoring press and public alike.

As the company's finances began to recover, the family felt that it was time to arrange a redistribution of capital shares. In order to achieve this, it was necessary for the company to lose its limited liability status for a few years. During this period the official designation of the company was 'Peter Morgan trading as The Morgan Motor Company'. In 1970 the business was re-formed as a limited company. It was whilst in this state of flux that the company took the decision to introduce the Plus 8.

Although the development and introduction of this model was ridiculously inexpensive in terms of the motor industry in general, for Morgan it was very expensive, and financed completely from its own reserves. The new model was well received by press and public alike, and proved an excellent investment.

Expansion

Another indication of the company's improved financial stability came in 1972 when it made its largest investment since the Second World War. Peter Morgan had become increasingly aware of overcrowding within the factory due, in the main, to the increased complexity of the cars and the greater time needed to complete each one. The adjoining factory site and building came on to the market, but was really far too large for the company's requirements. However, when, after a considerable time the site remained unsold, Peter Morgan realized he could purchase it at a very reasonable price. This he did and moved the trim, wiring, final finish and test departments into the largest building, situated furthest from the

Gas powered: Morgan Plus 8s imported to the USA are converted to run on propane gas by Isis Imports of San Francisco.

main entrance. The remainder of the site he rented out to provide the company with a steady income. Yet again the company was able to finance the whole deal without need of outside assistance.

Postwar depression

In the mid-1960s there was renewed interest in the marque. All through the 1970s the demand and orders increased, as did the wait for delivery. This trend has continued and, except for a period in the early 1980s when Britain and most of Europe went through an economic depression, delivery dates of about five years have been the norm. Had it not been for the American market, there can be little doubt that the company would have collapsed in the late 1950s, a time when nearly 85 per cent of production was exported to the USA. In 1961 there was a great recession in the Californian aircraft industry and not only did it hit the prosperity of those who were directly involved, it also spread throughout the whole country.

So dramatically did sports car sales drop in the USA that MG and Triumph agents were reduced to holding weekly auction sales in an attempt to reduce their stocks. Morgan's New York agent, Fergus Motors, cancelled an entire shipment. It just so happened that when the cancellation was received at the factory, Peter Morgan and his wife were taking their first ever holiday in America. The factory cabled him at his hotel and told him the bad news. Peter ordered the shipment be dispatched and immediately went to New York to see if his personal intervention could improve matters. His strategy worked; he persuaded the agent to accept delivery and whilst he was in the US met a large number of American enthusiasts.

Back in England, production had to be cut by one car a week. This does not sound very much but it did represent about 14 per cent of total output. Slowly, the market improved but it was almost two years before production and sales returned to their previous levels.

Several valuable lessons were learnt from this episode. The company needed to broaden its markets to avoid ever again being too dependent on sales to one country. Also, it was vital for Peter Morgan to be available to visit wherever possible enthusiasts of the marque around the world.

Emission and safety controls (USA)

Throughout the initial development of the Plus 8 everything had been done to incorporate all the foreseeable requirements of the new motor vehicle regulations introduced in the United States in the late 1960s. However, Morgan was forced to drop out of the US market by the end of 1971. The main

Country pursuits: American advertising photograph shows open-air Morgan motoring with rather more than three horsepower...

reason for this was the increasingly tough exhaust emission control standards which caused Rover to pull out of America at the same time, and automatically forced Morgan out as well. Other requirements such as 5mph 'no damage' bumpers and seatbelt warning buzzers also caused problems. These devices and necessary technology would eventually become available to Morgan as introduced by the major manufacturers; however, there was always a delay.

When the supplies of Morgans to the United States stopped, one American enthusiast, Bill Fink, who, from 1969, had traded in Morgan parts, repairs and the sale of second-hand Morgans, was determined to find a way to import Morgan cars. From the inquiries he made of the Department of Transport and all the other relevant governmental departments, he discovered that any car powered by propane gas was exempt from the emission control regulations. Having successfully overcome the most serious of all the legal requirements, he set about investigating ways to overcome the remainder. He worked out that if some of the modifications were carried out at the factory during construction, then he would be able to complete the work in the United States, making each car acceptable for resale. In 1973 he was in a position to approach Peter Morgan, who eventually agreed with his basic ideas, and promised to supply him with up to 24 cars a year, although bureaucratic delays meant it was not until 1976 that Fink sold his first car.

For these special US cars, some of the work was carried out at the factory - telescopic gas-filled shock absorbers, mounted behind reinforced front and rear bumpers, to comply with the 5mph head and tail end impact legislation; four-seater size windscreen; special steel reinforcing hoop bolted to the chassis behind the dashboard.

On the car's arrival in the United States, steel beams were fitted into the doors to meet the side impact regulations; in addition, a rear roll-over bar, inertia reel seat belts, padded sun visors and seatbelt warning alarm were all fitted. The interior was coated with a flameproof fluid.

The conversion of the engine required a special 18-gallon fuel tank with the filler valve neatly hidden under the spare wheel mounting, sealed fuel lines and a vaporizer and a lock-off to stop the supply when the engine was switched off. The inlet manifold and carburettors were changed. Finally, a 45-page dossier, with full written and photographic evidence of all the modifications carried out, had to be prepared for each car.

In 1986 another American enthusiast, Win Sharples from Virginia, successfully obtained permission to import Morgans into the United States, and was granted an agency by the Morgan Motor Company. All the modifications previously listed above also had to be carried out by his company.

In Canada the exhaust emission control laws are not so strict as in the United States, as a result of which the fuel injected Ford XR3 and Rover engines comply with regulations. All other modifications as required for the American market do apply and all new Morgans imported to Canada must be modified by Chris Charles, the authorized distributor. New Morgans can now be purchased in either the United States or Canada. However, the extensive modifications necessary always mean that the numbers sold will not approach those of the 1950s or 1960s.

Exports worldwide

Learning from the near-disaster of the 1960s, Peter Morgan has slowly broadened the company's overseas sales. The main European outlet is West Germany which has two distributors and now accounts for 30-35 per cent of the company's export sales. All the western European countries have a Morgan distributor but a problem is that in recent years Europe has also seen the introduction of tougher rules and regulations governing motor vehicles. The result of this is that it is now impossible to buy a new Morgan in Switzerland, Denmark and Sweden.

Dealing with the legislation around the world is a complicated matter. For example, in Japan, the Morgan distributor has to send all his imported cars to a specialist company for modifications so that they meet the emission control regulations. But before a 4/4 is exported to Japan the works has to fit the car with wider wings than normal to

HFS competes at Angle Bank in July 1923 in a wide-track, flat-sided Aero model, fitted with an 8 valve Anzani engine.

comply with the local wire wheel regulations. In all, at the time of writing there are a total of 15 European distributors; two in the USA; one in Canada and five in the rest of the world. Exports account for approximately 35 per cent of total production. It therefore follows that the home market must be buoyant to sustain production levels.

The welcome mat

The factory's 'open door' policy, whereby prospective owners, or even just the inquisitive, are shown around the works, and the friendly nature of all the employees, are important aspects in promoting sales. So is the buyer's personal choice of car specification when his or her turn on the four to five year waiting list eventually arrives.

Campaign agents

Throughout the three-wheeler era, many Morgan distributors proved the car in competition. Such men as Frank Boddington (Worcester), Freddie James (Sheffield), Jack Sylvester (Nottingham), FG Cox (Bristol), Horrocks (Bolton), Jim Hall (Stevenage), Roger Darmont (Paris), 'Joe' Huxham (Bournemouth), Billie Elce, Harold Holmes, Stanley McCarthy and Arthur Maskell (London) competed in all manner of competitions from trials to circuit racing.

With the introduction of the four-wheeler, much of the distributor competition participation faded away and motorsport was left mainly to the works. The most spectacular exception to this rule was Chris Lawrence who, although not an official distributor, was a works concessionaire. Then, around the late 60s, the distributors began to get involved again; John McDonald, Rob Wells, Bruce and John Stapleton, Mike Duncan, John Brittan, service agent Rick Bourne and other Morgan specialists have all been highly successful, particularly in racing. Their exploits are well covered in this and other books and it's as well to remember that by their skills and expertise they have promoted the marque wherever they have appeared.

Additional promotion for the marque has come from a large number of specialist companies who are able to offer the Morgan owner not only a spares service, but tuning, preparation and complete rebuilds to the highest possible standards.

Morgan Owners' Clubs around the world also help to create the reassurance and spirit of camaraderie so important to most Morgan owners.

Although he loved competing in sporting trials, HFS Morgan was basically of a nervous disposition. He dis-

Charles Morgan with Plus 8 at St Moritz for the 100th Anniversary of the Cresta Run in 1985. The Sotheby Cup races were part of the celebrations.

The next generation. Xan Morgan in a racing Three-wheeler with father Charles at Silverstone.

liked racing and speed events, but realized he must compete in them to prove his product. His most outstanding achievement in those early days must have been on November 23, 1912 when he came within yards of achieving the magic average of a mile a minute for an hour at the Brooklands race track, and won the Cyclecar Trophy.

HFS's nervousness was not inherited by either his son, Peter, or his grandson, Charles. Peter Morgan always maintains that the war spoilt his racing career; in the few races in which he did compete he showed great promise. Charles had no such interruptions, other than the different locations in the world where his work took him. He was very successful on the race track, as can be seen in the competition chapter, winning the 1978 British Production Sports Car Championship.

The next generation

Charles Peter Harry Morgan was born on July 29, 1951, schooled first at The Elms Preparatory School, Malvern, and then Oundle like his father, and went on to Sussex University, where he read the history of art, gaining a Bachelor of Arts Honours degree. On leaving university he worked for a year as a bookseller with Idea Books Limited, and then joined Independent Television News as a cameraman. This work took him to many remote and far-off places such as Vietnam, Kampuchea and Afghanistan, where he made an award-winning documentary programme about the guerrillas.

Although his father never placed any pressure on Charles to join the company, always insisting he should do whatever he was best at, he was obviously very pleased when Charles decided to join him full-time. In 1979 Charles married Vivien Lipschitz who, like the previous Morgan wives, supported her husband in all his motor sport activities. On January 8, 1985, Vivien gave birth to a son, named Xan, who already loves to visit the factory.

For relaxation, Charles loves motor racing, gardening, studying architecture and the Cresta Run. The last interest came about as a result of photography and directing a film, made for the St Moritz Tobogganing Club as part of the celebrations of 100 years of the Cresta Run in 1985. Whilst making the film he fell in love with the whole concept and was allowed by the Cresta Club to try it out a few times. They were so impressed by his skillful performances and keenness that they made him a member there and then, something which normally takes years to achieve.

Visiting the factory

Many authors, myself included, have tried to put into words the feelings and impressions they experience when they step into the visitors' waiting room at the factory for the very first time. One, in 1966, even went so far as to write 'it has a dinginess that could almost be considered studied. The manure-coloured paintwork is lustreless with age.' Things have improved since those days; the paintwork has been renewed, and a glass display counter installed, but the walls are still covered with display boards listing, in hand painted lettering, the competition achievements of the three-wheelers, plus various framed photographs. Entering, one gets the impression of stepping through a time warp, back to the 1920s.

The factory is an enigma. From initial impressions, you might expect the workforce and management to be of comparable age to the premises. Nothing could be further from the truth, with the average age in 1985 of the whole workforce being just 34. This clearly indicates that Peter Morgan and his board of directors are not afraid of

The Morgan factory at Malvern Link - mecca for all true sports car enthusiasts.

change or of young ideas.

There have been many subtle changes in recent years which have ensured an on-going supply of skilled staff on the shop floor. The biggest of these has been the removal of the foremen from the piecework bonuses scheme. They are no longer required to work full-time at their various trades and now spend a great deal of their time in on-the-job training and leadership. In this way their skills are handed down to an up-and-coming team of young men. Many of the skilled trades necessary to build a Morgan, such as tinsmithing and coachbuilding, are fast dying in outside industry. It is not that there is a lack of suitable young men to train, it is that these skills are no longer generally required, due to the continuing change to automation.

It is not just on the shop floor that the influx of young men is occurring. Obviously, the youngest member of the board of directors is Charles Morgan, but at the next level of senior management, Peter Morgan's assistant, Mark Aston, whose job description would be almost impossible to write due to his numerous and diverse responsibilities, has only just turned 30, while Derek Day's deputy in the sales department, Mark Read, was only 29 when this was written.

Litigation

In July 1980 an incident occurred in the USA which could have brought about the complete collapse of the Morgan company. A 1967 Morgan, which had been illegally imported into America from Canada, was stationary in traffic when it was rammed from behind by a small truck. The small truck was, in turn, rammed by a 55-ton gravel truck. As a result of the impact a fire ensued and the Morgan driver's hands were quite badly burnt. In 1983, solicitors acting on behalf of the injured driver included in their litigation lawsuit an allegation that the Morgan was partly responsible for the injury, due to the design and location of its petrol tank. It took the company's legal representatives almost 10 hours to present the technical evidence to the court, which in the end accepted that the claim was almost completely unjustified. No legal defence is cheap and American legal costs are notoriously high. The action cost the company almost £150,000, an amount, although large, was minute in comparison to what it could have been if the court had held that the company was wholly or partly responsible.

75 years young

1984 was to mark the 75th anniversary of the Morgan and also Peter Morgan's 65th birthday. The celebrations of the anniversary were to bring together more Morgans than had ever been seen in one place, with well over 1000 cars converging on Eastnor Castle on the final day of the 10-day of celebrations. For those of us who were involved in the organization in any way, there was a feeling of immense honour and satisfaction; to have taken part was to have gained a place in the unique history of the oldest one-family motor car manufacturer in the world. Media coverage was enormous, with over 120 magazines and newspapers worldwide publishing some form of tribute; then there was also coverage on local, national and international radio and television. Morganeers from all over the world made lasting friendships.

A few years ago a motoring magazine published an article on the Morgan which posed the question 'Morgan's Raison d'Etre?' If its author was at the 75th, his question was answered. The article ended, 'Dr Johnson once said "The ultimate end to all employments is to produce amusement". Morgans are sheer fun to drive, and they produce more amusement than a train load of Playboy bunnies.'

II

FOUR-WHEELERS IN COMPETITION

HFS knew from 25 years' experience that the only way to prove a new car was in competition, and so it would be with the four-wheeled version of the Morgan Sports. The two main considerations for the motoring public were reliability and competition achievement.

Although today's Morgan holds a particular attraction because it is in tune with the current interest in nostalgia, this has not always been the case. It was largely through success in competition that the 4/4, and later the Plus 4, were able to establish such strong positions in the four-wheeler sports car market.

HFS decided to make the public announcement of the new model to coincide with the date of its first competition appearance. He entered himself and the 4/4 for the London-Exeter Trial. On December 27, 1935, when the first magazine to publish details of the new Morgan 4/4 went on sale, HFS set off, in pouring rain, to prove that this new four-wheeler was the best in its class. It rained continuously throughout the two-day event, making it even more testing. But both car and driver excelled themselves, claiming a Premier award and earning special mention in the motoring press for the new car's hillclimbing ability. HFS could not have wished for a better launch.

Twenty-seven years later at an official reception, HFS defined the qualities required of drivers for success in trials and rallies: skill, stamina and the ability to use a stopwatch. But he added that good drivers also needed good, reliable cars, which could only be produced by skill in manufacture and preparation. He had certainly proved his point with the introduction of the new four-wheeler, for it was to be the start of a long and successful history in four-wheeled competition.

The Morgan family have always encouraged this aspect of Morgan ownership, firm believers that their cars should be enjoyed in their natural environment - motorsport. Furthermore, they have always been willing to spearhead these beliefs by participating themselves. Even in the 1930s it was rare for a manufacturer to actually drive one of his own vehicles in a competitive event. In modern times it is almost unheard of, but Charles Morgan still competes on a regular basis, albeit rather less now than previously.

Races, rallies, hillclimbing, reliability trials, off-road events, driving tests and many other forms of motorsport are all within the scope of the versatile Morgan.

In the following year-by-year record I have not attempted to list or mention every event that a Morgan has competed in but rather give a general overview of the sporting achievements of the four-wheeled Morgan. These follow no particular pattern, car or driver, and are selected on the basis of the insight they provide into the versatility of the Morgan as a true sports car.

1936: a slow but sure start

The second event for the new 4/4 was the MCC Lands End Trial, held in April, where yet again HFS claimed a Premier award. This was followed in May by the Edinburgh Trial, in which the works entered three cars. They gained a Premier and two silver awards. In June, E G Boutle took a first class award in the Junior Car Club's Evening Trial at Guildford, Surrey, and earned the distinction of becoming the first private owner of a 4/4 to win a motoring competition.

Harry Jones of the Morgan factory drives the prototype 4-4 in the Edinburgh Trial in May 1936 in which he won a Premier award.

1937: 4/4s go racing

The year started for Morgan in the Lands End Trial in March, in which all five Morgan starters gained Premier awards.

The major rally in Britain was initiated by the RAC and was held in early March. Three Morgans were entered in Group 1 (for open cars up to 10hp). The Group was won by George Goodall's 4/4, earning him a Premier award. HFS and E D Bowman finished eighth and tenth respectively and took silver awards.

In May, Peter Morgan made his competition debut as an official entrant/driver in the Edinburgh Trial. He succeeded in winning a silver award, whilst the other Morgan entries won two silver awards and four Premier awards.

The 4/4's first race was in the Ulster Trophy, held in June on the 4-mile Ballyclare road course. Belfast mechanic Robert Campbell was the surprise winner after a series of accidents and mechanical breakdowns sidelined 12 of the 21 starters in this handicap event.

The following month Morgan was to win yet another Irish Race, this time the Leinster Trophy, which was won by D C McCracken in his 4/4, with Robert Campbell finishing fifth in the same car that had won the Ulster Trophy. McCracken had been very confident of winning; he had told a representative of *The Motor* that he would win easily, a full month before the race. However, he did so only after W F O'Riordan in a Riley collided with a slower car and somersaulted out of the race.

The first English venue for the RAC Tourist Trophy Race was the Donington Park circuit. The only Morgan entry was a works-entered 1098cc 4/4 driven by the famous Morgan three-wheeler racing driver, Henry Laird. A broken stub-axle put him out of the race on the 84th lap when he was well placed.

1938: Morgan at Le Mans

This year's Morgan competition successes were dominated by two major events, the first of which was the RAC Rally, in which George Goodall won the Group 1 prize with HFS finishing third and J F Heaton seventh. Miss D L Bean won her class in the coachwork competition; the Morgan looked good as well.

The second big affair was the world famous 24 hour race at Le Mans. Such were the rules at that time that, providing you had a car, money and were impetuous enough, there was nothing in the regulations which said that you could not make Le Mans your first competitive event. One of the competitors who did just that was the 25-year old daughter of a Sheffield solicitor,

One of the two early 4-4s which did so well in the Irish Leinster Trophy and the Ulster Trophy races in the summer of 1937.

Harry Jones competing in the Lands End Trial in April 1938.

Prudence Fawcett during practice in the 1938 Le Mans.

Miss Prudence Fawcett. Her co-driver was Geoffrey White who was chief sales manager for Winter Garden Garages. The car was prepared for her by Dick Anthony, Winter Garden Garages' chief racing mechanic. Dick was a very experienced competitor, having earned himself a considerable reputation on the track driving Aston Martins. The pit was manned by Winter Garden Garages owner, Lancelot Prixeaux-Brune.

The race was dominated by the big Delahayes, Talbots and Alfa Romeos and, as usual took a heavy toll of cars; by 7am on Sunday only 22 of the original 42 starters were still running. By the time the finishing flag was held out at 4pm, only 15 were left. Miss Fawcett's Morgan might well have finished a lot higher than its 13th place if it had not been for a radiator leak which developed in mid-morning and necessitated the car being nursed to the finish. When the cylinder head was removed the following morning it was discovered that three of the exhaust valves were burnt-out due to the overheating. The car finished second in its class, covering 1372.98 miles at an average speed of 57.2mph. It was classified 14th in the Index of Performance.

As for Miss Fawcett this turned out to be her first and only race. Later that year she became engaged, and made a pact with her fiancé that she would give up motor racing if he would give up aeroplane racing. It was a pact which was to last all their married life.

Her car was eventually returned to the works and was entered by them in

Prudence Fawcett driving in the 1938 Le Mans 24-hour race.

Left: Henry Laird driving car 29 in the 1938 Tourist Trophy race.

Below: Henry Laird makes a pit stop during the 1938 Tourist Trophy race.

the second Tourist Trophy race at Donington Park. The car was driven by Henry Laird and, in treacherously wet weather, earned praise from the motoring press for its stability. One report said, *'H Laird's was probably the steadiest car in the race, with not a trace of skidding on the wet roads'*. Yet again, the car was to suffer from a leaking radiator which brought it to the pits six times, and Laird was eventually forced to retire on the 94th lap.

One of the most coveted of all trials awards is the Gold Medal Triple Award, acquired by gaining a Premier award in the Lands End, Edinburgh and London-to-Exeter Trials in the same season. Two such awards were presented at the end of the 'Exeter' - one went to HFS and the other to Peter Morgan.

'Uncle' George Goodall and 'Sonny' Jim Goodall competing in one of the 1122cc 4-4s which had been modified to cycle-type front wings. They finished first in class and took a Premier award.

Despite the many trophies which have come his way since, Peter still looks upon his 1938 Triple Award, won in his first full year of competition, with special pride.

1939: back to Le Mans

This year was dominated by the RAC Rally and the Le Mans race. In the premier British rally, George Goodall was out to complete the hat-trick for Morgan and himself. Only two Morgans were entered, the second being HFS with a 4/4 Coupé. In both cases it was a close-run thing: HFS won his class by a narrow margin and, when all the marks had been added up, George Goodall had won Group 1 by just a single point from A E Curtis' HRG.

This success continued in the Le Mans 24 Hour race. Recently discovered photographic evidence appears to show that a different car from the 1938 race was used, it was obviously a works car which had been rebodied. The shape, with sloping rear and the insert spare wheel, was subsequently adopted for the 4/4 Le Mans Replica model. The engine of the car had been bored out to 1104cc in order to move it up a class, and the carburation altered, with twin downdraught SU carburettors being fitted, one at each end of the cylinder head. In addition, the dynamo was moved away from the top water manifold and now rested in a special cradle. The car was driven by Dick Anthony and Geoffrey White, with the pit being manned by Charlie Curtis, the Morgan Motor Company's chief tester. It was intended he accompany the race car in an identical spare car in case it was

Dick Anthony and Geoffrey White seen here soon after arriving at Le Mans in 1939, having driven the car there!

Dick Anthony and Geoffrey White enjoy a well-earned rest after the 1939 Le Mans race.

A special-bodied Series I 4-4 single-seater, with Coventry Climax engine, belonging to Jim Boughton, is seen here on April 10th, 1939, at the 150-mile race at Bathurst, Australia.

needed but, although this car was prepared, it did not actually travel to Le Mans but was sold after the war as a Le Mans Replica.

In spite of fuel troubles the Le Mans car performed well. The first problem came with only four hours to go, when the engine continued running at high speed even after the driver's foot was lifted from the accelerator. White drove the car very carefully back to the pits, where it was discovered that the needle valves on both carburettors had collected a tar deposit which caused them to stick. It was useless trying to remove the deposit with petrol so the car rejoined the race, but the driver had to lift off much earlier for every bend.

Then, with less than an hour to run, the engine stopped, starved of fuel. Because of the sticking needles it had been necessary to schedule an extra refuelling stop, but this had not been due for another two laps. Anthony climbed out of the car, looked down the filler pipes and discovered that although there was still fuel in the tank, because of the considerable camber of the track (which also happened to be mainly right-handed) the fuel surface was tilted away from the outlet pipe, which was now exposed. The regulations stated that in the event of a car breaking down on the circuit the driver alone could work on it, but to push the car back to the pits, single-handed, a distance of nearly a mile, was out of the question. However, by cutting across country it was only about half a mile, so Dick ran back to the pits, collected a can of fuel, ran back to the car and poured it in, whereupon the car restarted immediately.

It was a fine effort which enabled the Morgan to become the highest-placed British car in the Rudge Whiteworth Biennial Cup and to finish 15th overall, having covered 1548.62 miles at 64.53mph. This compared with the race-winning average of 86.85mph set by Wimille and Veyron in their 3.3-litre Bugatti, the first of 22 survivors out of the 42 starters.

Recounting his experiences that year, Dick Anthony said, *'The car went very well except for the fuel trouble. As you know, I have competed in many makes of cars (though mainly Aston Martins) in many types of events, and my only criticism of the Morgan is that instead of building a car to last me 24 hours at Le Mans, HFS, as usual, built a car to last 24 years or more.'*

The works entered the Le Mans car for Laird in the Tourist Trophy Race. Also entered was Stanley Wood, another famous Morgan three-wheeler racing exponent, who used a supercharged car. However, just a few days before the race, war was declared and all motorsport curtailed, though not forgotten, for six long years.

1945: immediate postwar years

The distinction of winning the first postwar event in a Morgan must go to Ken Burgess who won his class in the Bristol Motor Cycle & Light Car Club's hillclimb at Naish House in October.

1946

Motorsport events were very few and far between due to fuel restrictions. Those clubs which did manage to stage an event found that they were overrun with entries. Morgan drivers did manage to gain entry to some events, but did not perform all that well.

1947

As the country slowly thawed out after one of the coldest winters ever recorded, the first event of any note was the Lands End Trial in April. Seven 4/4s were entered, of which five survived to collect Premier awards.

1948

With petrol and other restrictions still in force, this proved to be an abbreviated sporting season, but the MCC's Devon Trial in July was a very well organized and testing event, in which three Morgans were entered, with W A G Goodall and R Hellier scoring 'firsts' in two of them.

George Goodall completes the Kerb Test at Pitlochry in the eighth RSAC Scottish Rally, May 1939. He finished first overall.

In the Cheltenham Car Club's trial in October Peter Morgan and 'WAG' collected two more First Class awards and won the team prize.

1949

The first day of the year brought a Premier and two Second Class awards for the three Morgans entered in the MCC's Exeter Trial while, by the time of the Lands End Trial in April, fuel restrictions had been eased and there were six Morgans amongst a much larger entry. Four claimed First Class and two Third Class awards, the team prize going to Peter Morgan, W A G Goodall and C J McCann. The same team repeated their success in the MCC's Edinburgh Trial in June, when all four Morgans entered took First Class awards, but there were to be no Triple Awards for Morgan drivers that year, the MCC's 'Exeter' providing them with just one Second and two Third Class wins.

1950: Plus 4 enters the competition scene

Jeff Sparrowe, who later became the first president of the Morgan 4/4 club, won the April Woodberry Chilcott Cup in the Bristol Motor Cycle & Light Car Club's race meeting at Lulsgate, driving a very modified 1098cc Le Mans Replica.

A Verkey upheld the Morgan name on the continent, taking a second in class in the Tulip Rally.

The MCC's Exeter Trial in December was the first full-length 'Exeter' to have been staged since 1938. It was also the event in which the latest Morgan model, the Plus 4, made its debut. The works had entered two examples, driven by Peter Morgan and W A G (Jim) Goodall (son of G H Goodall), who won First Class and Second Class awards respectively.

1951: rallying returns

The year's most important event was the RAC's first International Rally of Great Britain (prior to this, the event had national status). The event, in June, was divided into four classes: open and closed cars up to and over 1500cc, with

Peter Morgan competes with his short-wheelbase trials car at Brooklands race track in 1939.

Rob Lowrie was an experienced Le Mans competitor who often entered his own cars; in 1949 an Aston Martin, 1950 a Riley and 1951 a Jaguar XK120. In 1952 he entered the race with a Plus 4 supplied by the Morgan Motor Company. This car was not new and had been used in numerous rallies and trials. He was forced to retire after about six hours when a valve and piston made contact.

each class qualifying for the same amount of awards and prize money.

The huge entry of 267 cars included four Plus 4s, three of which were works-entered, as well as the works team of Jaguar XK 120s, which was led by Ian Appleyard in his soon-to-be-famous white two-seater, NUB 120. From starting points at Harrogate and Cheltenham the routes converged at Silverstone for a special test, then went through Northern England into Scotland, back down to Blackpool for an overnight stop, then on through Wales and back to the final test and finish at Bournemouth, four days after the start.

When all the road penalties and test marks had been assembled, it was found that the small open-car class had been won by J V S Brown in an HRG, closely followed by Nancy Mitchell in a similar car.

J Readings, driving an MG, had taken the small closed-car class and P Harper, with a Vauxhall, claimed the larger saloon class. That left the over 1500cc open cars, which had been a needle-match all the way through between the Plus 4s and the XK 120s, and the destination of the coveted team prize.

Peter Morgan well remembers the tension he felt before that final test at Bournemouth: *'At that stage I was still fractionally ahead of Appleyard but, unfortunately, I managed to lose time on the final test. It was a great disappointment, although we were still in with a chance for the team prize.'*

Appleyard tackled the vital manoeuvrability test with his usual skill to bring his aggregate score to 109.61 penalty points which, when added to those of his team-mates Bradnack and Grounds, gave the Jaguar team an aggregate score of 353.67 marks. Then it was the turn of the Morgan team of Morgan, Goodall and Dr W D Steel. Goodall was the first to go through the sandbag-lined course and his time of 74.42sec gave him an aggregate of 114.55 marks - second in class at this stage. Peter Morgan was next. He had been faster than Goodall in the test at Blackpool, but this time his blue coupé was a fraction slower. Nevertheless, 75.41sec gave him an aggregate of 112.99 marks, which meant that he would be second in class and Goodall third. But what about the team contest? Pencils and paper came out as Steel prepared to make his run; if he could keep his aggregate score below 126.13 marks then the Morgan team would do it. He did his best and it was good enough; his total of 123.98 marks gave the Morgans a winning margin of just 2.15 marks! It was a popular victory, for it was one against the odds with no fewer than 37 XK 120s in the rally. In the end, eight Jaguars and two Morgans formed the top ten in the class.

Immediately after the RAC, Dave Price and his daughter took their Plus 4 across the Channel to compete in the 950-mile Evian-Mont Blanc Rally and won their class. They were the highest placed British entry in the event.

1952: triple success

The Morgans of Morgan, Goodall and Steel won the RAC Rally team prize for the second year running. To mark their achievements the Malvern Urban District Council held a dinner at the Winter Gardens, Malvern.

For the first time since 1939, a Morgan was entered at Le Mans, but the specially prepared Plus 4 of Lawrie and Isherwood dropped a valve quite early in the race. It was another 10 years before Morgan was represented again in the Le Mans 24 Hour race.

Morgan successes included an outright win in the London Rally, with Jimmy Ray taking his revenge for Morgan over Ian Appleyard in his XK120, who was regarded at the time as Britain's leading rally driver.

Peter Morgan collected two Premier awards in the Lands End and Edinburgh Trials and, if he could manage another in the 'Exeter', would win his second 'Triple'. Morgans were also well placed to win the Team Championship. As usual, the Exeter took its toll on cars and drivers, but Peter won his third Premier of the season and his Triple Award. With the two other members of the team, Goodall and M Cann, only managing Second Class awards, a good deal of tension was generated whilst the officials did their arithmetic. This proved that the team had done enough to win the Team Championship for the first time since the introduction of the four-wheeled Morgan.

1953

There was to be no repeat of the Morgans' successes in the RAC Rally this year. However, Jimmy Ray recorded another fine victory in the London Rally to retain the Wick Trophy for another year.

At the end of another successful year for Morgan drivers, Jimmy Ray won the British Trials Drivers' Championships in his Plus 4. Lady driver Anne Neil completed a fine Morgan double by

Dixon Smith also used the prototype Series II 4/4 in competition, this time in a Curborough Sprint meeting.

Dixon Smith tackles Prescott Hill in a TT Replica in 1955. (Note the exhaust system on the offside, indicating that the car, by this time, was powered by a Standard Special engine).

winning the British Trials Lady Drivers' Championship.

1954: mud and Macau

This was a year of very mixed fortunes for Morgans, with the works drivers not gaining many major awards.

Private owner/drivers had some notable successes in hillclimbs, autocross and rally events. In particular, Jimmy Ray, who won his class in the RSAC Scottish Rally, received praise from the motoring press for the handling of his Morgan. He hoped to repeat his performance in the London Rally and make it a hat-trick of wins, but he became stuck in mud near one of the controls and lost too many points to be able to recover.

On the other side of the world, the first Macau Grand Prix was held on October 31 and it attracted the entry of a Plus 4 four-seater owned by 'Dinger' Bell, later the director of the Hong Kong Royal Observatory. He had shipped the car out from England and it now sported a TR2 cylinder head, twin SU carburettors and a specially-made aluminium bonnet. Dinger was first away from the Le Mans-type start, soon set the fastest race lap and led the field for five laps. Then disaster struck as he was beginning to lap some tailenders; the front nearside brake drum sheared and the whole unit, complete with wheel, came adrift, so ending the racing career of 'Betsy', as the car has always been known.

The year ended with the British Trials and Rally Drivers' Association annual dinner. A new prize, the Autosport Rally Navigators' Trophy, was presented to J C Dixon, Jimmy Ray's navigator, who won the Association's Gold Star rally competition (which excluded works drivers). Les Yarranton finished in third place. In the Silver Star competition, Dr J T Spare was first, Jimmy Ray fifth, Peter Reece seventh and K N Lee eighth, while Anne Neil was joint fourth in the Ladies' Silver Garter competition.

1955: success in America

This year's Exeter Trial saw Morgan drivers D L Lewis, H H Gould and Dr J T Spare win the team award and W A G (Jim) Goodall secure one of the only two triple awards won that season.

The RAC International Rally became a farce due to weather conditions. Competitors were repeatedly held up behind stuck cars, some losing up to two hours. Crashes were numerous and Peter Morgan had to abandon his car after one. In all, only three Morgans finished and none was among the prize winners.

In marked contrast, the Sebring

Morgan

12 Hour race in scorching Florida sunshine on March 13 included two Morgans amongst the 80 entries, both of which were amongst the 49 finishers; that of Rothschild and Kunz finishing third in class behind a pair of Arnolt-Bristols, with the Morgan of Weitz and MacKenzie narrowly failing to take fourth place.

Also in America, Gaston Andre took his Morgan to a class win at a race meeting at Beverley Airport, New England on July 4, and Mike Rothschild ran him a close second in the car which had gone so well at Sebring. Two months later, at Watkins Glen, the weather was so hot that the track had started to melt by the time the production sports car race - supporting the Grand Prix - got underway. However, Morgan stability was well demonstrated in the tricky conditions, and Gunnard Rubini and Gaston Andre finished first and third in their class, separated by an AC Ace.

Peter Reece won a 30-lap Production Car Race at the Daily Dispatch Gold Cup Meeting at Oulton Park in September. Having leapt into the lead at the start, he was overtaken by one Austin Healey on the first lap and by another on the second. He sat back, holding off the remainder of the field for five laps while a hectic battle developed ahead of him. Suddenly the two Austin Healeys touched and overturned, leaving Reece to go through and win by 38 seconds.

Back in America, Gunnard Rubini and Mike Rothschild continued to do well with their Plus 4s, finishing first and second overall in an inaugural race at Fairchild Airport, Maryland, while Gaston Andre was third in his class.

The year ended with Dr J T Spare winning his class in the MCC's National Rally, but the Morgan team just missed the team award when WAG (Jim) Goodall's car broke a halfshaft on the final test.

Dixon Smith using his TT replica, now powered by a Standard Special engine, in an Inter Club Driving Test meeting at Honeybourne in 1955.

1956: Suez crisis

After the criticisms of the previous year's event, the RAC Rally turned out to be a first-class affair. Dr J T Spare was placed third overall and won his class, while Les Yarranton managed third place in the same class.

This was to be the year of team awards for Morgans with wins in the Highland Three-Day Rally; Midland AC's Birmingham Post Rally; Morecambe National Rally; Worcestershire AC's Autocross and the MCC National Rally.

Another team award for Morgan was won in the Aston Martin OC's 21st Birthday race meeting in July. The most exciting event of the day was the 21-lap handicap relay race for the David Brown Trophy, with 24 teams of three cars entered. Every driver had to complete seven laps before handing over the sash to the next team member. Therefore, it was not too long before the pits became utter confusion. The team of Ray Meredith, Barry Phipps and John McKechnie, all in Plus 4s, emerged as overall winner.

This was the year of the Suez Crisis and on November 20 the RAC announced that no more motorsport permits would be issued until fuel supplies were re-established. On December 17, petrol rationing was introduced and motorsport resumed on a limited scale.

1957: explosive racing

There seems to have been plenty of fuel in the USA, and the Sebring 12 Hour race on March 23 was attended by 30,000 spectators, who saw the lone Morgan entry of Bob Grier and Bob Kennedy run like clockwork for 149 laps to be classified 28th. The 1.5-mile circuit at Lime Rock, Connecticut, was opened on April 28, when Bill Shade and his Plus 4 just failed to split the two AC Bristols leading the sports car race.

Back in England, Morgans competed well in race meetings up and down the country. In one race at Goodwood, John McKechnie had a race-long battle with an MGA, the latter benefiting from having disc brakes. The two cars entered the chicane side-by-side on the last lap, but it was the MGA that emerged on the right line, and McKechnie's Morgan exploded into view through the middle of the chicane wall in a shower of planks!

On the same day Peter Morgan and Jim Goodall competed at Silverstone in several races at the MCC race meeting. The best was a 10-lap handicap battle with an Austin Healey, which ended with Goodall just fractionally ahead of Peter.

On the rally side, it was a very mixed year for results for Morgans, the best of which came in the London Rally, with Pauline Mayman and Valerie Domleo winning the Coupe des Dames, and in the Guy Fawkes 200 Trial they gained a First Class award. F A Free-

Twenty four teams of three cars were entered in the Aston Martin 21st birthday race meeting at Silverstone in July 1956. One of the Morgan teams entered consisted of John Hooker (OWP 38), Alan Smith (KUY 714) and Robin Brown (19c), who are seen here just before the start. The spectator in the foreground with his hands in his pockets is Barrie Phipps, the well-known Morgan competition driver who was a member of the other Morgan team, which won the event.

man won the Guido Vase for the best performance on the test hills.

1958: the star and garter

This year the Pauline Mayman/Valerie Domleo combination won the Coupe des Dames in virtually every event they entered. It was therefore not surprising that Pauline was the runaway winner of the BTRDA Silver Garter and was placed third overall in the Gold Star table.

Morgans also did well in driving test events this year, with Peter Morgan the overall winner from an entry of 80 in the Taunton MC's annual driving tests and J F Livingstone finishing as runner-up in the Flather Star driving test competition series held throughout the season.

Alan Smith sets off from the pits on his seven laps before handing over to the last man in the same event.

In Chris Lawrence's first full season of racing in 1959 he was to win nine marque races and one handicap race. He also came second four times and third once and finished every race. Here he leads the field in the BRSCC Brands Hatch marque race on May 10, where he achieved his first win in a Morgan.

Lyndon Simms partnered by R E Stokes took a highly creditable second place in the 1601cc-2000cc GT class in his Plus 4 Drop Head Coupé, splitting two Triumph TR3As. The rally concluded with a race around the Zandvoort circuit, in which Simms had a race-long battle with Annie Soisbault in a TR3A. He held off her final challenge to win.

1959: newcomers' success

The season started with the Mayman/Domleo partnership looking as if it were to enjoy another year of domination. In the Scottish Sporting CC's Highland Rally they ran away with the ladies' award and finished well ahead of the majority of male competitors in this 1000-mile event. A week later they repeated their success in the Midland

Chris Lawrence laps an MGA on his way to victory in the BARC members meeting marque race at Goodwood on June 27 1959.

In 1961 Pip Arnold competed in the Coppa Inter-Europa race at Monza in XRX 1, which was fitted with an experimental Lawrence-tune hardtop that was eventually marketed to Morgan owners.

AC's Birmingham Post Rally. But shortly afterwards, while competing in a hillclimb at Prescott, Pauline Mayman crashed, sustaining injuries which kept her out of competition for several weeks. She was back in action again by August and had clearly not lost any of her earlier form. Between August and October the Mayman/Domleo duo won four Coupe des Dames in the Plus 4.

Chris Lawrence's first full season of racing a Morgan proved to be spectacularly successful. He took part in virtually every marque race and most of his wins were runaway successes. His final tally was: wins in nine marque races and one handicap race; four second places and one third, a record which made him a comfortable winner of the Freddie Dixon Trophy.

On the RAC Rally, Peter Morgan earned the title 'King of the Tests', having won his race and losing only two marks from all the other tests. The final positions were not decided for several months due to a protest lodged by the potential winner, Wolfgang Levy in a DKW, who also stood to gain second place overall in the European Championship. When the results were eventually announced - the protest having been turned down - Peter Morgan was class winner and sixth overall.

1960: enter the big works teams

Lyndon Simms with Roger Jones started from Glasgow in the Monte Carlo Rally in a Plus 4. It was not a good event for them and, although they reached Monte Carlo, out of an original 311 starters they only managed to finish in 125th place.

The 1960 'Monte' is on record as being the first in which the works teams really came to the fore, taking the first three places overall as well as the team prize and signalling the start of an era during which highly-financed works teams dominated rallying at international level. The costs involved were huge squeezing out the smaller concerns - the Morgan Motor Company amongst them.

Undaunted, Lyndon Simms, this time partnered by Barry Hercock, once again competed in Europe in the Tulip Rally. They finished 14th overall but not without an alarming incident along the way. After leaving Eindhoven they were on a wide level crossing when suddenly the gate dropped down, trapping them. They looked around to see an express train approaching, so Lyndon edged the hardtop Plus 4 as far off the tracks as possible and waited anxiously. After what must have seemed an eternity the express rushed by and went on its way, having missed the back of the Morgan by about two feet!

On the other side of the world, the Warwick Farm race circuit was opened at Sydney, Australia, on December 18, and the very first race was won easily by David McKay driving a Plus 4 prepared by Ken Ward who, with his brother had just formed a Morgan distributorship and also founded the Morgan Owners' Club of Australia.

1961: too 'old fashioned'

The beginning of the year provided little reward for Morgan drivers, the best result being that of Brian Harper who was first overall in the *Express & Star* Rally, but then only after a protest.

Later in the year three Morgans, driven by Richard Shepherd-Barron, Chris Lawrence and Peter Marten, were entered for the Peco Trophy race at the BRSCC Guards Trophy race meeting at Brands Hatch on August Bank Holiday Monday. Although Marten's car ran its bearings on the second lap, the other two Morgans went extremely well, Shepherd-Barron's winning its class at 77.09mph and setting a new class record at 78.45mph, while Lawrence kept ahead of an AC-Bristol to finish in second place.

In September, Chris Lawrence, who earlier in the year had had his application to race at Le Mans turned down because his car was *'too old-fashioned'*, travelled to Monza with Pip Arnold for the Coppa Inter Europa. The 3-hour race was held prior to the Italian Grand Prix and the Morgan drivers finished a very impressive third in class with their 'old' car, and might have been even higher had they and the car not been so troubled by the extremely hot weather.

Brian Harper continued to perform well in rallies throughout the season and was declared outright winner of the BTRDA Silver Star rally award.

1962: 93-97mph for a day

The international season got off to an encouraging start for Morgan when a Plus 4 shared by A Rogers and J Bailey

Peter Morgan, with Mrs Jane Morgan as passenger, strives to maintain adhesion and motion in his 4/4 in the MMEC/Shenstone/Morgan Clubs Production Car Trial in March 1963.

won the 2000cc GT class at the Sebring 12 Hour race.

Pip Arnold, Richard Shepherd-Barron and Hugh Braithwaite entered the Spa Grand Prix, contesting the 1301-2000cc class; Arnold finished in fifth place. The following day, Shepherd-Barron finished an excellent second to a Porsche. This was with a replacement crank from an old TR2 engine lying in a village about 100 miles away from the circuit, his own having broken in one of the practice sessions.

Braithwaite was lucky to walk away unscathed from a high-speed crash at Spa, which severely damaged his car. He teamed up with Pip Arnold for the Nürburgring 1000 Kilometre race, while Shepherd-Barron and Lawrence shared their Le Mans car and the Savoye brothers entered a third Morgan. This was to prove the race of starter motor troubles and both the Lawrence-tuned cars were afflicted; between them, they exchanged motors 12 times during the race. The Le Mans car succumbed with engine trouble on the 25th lap after being timed at 10min 34.9sec for a lap and, at the end of the day, the only thing to celebrate was third in class for the Savoye car.

The most important race of the year, of course, was the Le Mans 24 Hours, on June 23-24 which, as usual, had been preceded by a special practice weekend in April, at which, on the only dry day, the Chris Lawrence/Richard Shepherd-Barron Morgan had lapped in 4min 56sec, making it the eighth fastest of the cars present.

During practice on race week the Morgan drivers were more concerned with track familiarization and minor adjustments than in setting any records, and when the time came to line up for the 4pm start on the Saturday, both drivers were quietly confident that their 'old-fashioned' car would do well. Theirs was one of 16 British cars amongst the 55 starters, and throughout the race they were never forced into an unscheduled pit stop. Their regular stops were made every three hours to change drivers, refuel and top up with oil, and their only drama came when an ex-

Cyril Charlesworth tackles Tillerton in the 1964 MCC Exeter Trial.

haust pipe broke close to the manifold. An attempt was made to secure the pipe during the 7am pit stop, but rather than waste a lot of time the car was sent back into the race sounding more like a Grand Prix car than a production sports car as it continued to lap at close to 100mph. For every Morgan enthusiast present that year the last hour of the race was a nail-biting time, but the Morgan was still running well and it crossed the line in 13th place (the same position as the Fawcett/White car had managed in 1938) amongst only 18 finishers. Most important of all, it had won the 2-litre class at an average speed of 93.97mph, having covered over 2255 miles in the 24 hours. This was the most important Le Mans result in the history of Morgan.

Back home, Morgans continued to do well in all forms of the sport, but 1962 was the first year when government legislation all but finished road rallying, because of a public outcry against night driving at high speeds and the continuing disturbances such events caused.

1963: Spa(rring) with the opposition

As competitors and spectators streamed away from Brands Hatch on Boxing Day 1962, a blizzard began in south-east England, heralding the start of the 'Big Freeze', which was to last almost to the end of March. Inevitably, many motor sporting events were cancelled, but as Britain froze, in Sydney, Australia, drivers were suffering from heat exhaustion. This was the main hazard at the Australian Grand Prix meeting at Warwick Farm in February when A J Reynolds was placed second in his Plus 4 in the marque race supporting the main event.

Once again, Chris Lawrence and Pip Arnold with, for the first time, Bill Blydenstein, sallied forth into Europe. At Spa, the Grand Prix became a marque championship event with the three Morgans entered in the 2001-2500cc class, which they dominated. Lawrence and Arnold finished first and second respectively, Lawrence having averaged 101.69mph for the 36-lap, 315-mile race, and set a fastest lap time of 105.14mph. Blydenstein failed to finish.

Arnold and Blydenstein were joined by Hugh Braithwaite when they travelled to the Nürburgring for the 1000 kilometre race. Despite many mishaps and breakdowns in practice and in the early part of the race, Braithwaite won his class with Arnold runner-up.

In the tiny Republic of Panama the major motorsport event was the 100-mile race on the Chane circuit. Here, BARC member Pat Kennett, the service manager for the local Morgan agency, entered a Plus 4 Super Sports powered by a Triumph TR4 engine. Kennett got away well and soon built up a useful lead, but the very hot weather was taking its toll on cars and drivers, as was the track condition. After 35 laps, Kennett made a stop to check tyre wear and to pick up a drinking bottle. Forced to slow in the later stages of the race due to strange noises emanating from the rear wheels, he judged his pace carefully to finish 38 seconds ahead of the field.

1964: road racing

The start of the year in England saw Peter Morgan driving the Plus 4 Plus perform well in the Exeter Trial.

On the continent, Paris Morgan main agents, Savoye, entered a Morgan in the Monte Carlo Rally. (The regulations having been changed once again so that Morgans were eligible for entry). The car was driven by Claude Savoye and his brother-in-law Etienne Girard. In spite of the extreme cold, in true

35
B9

Club racing is obviously not confined to the UK. Here, members of the Swedish Morgan Owners Group compete at Skarpnack Airfield in 1966. A 1965 Plus 4 Super Sports sorts out the opposition.

Morgan tradition the hood remained firmly down for the whole event. The car and crew performed very well and finished second in the 1300-1600cc class.

Back home, production car trials, circuit racing and hillclimbs now became the battle ground for the private Morgan owners, with such drivers as Harry Rose, Amie Lefevre and Ray Meredith doing very well.

1964 witnessed the inauguration of the American Road Race of Champions. The Sports Car Club of America stage races throughout the year in six regions from which emerge regional champions in many classes, who then go forward to compete in this final event for the national championships. With classes C and D being combined into one race, Earl Jones' Plus 4 Super Sports was up against such cars as Cooper-Chevrolets and big-engined Specials, so an overall win was out of the question, but Jones set about his task very well to emerge production car champion in Class C. The entries in class G included a 4/4 entered by 24 year-old Pat Menone of Washington DC - the only woman driver to have won through from the regions - who took the big race seriously enough to arrive at Riverside several days early and take a short course at the Carroll Shelby Racing Drivers' School operated from the track. However, it was all to no avail as mechanical trouble put her out on the fifth lap.

1965: going the distance

Chris Lawrence entered one of his SLRs in the Guards 1000 race at Brands Hatch - the longest race staged in Britain since the war - which was divided into two 500-mile races. After an American-style rolling start from behind a pace car, the lead was taken by a Sunbeam

Tiger, but as the race reached Bottom Straight the SLR shot ahead and led for 66 laps, averaging 77.18mph. As the car was fitted with long range tanks, it looked every inch a winner. But this was not to be, for on the 67th lap Lawrence pulled into the pits with a broken kingpin post. The repair cost 21 laps before he could rejoin the race, only to be further delayed by brake trouble. The next day, starting from the back of the field, Lawrence worked his way up through the field and into the lead when the leading car, a Jaguar E-Type, broke a camshaft. From there on they were never challenged and completed the 189 laps of the second heat in 6hr 36min 20sec, 64 seconds ahead of the second-placed car.

Alan House continued to perform well with his Plus 4. In the final of the Autosport Championship at Snetterton he succeeded in winning his class and in so doing became 1965 Autosport Champion in Class B.

1966

At the Exeter Trial held at the beginning of January, Amie Lefevre won a First Class award and one of the MCC's coveted Gold triple awards.

Chris Lawrence's SLR competed in the BARC Goodwood meeting in May. In the marque race he was joined by the Triumph-chassised SLR of Neil Dangerfield. Lawrence's car enjoyed an easy victory, finishing over 300 yards ahead of the second-placed car. Dangerfield's car lost power halfway through the race and finished well down the field.

Shortly after this race, a letter appeared in *Autosport* protesting about the SLR's eligibility for marque races. However, as Neil Dangerfield pointed out in his reply, Chris Lawrence was in business to make Morgans go faster and this he was doing very successfully.

Amie Lefevre rounded off a most competitive year by taking a First Class award in October in the S Rodney Whiteley Trophy Trial, and another in the Worcestershire Production Car Trial at the end of November.

1967: a varied success rate

In the Exeter Trial Peter Morgan chalked up another class win, this time with a Plus 4 Plus.

The special Morgan-only five-lap handicap race at the Bentley DC's Silverstone meeting in August attracted 23 entries. The winner was Ray Meredith, whose TR4-engined Plus 4 averaged 74.95mph and finished ahead of the Vanguard-engined Plus 4 of Harvey Postlethwaite - later to become well known as a designer of Formula One racing cars. Ken Davies was third in his Ford-powered Series 5 4/4, while G Griffiths' Plus 4 was a class winner. Although they were not amongst the winners, two of the more interesting cars were Andy Duncan's Lotus-Ford-engined 4/4 and the Hon Basil Fielding's Daimler SP250-engined Plus 4.

Alan House could only finish fifth overall and second in his class in the final Amasco Championship race at Brands Hatch on Boxing Day, but with four class wins and one second place in five races he was a worthy 1967 Champion in the 3000cc class.

So ended another varied year of Morgan successes in motorsport. Although the cars were no longer fully competitive in certain areas in which they had shone in the past, Morgans continued to do well in selected events, notably marque racing, sprints and hillclimbs, against cars of much younger pedigrees.

1968: triple effrontery

When the Daytona Continental 24 Hour race came to an end the afternoon of February 4, the applause for the winning Porsche 907 was little more than that which erupted when a Morgan Plus 4 crossed the finishing line at the other end of the field. George Waltman had had the effrontery to enter the Morgan in the GT class at Daytona, where its main rivals were 7.3-litre Chevrolet Corvette Stingrays. He caused quite a stir as, wearing regulation racing kit but with a red scarf flying in the wind, he plodded on steadily as one-by-one faster cars dropped out, and was eventually classified last of the 30 surviors out of 63 entries. The winning Porsche had covered 673 laps at an average speed of 106.97mph, whereas the Morgan completed just 339 laps at 53.69mph. Nevertheless, bearing in mind that the little car was being lapped every other lap, it was a remarkable piece of endurance driving.

In a very different environment, David Way set a new class record of 37.69sec with his 4/4 to win his class in the BARC South Wales Centre's hillclimb at Pontypool Park on September 29. To round off another interesting year for Morgan drivers, the MCC announced that its coveted Triple Baddeley award, which is only presented to a driver who has cleared every section in the classic Lands End, Exeter and Edinburgh Trials, had been won by Roger Bricknall and his Plus 4.

Morgan

III

4-4s

Flat Radiator 4/4s (1936-1951)
Coventry Climax 1122cc engines.
(March 1936-September 1939)

By the early 1930s HFS realized that if his company was to survive he must alter his thinking. The warning signs were there in the failure of so many of Morgan's three-wheel manufacturing competitors. He had to move to four-wheelers.

A very basic prototype was constructed using the Z-section chassis, and the Ford 8hp, 993cc side-valve engine of the then new F-type three-wheeler. The front half of the car was really nothing more than a F-type roll-

The experimental four-wheeled Morgan was based on an F-type three-wheeler chassis, as comparison of these 'before and after' photographs shows.

4-4 chassis no 2 was extensively tested at Brooklands. It is shown here at the track in July 1935.

ing chassis, complete to, and including, the windscreen. From here, the chassis rails continued parallel and the chassis was fitted with a two-wheeled rear axle. Two bucket seats were bolted to the chassis. Each rear wheel was covered by a piece of flat steel plate, very slightly bent to follow the wheel-line mounted on outriggers acting as rear wings. Assigned chassis number 1, this experimental car was never registered for the road and was always driven on the company's trade plates.

Experiments with this test-bed showed that the Ford engine would not be powerful enough for use in a production car, especially with the extra weight when the car was fully bodied and fitted with twin spare wheels. HFS looked around for an alternative engine of comparable size and decided to experiment with a Coventry Climax power unit. Thousands of these engines had been produced in various forms and had a good reputation for durability, having been used in such cars as the Crossley Regis and Triumph Gloria. Two more experimental cars were built, with full roadgoing bodywork, and were subjected to exhaustive road and track trials throughout 1935. In July of that year a member of the staff of *The Light Car & Cyclecar* magazine was invited to test drive the car with chassis number 3 at the Brooklands race track. The test car was driven at speeds *'in excess of 70 mph'*, but the report was not published until Christmas 1935, when the model was officially announced. However, the first production car was not delivered until March 1936.

In spite of its complicated layout of overhead inlet and side exhaust valves and thermo-syphon cooling. the 1122cc version of the Coventry Climax engine was found to have a much better performance than the Ford and was far more in keeping with the sporting image that the marque had built up over the years.

The new model retained the well-tried and proven Morgan front suspension as used on the three-wheeler. The Z-section chassis frame was arranged so that a light but strong plywood floor could be supported on the inward-facing bottom flanges. The floor extended from the bulkhead to just in front of the conventional live rear axle, with nothing protuding below it, which meant that the ground clearance could be kept to a minimum (6in) and the car's centre of gravity as low as possible. The exhaust system passed through the side wall of the chassis and along the outside of it; in this way it was protected from any rough ground and, at the same time, passengers were shielded from it by the running board.

The top flange of the side members was used as the platform for mounting the bodywork which consisted of a wooden ash frame on which were fixed steel panels. The tops of the two rear-hinged doors were cut away to allow the driver and passenger extra external elbow room. The overall width was just 41 inches.

The seating consisted of a bench seat back with inflatable, individual seat cushions. Between them was a hump covering the end of the gearbox which was connected to the engine via a short shaft to the flywheel and Borg & Beck clutch.

The windscreen, which could be folded flat, was fitted with a single wiper driven by an electric motor. It was not until April 1937 that two wipers linked by a tandem bar were fitted. Aeroscreens could be supplied as an option.

In the luggage space behind the seats was stored the canvas hood, the supporting framework and sidescreens, leaving just enough room for a small bag. Behind that, the fuel tank led to an elegant sloping tail that followed the contours of the rear wings which, in turn, housed twin spare wheels mounted vertically.

The 16 inch wheels were fitted with 5 inch wide tyres. On the first two prototype cars wire wheels were fitted, but the road-going prototype had disc wheels. These were changed to 'spoked-disc' wheels (Easyclean) around July/August 1937. By mid-1938 disc wheels were only being fitted to order and by early 1939 were only used on the Drop Head Coupés.

Instrumentation was set at the centre of a polished wood dashboard, which had a small cubbyhole on each side. In the middle a 90mph speedometer containing an odometer and a tripmeter was sited, with a small mechanical clock in the centre bottom of the face, Below this was a large horn push with the various push/pull buttons for controlling the panel lights, fog, lamp, choke and starter arranged along the bottom of the panel. To the left were ammeter and oil pressure gauges in a single unit and below this

Publicity photo of the definitive first series 4-4 with the Easyclean spoked-disc wheels that became standard fitting for the two-seater from 1937.

All pre-war 4-4s were fitted with black-faced instruments and a Brooklands steering wheel. Instruments are, left to right: ammeter/oil pressure; ignition switch/lights/charging; speedometer/clock; fuel/temperature. The large button beneath the speedometer is the horn push.

Lucas ignition and lighting switch. To the right the fuel and temperature gauges shared a single unit. The steering wheel was a 16 inch bakelite-rimmed four-spoke affair.

The resulting overall appearance of the new car was one of compact sporting elegance. Not only was it pleasing to the eye, it was also a car which was to prove itself, in true Morgan tradition, to be as good in competition as on the road. It was never intended as a replacement for the three-wheeler, but rather a supplement to it. By 1937 demand was already outstripping the company's production capacity. The four-wheeler was a success and Morgan sought to build on this by introducing additional models.

To meet the popular demand for a family sports car, the first production 4-4 four-seater left the works in December 1937. The extra space necessary was achieved without lengthening the chassis: the fuel tank was repositioned and the two 6-volt batteries mounted in the space occupied by the forward of the two spare wheels. A 24.5 inch wide bench rear seat was fitted and an 18 inch back rest positioned over the rear axle.

The designation '4-4' had been decided upon because the car had four

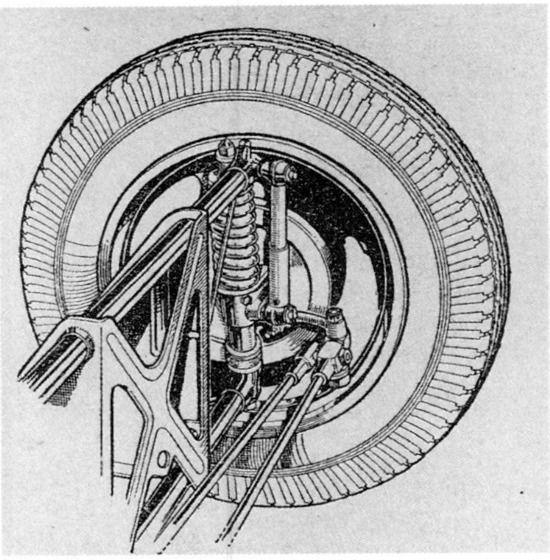

Morgan four-wheeler front suspension is the familiar sliding pillar type, but the springs are inclined slightly rearwards to provide better castor action.

wheels and four cylinders. This simple formula has been used for model names for all four-wheeled Morgans since. However, after the war, the company, for some obscure reason, changed the actual style of the name from 4-4 to 4/4, which has been the norm ever since.

Flat radiator 4-4s or 4/4s are often nowadays referred to as Series I, but they are strictly speaking just 4-4s or 4/4s. Coincidence or not, the first production car was chassis number 4, and the first one to be exported was chassis number 44, which went to Stewart Sandford in Paris - this was also the first 4-4 to be painted a colour other than green.

Drop Head Coupés

The first experimental Drop Head Coupé was powered by the prototype Standard Special engine, which later replaced the Coventry Climax unit.

As Morgan had no experience of making this style of body, it was decided to send a rolling two-seater chassis to the Avon Coach Building Company at Wall Street, Warwick. The body they constructed was so attractive that HFS immediately took the car for his own personal transport. Later it was used by George Goodall and became known at the factory as *'Uncle George's Winter Carriage'*.

Using this design as a basis, the company produced their own, even more attractive, bodywork. It had 26 inch wide rear-hinged doors with no cutaway at the top, elbow room being obtained by recesses cut in the inner panel of the doors. Detachable windows were fitted, consisting of two panes which slid in either direction sideways in a metal frame, so permitting front or

Two views of the first Drop Head Coupé bodied by the Avon Coach Building Company which was used as the basis of the Morgan factory's production car. Note the spare wheel cover with petrol filler cap at its centre.

The first factory produced Drop Head Coupé was this 1939 car rallied by George Goodall. It now lives in the United States.

The works 4-4 two-seater that was loaned to Prudence Fawcett for the 1938 Le Mans 24-hour race, photographed at Madresfield Court with a friend of George Goodall at the wheel. The car's bodywork was extensively modified before the race.

rear ventilation. Further ventilation when the hood was erected was provided by opening flaps fitted in the side walls of the scuttle.

The three-position hood could be used fully closed or fully open, or rolled back and fastened immediately behind the occupants' heads in the 'Coupe de Ville' style. Other features were wider section 16 inch steel disc wheels; slatted radiator grille; leather upholstery; wood-capped interior; chromed body trim; twin upright spare wheels and optional two-tone paintwork. It made a very acceptable alternative version of the 4-4. A few were fitted with the smaller but more powerful 1098cc Coventry Climax competition engine and, later, the 1267cc Standard Special unit; these won many competition honours, both as works and private entries.

1098cc competition engines

For some years the Triumph Motor Company had been producing Coventry Climax engines under licence for their Gloria and Southern Cross models and also for the Crossley Regis. These engines were rated at only 1087cc and the competition version had slightly increased capacity. Coventry Climax also wanted to add a similar competition engine to their range, but they were too small to produce them in volume and therefore arranged for Triumph to produce them under licence. But Triumph did not really have the manufacturing capacity or manpower to produce a full competition-tuned engine in large quantities either, so they compromised by producing some bolt-on accessories to further improve the engine's performance.

The basis of the competition engine was still the inlet-over-exhaust valve unit that was already being supplied to Morgan. By under-boring the engine to 63.2mm, a 1098cc displacement was achieved, thus bringing the engine into the very popular international under 1100cc competition class. The engines were balanced throughout, including the flywheel, but did not have polished ports, etc. These partially-tuned competition engines, delivering 46bhp at 4500rpm, were fitted as an option by Morgan to standard-chassis cars with normal bodywork, including four-seaters and Drop Head Coupés, for those owners who wanted the extra performance.

The bolt-on extras which improved the performance even more were supplied by Triumph only to order. They consisted of a four-branch internal and external tapering exhaust manifold which produced a venturi effect. This, in turn, was fitted directly into a Burgess straight-through silencer. Carburation was either by Solex or Zenith, with a larger than normal choke tube. Adding these extras increased the power to about 54bhp at 4500rpm.

A small number of these engines were tuned further by the Morgan factory for specially-ordered cars. These had polished cylinder head ports and were fitted with domed pistons, plus a camshaft of a different profile.

Any Morgan which was to be fitted with either of these latter engines had to have its bodywork altered to accommodate the exhaust manifold. This consisted of removing the normal swept front wings and running boards and replacing them with cycle-type front wings, which were fixed and did not turn with the front wheels, unlike many British cars of the period. Also, the nearside of the bonnet was altered with a cutaway for the exhaust downpipe.

Le Mans Specials - TT Replicas and Le Mans Replicas (1098cc Coventry Climax engine - complete competition or fully tuned)

A certain Miss Prudence Fawcett decided she would like to compete in the 1938 Le Mans 24 Hour race. After long discussion with friends of the family, namely Rivers Fletcher and Lancelot Prideaux-Brune (who owned Winter Garden Garages, one of Morgan's main distributors), it was decided to

The body alterations to the first Le Mans Morgan are clear from this pit shot: twin fuel tank fillers, aero-screens behind the main windscreen and cycle-type wings.

Right: 'Competition success sells cars'. The factory continued its Three-Wheeler advertising practice with the Four-Wheeler.

approach several different British motor manufacturers to see if any would be willing to loan a car. At least three replied that they would. It was decided to accept the Morgan offer, which seemed to have the best chance of success because the 1098cc engine had the best potential for extra tuning. A perfectly standard two-seater was fitted with a 1098cc engine and was collected from the works, along with the set of bolt-on extras, in March or April. (At this time the 1098cc engine had not as yet been fitted to a production car, although an experimental one had been used in the works-entered 4-4 in the 1937 RAC Tourist Trophy race, held at Donington Park).

When the car arrived at Winter Garden Garages in Holborn, London, it was completely stripped down and reconstructed to reduce weight as far as possible within safety limits set by the Morgan works. The alterations included the removal of the front wings and running boards and their replacement with cycle-type wings, as previously described. The rear panel which housed the two upright spare wheels was removed from immediately below the luggage compartment lip where the tonneau cover was fitted. The nine-gallon fuel tank was replaced by a 24-gallon tank which was shaped to fit the space above and behind the rear axle. Two quick-fill, racing filler pipes and caps were placed through the tonneau cover where the hood and sidescreen storage compartment was normally located. The rear panel was replaced with a solid flat panel and positioned so that it followed the sweep of the wings. Two spare wheels were mounted one on top of the other, on top of the panel and at the same angle. Only one spare was carried in competition.

The engine was also completely stripped, and race-tuned by the garages' chief racing mechanic, Dick Anthony, who was to be Miss Fawcett's co-driver in the race. In addition, 17 inch spoked disc wheels replaced the normal 16 inch versions. The gearbox and rear axle ratios were also changed. Chronometric instruments, including a tachometer, were fitted.

The car was tested several times at the Brooklands track before being driven to Le Mans. The race entry was a completely independent venture which received no other factory support. After the race the car was eventually returned to the works and was then entered in the Tourist Trophy but failed to finish, due to overheating caused by a leaking radiator. In interviews that I have had with both Miss Fawcett and Dick Anthony, they have stated that the same car was used in both the 1938 and 1939 Le Mans and TT races. Photographic evidence has recently come to light appearing to reveal that unless two registration numbers were used on the same car, they were in fact two different cars. However, since the works had a history of moving registration numbers from car to car on three-wheelers, and there is no reason to suppose the four-wheelers were any different, I feel it is still quite possible that only one car was used.

In the 1939 Le Mans 24 Hours, the car had the spare wheel lowered into a recess in the sloping tail. One or two other works competition cars were rebodied in the same style, but not necessarily fitted with the competition engine, with the bolt-on extras or the 'fully-tuned' engine. This competition car gave rise to three very distinct and different limited production models, although only one of them was ever officially announced.

The first of the three was produced immediately after this latest race success: the Le Mans Special, as designated by the factory. This car had a normal production body from the scuttle back, including the twin upright spare wheels. In front of the scuttle it was fitted with cycle-type front wings and powered by the 1098cc engine fitted with the four-branch exhaust manifold. In all, seven of these Specials were produced between June 1938 and September 1939; six were two-seaters and one a four-seater.

The second was known as a 'TT Replica', although it was never officially designated or announced as a production model. This had the sloping rear panel with a single spare wheel mounted

on top of it following the same sweep; cycle-type front wings; full competition engine; a nine-gallon fuel tank with twin filler caps in the sloping back panel immediately above the spare wheel and a normal cover over the luggage compartment. This was the same configuration as the 1937 Le Mans and Tourist Trophy race car. In all, four cars were made or converted to this body design.

The third was officially announced in February 1939 as the 'Le Mans Replica' and was priced at £250. This model had the spare wheels mounted so that the bottom one was partly inserted in a recess cut in the rear sloping tail (this being the same as the modification made to the 1939 Le Mans race car). The Le Mans Replica was fitted with the 'fully-tuned' 1098cc competition engine, four-branch exhaust and cycle-type front wings.

One of these cars had been specially built as the back-up car for the 1939 Le Mans race but, for various

The only known TT Replica 4-4 which, today, lives in America, is the ex-works car used by Peter Morgan in competition.

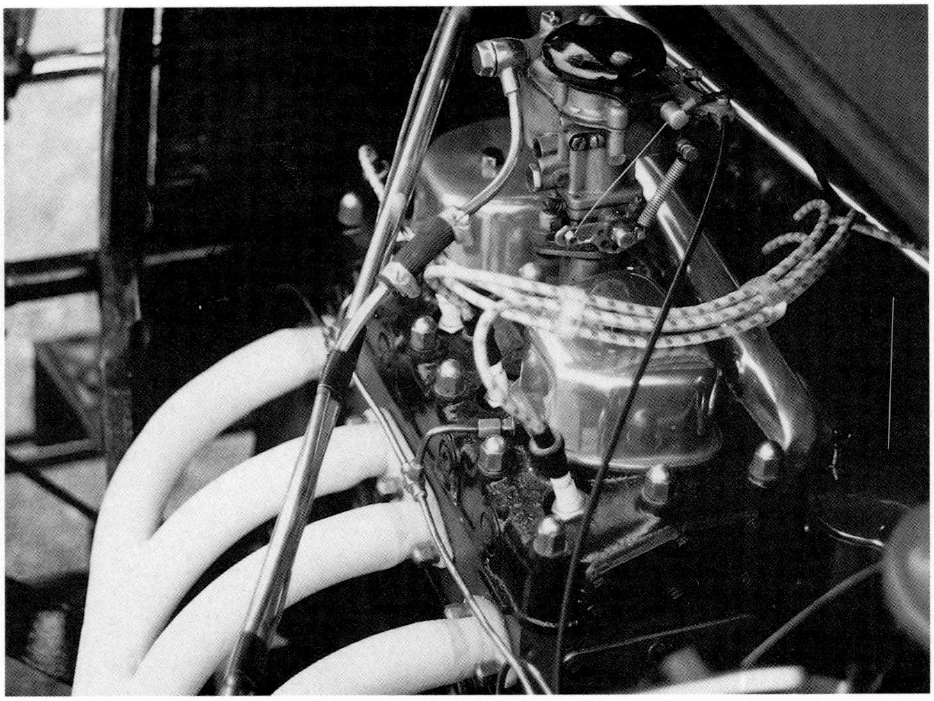

These pictures of the author's Le Mans Replica show the model's characteristics; also spare wheel, dual filler caps and the 'Fully Tuned' engine with its four-branch exhaust.

reasons, did not actually attend. Modifications included the engine being bored out to 1104cc and the storage compartment tonneau cover replaced by a hinged metal lid, which provided a small boot with a removable floor to give access to the rear axle. This hinged lid followed the curve of the wings and back panel from immediately behind the seats, giving a more streamlined appearance. A nine-gallon fuel tank and twin fillers were fitted, as for the TT Replicas. The car used for press photographs at the official announcement was the specially-constructed Le Mans back-up.

The 1939 Le Mans car and the back-up car were entered in the Tourist Trophy again, but with the imminent onset of war the race was cancelled. According to Miss Fawcett, the works accepted an offer for the 1938 race car and it went to America; it has not been traced since.

It had been intended to use the back-up car as a works competition and road test car but when war was declared, the two cars were placed in store along with two rolling chassis fitted with 1098cc competition engines and two or three of the works competition cars, including those of HFS and Peter Morgan.

After the war the two finished cars were overhauled and repainted. Morgan had planned to use them as competition cars, but the back-up car was sold by mistake in 1946 (they attempted, unsuccessfully, to buy it back a week after its sale). Because of this, and the fact that Coventry Climax would not be resuming production of the prewar engine, it was decided to sell the second car as well.

The other two chassis were bodied and sold in 1947, one as a Le Mans Replica and the other as a normal two-seater 4-4, which was exported to France. All four were allocated post-war chassis numbers, to correspond with the sales dates and, with the exception of the export car, recorded as Le Mans Replicas.

After nearly 20 years of extensive research, I have recently been able to establish exactly how these special 4-4s were designated and the total production figures for each example. Comparing other known examples and recent photographic evidence from the private collections of Miss Prudence Fawcett and other prominent Morgan celebrities, plus many hours of studying the factory records, I can clarify the production details of these special models as follows:

Seven **Le Mans Specials**, six of which were fitted with twin upright spare wheels, 1098cc engines, four-branch exhausts and cycle-type front wings. One four-seater with the normal single upright spare wheel was produced to the same specifications. These were not fitted with the 'fully-tuned' engine.

Four **TT Replicas**, with the spare wheel mounted on top of the rear panel, 1098cc engine, four-branch exhaust, cycle-type front wings. This number includes the actual race car and those converted to this specification. All were fitted with the 'fully-tuned' engine.

Three **Le Mans Replicas** with sloping rear and inset spare wheel. Fitted with 1098cc (1104cc back-up car), 'fully-tuned' engine, four-branch exhaust, cycle-type front wings.
Note: two or three works trials cars were rebodied in the Le Mans Replica style, including the 1939 race car, but were never production examples of that model.

Experimental 4-4s

In 1937 Morgan experimented with supercharging a 4-4. The standard Coventry Climax engine was a Roots-type blower by Centric. Peter Morgan remembers: 'After preliminary testing on short journeys around Malvern, my father (HFS) said I could drive it back to our home at Cannon Hill. Everything went well and the car appeared considerably faster than the standard

This is the only four-seater Le Mans Special, recently rediscovered. Basically, it's a regular-bodied four-seater with cycle-type wings and the 1098cc competition engine.

was a closely guarded secret - even though it was tested over quite a long period - for externally it looked very little different from any other 4-4, except that it was finished in battleship grey primer.

Peter Morgan, who drove the car extensively, recalls it was very successful in tests and that he was delighted with the ease with which he could 'sort out' other cars, whose drivers thought they had the measure of any 4-4.

The Morgan V8 might have found its way into the Company's prewar catalogue, but its development was halted by the Government's decision to raise the road fund tax.

The 1122cc Coventry Climax engine was the mainstay of pre-war 4-4 production. It is shown here fitted to a 1938 four-seater.

4-4 and on getting on the Oxford by-pass I decided to have a go, but after getting up to 85-90mph a main bearing failed. It is possible that the boost was too high, I think it was about 8-10psi, but in any case the crankshaft had a tendency to whip on the standard engine so my father abandoned supercharging the 1122cc Climax engine. The 1098cc engine would have been more suited to this because of its fully balanced crankshaft, but the factory never 'blew' the Le Mans or TT cars because of the class structure in racing and the higher performance that would be needed.'

The overhead valve gear of the Coventry-Climax. A cup is fitted to the top of each pushrod to hold the lubricant.

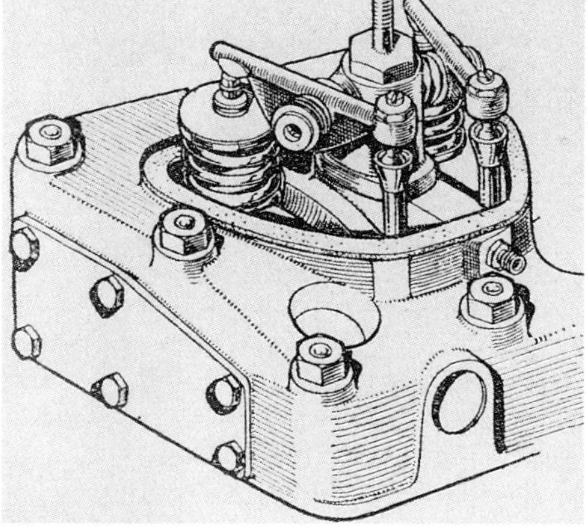

V8 two-seater

A second experimental car was commissioned in April 1937 which was equiped with the 2228cc Ford V8 '60' sidevalve engine. The car's existence

This short wheelbase trials car was campaigned with success by Peter Morgan. It started with a Ford engine but for most of its life was powered by a Standard Special.

Supercharged 4-4

After the V8 engine had been removed from the chassis the car was fitted with a Ford four-cylinder sidevalve engine. The installation, which was carried out in collaboration with Carburettors Limited, involved fitting the engine with a belt-driven Arnott supercharger which operated at the relatively low pressure of 4-5psi. The blower, which had its own oil supply from a small tank located between the engine and the scuttle floorboard, was mounted high up at the front of the special high compression cylinder head. The cylinder dimensions had also been altered, raising the displacement from 933cc to 1021cc.

No performance figures for this experimental car seem to have survived, but having driven a Le Mans for several years that was fitted with an identical supercharger, I am sure they were impressive.

This car is reported to have been used in two rallies - although no record of this can be found. Unfortunately, World War II brought its development to an abrupt halt and, although at one time there was talk of it being put into production after the war ended, this did not happen.

There is no doubt that low pressure supercharging has certain attractions; I have found that markedly improved acceleration and top speeds are obtainable at not unreasonable increases in fuel consumption - 4-5mpg overall has been my experience. Peter Morgan says 'This project was undertaken by my father to achieve an economy sports car with a really good performance, but the war prevented it from being productionised. In any case, we had started using the ohv pushrod Standard Special 1267cc engine which was used for all four-wheeler models after the war'.

Standard Special engine 1267cc (May 1938-February 1951)

At the beginning of 1938 it was becoming evident that the Triumph Motor Company was getting into serious financial difficulties, and the Morgan factory could see that it might not be too long before its supply of Coventry Climax engines would come to an end. (Triumph actually went bankrupt in June 1939). With this possibility in mind, HFS approached his old friend, John Black, who was by then head of the Standard Motor Company, to see if he could supply a suitable engine. (It had been John Black who, as a bright young man in 1909, had actually produced the patent drawings for the Morgan Runabout for HFS).

The Standard Motor Company was

Post-war 4/4s used the 1267cc overhead valve Standard Special engine specially designed and produced for Morgan by the Standard Motor Company. Production continued until 1951.

The neat little four-cylinder Standard Special overhead valve engine was based on a 10hp sidevalve Standard engine.

Below: This advertisement, which appeared in The Autocar *in December 1939, reflects the optimism of the British people that the war would be over in a few months.*

already producing sports car engines for Jaguars, although its own cars were anything but sporting. Sir John Black (as he was to become) agreed to design and produce an overhead-valve version of his 1267cc, 10hp sidevalve engine just for Morgan. The resulting engine produced more power than the 1122cc Coventry Climax and without moving it into a higher taxation class. (In those days the amount of yearly road tax was determined by a complicated formula based on the horsepower rating of the engine).

The first experimental Standard engine was installed into the first Drop Head Coupé in about May 1938, with the second and third test engines being fitted in two works trials cars, a two-seater and another Drop Head Coupé, in December the same year. The first production car fitted with the 1267cc engine left the works in July 1939. In all, 38 cars were fitted with this engine before war commenced and the Morgan factory was turned over to the manufacture of parts for the war effort.

No Coventry Climax engines as supplied to Morgan were produced after the war. With the exception of the four cars stored throughout the conflict, all postwar production of the 4/4 was with Standard Special engines.

Because of the postwar restrictions on materials available for the motor industry, production was very erratic, with orders far outstripping the number of cars that could be produced. When materials were available, production was permitted on the strict understanding that at least 90 per cent of output was exported. Supplies of sheet steel were so short that a few of the first postwar production cars were sent away to outside companies for panelling. These were companies who received an allocation of sheet metal, but did not have the export orders to use up their quota. However, all other work on the cars, including the wood work, was carried out at the factory. It was Janu-

Left, top & middle: Two views of the prototype that put the improved side-valve Ford engine into the cowled radiator Plus 4 to create the Series II 4/4. Note the white-faced instruments.

Bottom left & top right: In 1947 the factory issued these black and white photographs in the form of postcards. The reverse listed specifications and prices of the various models. Shown here is a 4/4 Tourer and a Drop Head Coupé.

ary 1948 before the factory managed to return to full production again.

4/4 Series II
October 1955 - November 1960

HFS periodically experimented with various sizes of Ford engines. He was attracted to the 10hp version, but this was not robust enough, nor did it produce enough power for a sports car.

Towards the mid-1950s it became obvious to the Morgan Motor Company that there was a gap in the market for a low-priced production sports car. It so happened that this coincided with Ford's redesign of its 10hp engine: without altering the outside dimensions, the engine was now stronger and more powerful.

HFS was delighted to discover that, not only was Ford willing to supply him with the new engine, but it could be purchased complete with Ford's gearbox. By using this transmission, Morgan could achieve a considerable saving against the Moss gearbox which was used in the Plus 4 model. In addition, savings could be made in production costs, as it was found that even with the bolted-on gearbox supplied direct from Ford, the engine would, with only very minor alterations, fit neatly into the Plus 4 chassis. This did mean, however, that a remote control gearchange would have to be designed, which was achieved by extending the normal lever vertically and connecting it to a rod running on a pivot across the scuttle,

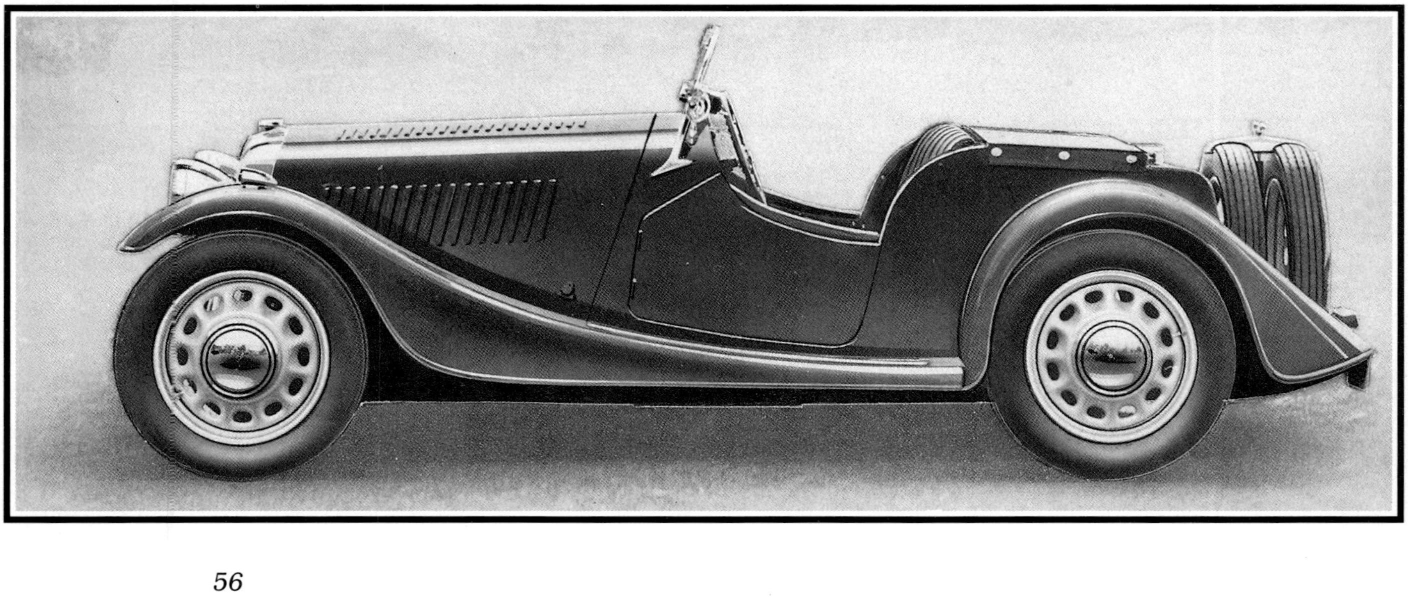

through the firewall to emerge under the dashboard. The main disadvantage of using the complete unit was that the Ford gearbox had only three forward gears, but the cost saving far outweighed the fact that the ratio of the second gear was so low it sorely restricted acceleration. The arrangement of the remote gearchange allowed for the use of the umbrella-type handbrake lever which was also mounted under the dashboard. All this meant that the new model had more foot room than the Plus 4. The savings were such that, in the end, the Morgan Company was able to offer the car at the amazingly low price of £714, compared to £894 for the Plus 4, £939 for a Triumph TR2, or £961 for the MGA.

The new car was introduced at the London Motor Show in October 1955 and was listed by the factory as a Tourer. Modest by sports car standards, its performance reflected the fact that it had less than half the power of the TR2-engined Plus 4. For those owners who wanted more power there existed a wide range of moderately-priced tuning kits for the 1172cc Ford Anglia 100E engine.

The Competition version of the Series II was introduced in September 1957 and was fitted with this Aquaplane aluminium alloy cylinder head and inlet manifold.

In September 1957 the Company introduced a 'Competition' version of the 4/4. The main mechanical changes were aquaplane aluminium alloy cylinder head and inlet manifold; four-branch exhaust manifold and a copper/asbestos cylinder head gasket. The compression ratio was raised to 8:1. Twin SU carburettors were fitted in place of the single Solex and stronger valve springs were fitted. All this increased the power output to 40bhp at 5100rpm. There were no body changes made for this model. In all, 43 Competition models were produced.

Although only produced as a two-seater, one experimental four-seater was built on chassis number A296. In addition, a Drop Head Coupé was specially built for Mrs Stella Skinner, a daughter of HFS Morgan, on chassis number A553. This car was finished in yellow and was delivered on February 1, 1960.

4/4 Series III
October 1960 - November 1961

By deciding to adopt the Ford 100E Anglia engine for the Series II, the company made a far shrewder move than it realized. Not only was it now using a readily-available, mass-production power unit, it also benefited from Ford's own intensive (and

Only one Series II Drop Head Coupé was made by the factory. This car was originally finished in yellow and was produced for one of HFS's daughters. It is now in America and has been slightly modified.

Below: The prototype of the Series III was the first to have the 997cc overhead-valve engine from the new Ford Anglia in place of the old side-valve unit. With the new engine came a four-speed gearbox.

expensive) future engine development.

The first of these was the introduction of a completely new 105E overhead-valve engine and four-speed gearbox for the new design Anglia in October 1959. Morgan negotiated the supply of this new engine, although it was a year before it could be offered in the 4/4, to create the Series III.

This lightweight (212lb) 997cc engine produced far more power and had enormous competition potential in the

A 1967 Series IV with an interesting chassis number. B690 was found under a pile of rubble in a contractor's yard in Nigeria. It has been fully restored and is now owned by the President of the Nigeria Motor Sports Club.

1-litre class. The increase in power was linked to Ford's first four-speed gearbox, which had synchromesh on the top three gears. Once again the complete assembly fitted neatly into the Morgan chassis, with only a crossmember requiring some relocation. The Series II remote gearchange remained.

Only a few external changes were made: the main identification points are the different style rear lights and the transfer of the fasteners for the side-screens to the outside of the doors. The wider body was used on the last 70 production Series IIs.

The Series III is sometimes - unofficially - referred to as the 50th Anniversary Model because its production coincided with this milestone in the company's history.

4/4 SERIES IV
October 1961 - March 1963

The 4/4 Series III was not produced for very long. In May 1961 Ford upgraded the capacity of the 105E engine to 1340cc for use in the new Consul Classic 315 saloon. It became available to Morgan by October the same year, giving birth to the 4/4 Series IV.

Designated by Ford as the 109E, the upgraded engine developed almost 40 per cent more power than its predecessor. The use of this power unit necessitated several modifications to the Morgan; more power meant greater performance which, in turn, meant better braking was required. The company therefore adopted the 11 inch disc brakes as used on the Plus 4 as standard for the 4/4 Series IV. In standard form the increase in performance over the Series III was substantial - the 0-60mph acceleration time was reduced from 26 seconds to 16.5 seconds, while the top speed went up from 80mph to 92mph.

The gearbox operation had been changed by Ford to provide the new Classic 315 with a column change. This necessitated the Morgan Company modifying the mechanism so that its same system of remote operation could still be used.

Produced for only 18 months, the Series IV's demise was induced by the

Outwardly, not much has changed, but the Series V was advertised with a Ford 1.5-litre engine - the 116E unit from the Consul Classic and Cortina. It remained in production for five years.

availability of engines. Ford once again uprated the engine, so Morgan had to introduce yet another new model.

Chassis number 664 is listed in the factory records with the prefix letter 'A'. The car is, in fact, a Series III Competition model which was specially made for the New York Morgan agent, Fergus Cars. The car was fitted with a Ford 105E Competition engine, with an Alexander cylinder head, special manifold, twin SU carburettors, disc brakes and wire wheels.

The Series V Competition model used the Ford Cortina GT's engine with a single twin-choke Weber carburettor. Several owners undertook further modifications - like adding twin Webers, as shown here - in the quest for more power.

Although the company's advertising themes had changed by the mid-fifties, once in a while the motoring public was reminded of the Morgan's competitive potential.

4/4 SERIES V
February 1963 - March 1968

The new engine destined for the Series V was designated 116E and had been developed by Ford to power the Cortina saloon, and also to improve the performance of the Classic 315 and the Capri coupé. The change was most fortunate for Morgan; it meant the company was able to offer a 4/4 which, even in standard form, could outpace its rivals - Austin-Healey Sprites, MG Midgets and Triumph Spitfires.

Beside developing the new engine, Ford also improved the gearbox. The new unit was fitted with synchromesh on all four gears and, although slightly larger than the previous gearbox, it presented no real problems in fitting to the Morgan chassis.

Within a very short time of the start of production, Morgan was able to offer a Competition version of the Series V for only £133 extra. This was fitted with the Ford Cortina GT engine; basically the same as the normal 116E but with a twin-choke Weber carburettor, high-compression cylinder head and a high-lift camshaft. This increased the power from 65bhp to 78bhp and provided 0-60mph acceleration in 11.9 seconds and a top speed of 95mph. Towards the end of its production run, the Competition model was fitted with an uprated gearbox with remote control change, which Ford had developed for the Lotus Cortina. Before this the model was fitted with a Wooler remote control gearchange.

One two-seater (B1384) was built to special specification for Mike Duncan, including aluminium body and modified bonnet. The car was supplied as a rolling chassis, less engine and gearbox. Mike Duncan then fitted it with a highly-tuned Lotus twin cam engine complete with Lotus Cortina gearbox and the car was returned to the factory to be fitted with the body and wings. The registration number of this car was OUY 2E (first registered 16.3.67) and it was used very successfully in competition by Mike Duncan. A few years ago it was exported to the USA.

Another two-seater (B996) was supplied, less engine and gearbox, to John McKechnie, the Morgan agent in Cheltenham. He fitted the car with a Cosworth-Ford single overhead-cam engine. This car was another which was very successful in racing.

Six cars were supplied from the factory fitted with hard tops. The chassis numbers of these cars are: B934, B971, B974, B984, B995, B996.

Morgan

IV

PLUS 4s

To launch the Plus 4, Morgan produced a gleaming exhibition chassis for the Earls Court Motor Show in October 1950. The two photographs (below & right) show the Vanguard engine, remote gearbox and the Salisbury rear axle.

The Flat Radiator Plus 4
February 1951-June 1954

Introduced at the Earls Court Motor Show in October 1950, the Plus 4 model became the flagship of the company. Its introduction heralded the start of an era: this model, in all its various forms, changed the face of Morgan and was to stamp the marque's authority on motorsport worldwide, recreating the phenomenal successes of the three-wheelers.

When, in 1947, the chief executive of the Standard Motor Company, Sir John Black, decided to adopt a 'one engine' policy for Standard's 1950 passenger car range, he settled on a four-cylinder overhead valve engine of 2088cc. This powered the new Standard Vanguard and its introduction meant that no 1267cc Standard engines would be manufactured after 1949. The adoption of the enlarged engine for the 4/4 was not, however, an automatic choice. Although its use seemed, perhaps, a logical progression, HFS was not convinced. He had envisaged that an engine of around 1750cc would be used for the Vanguard, but when he heard of its much larger displacement, he sent Peter Morgan out shopping for an alternative engine of not more than 1.5 litres capacity. He returned empty-handed; there was no suitable engine available. So they decided to use the Vanguard engine after all. HFS's hesitancy was because he thought it possible the Government would re-impose the prewar system of car taxation. This had made the licence fee proportionate to the number and bore of the cylinders, and would have been unfavourable for an over 2-litre four-cylinder Morgan. In the event, the Government adopted flat rate taxation.

The new Plus 4 was assured of a

great reception. Intended as a replacement for, rather than a supplement to, the 4/4, it was offered in two-seater, four-seater and Drop Head Coupé forms. Quite early on, an experimental two-plus-two Drop Head Coupé was built for HFS's personal use (chassis number 2227; engine number V122ME; registration number KNP 152), but it was another three years before it was put into production and then not until the change had been made to cowled radiators. Retaining the same basic shape of the 4/4, the Plus 4 body was 2 inches wider across the cockpit, gave 2 inches more legroom and 6 inches more width across the pedal boards. Important changes to the technical specifications had to be made to augment the new

The Vanguard-engined Plus 4 was a logical progression from the 4/4. The Plus 4 was offered with all three body styles, and there were few external differences to the less powerful models.

Smaller sidelights and block lenses on the headlamps were introduced from this car, chassis number P2360.

engine, notably softer front suspension, made possible by inclining the swivels very slightly inwards at the top, thus enabling the vertical length of the pins to be increased, and longer, softer springs to be fitted. Another innovation was the semi-automatic lubrication of the sliding sleeves of the swivels. This was actuated by the driver depressing a pedal valve in the cockpit which allowed engine oil to be fed under pressure to the bearing surfaces.

The Newton shock absorbers used on the front of the early 4/4s were replaced by Girling double-acting telescopic hydraulic dampers, while Andre frictional rear shock absorbers gave way to Girling piston-type dampers. The Girling cable brakes were changed to a full hydraulic system incorporating 9 inch diameter drums with two leading shoes, with the brake pedal acting directly on the master cylinder.

When introduced, the Plus 4 two-seater was listed at £510 (plus £142.42 purchase tax), while the top of the range Drop Head Coupé was priced at £565 (plus £157.69 purchase tax). In spite of the considerable increase in power and performance, Morgans were still competitively priced in the market.

This saloon body - now fully restored by Gerry Willburn of California - was produced by Cooper of Putney and fitted to one of the 50 flat radiator Plus 4s sold in chassis form.

Plus 4 interim and cowled radiator models
December 1953 - May 1958

The transformation from the prewar upright radiator and separate, free-mounted headlamps to the cowled radiator style as we know it today did not occur overnight but developed through three distinct stages.

By 1953 motor vehicle designs were beginning to change, more and more emphasis was placed on streamlining and simplicity of construction. The days of the free-standing chromium-plated header tanks and sides of integral radiators, and large, separate headlamps were numbered. More and more manufacturers 'modernized' their designs and manufacturing processes, and the component manufacturers naturally followed the market. All said they would continue to supply Morgan with traditional parts but, the question was, for how long and at what price? HFS was only too aware it would not be very long before the costs became prohibitive. It was therefore good business practice to

The introduction of the 'Interim Radiator' Plus 4 was forced on the company by diminishing supplies of integral radiators and free-standing headlamps. This is a 'high lamp' version, required to meet the new UK lighting regulations introduced in 1954.

adapt the design of his cars to accommodate the new mass-produced components and so remain competitively priced in the market place.

The first change was the introduction of a small pressurized radiator mounted beneath a curved cowling, with a flat grille/stoneguard set at an angle into it. Pre-focus headlamp units were fitted into semi-recessed nacelles on an inner valance between the cowl and the more generously proportioned wings. Initially, these headlamps were set quite low, but a change in the lighting regulations from January 1, 1954, which required a new minimum height above ground to the centre of the lamp, meant that the second change had to take place. A 'high lamp' version of what is now known as the 'Interim Model' Plus 4 was created. The exact production figures of the 'low lamp' and the 'high lamp' version of this model are unknown; neither, for that matter, is the total number of the 'Interim' model cars produced. It has been suggested by another author that the number was only 19 and there is no reason to doubt this figure. There is also no reason to believe that there was any change in the factory's practice of overlapping design changes, which means it is quite possible that a 'low lamp' example may have been produced after the introduction of the 'high lamp' version and sent to an export market.

The change started at chassis number 3000. It is a fact that chassis number 3128 is a 'high lamp' version and, according to Mr Day of the factory, the last car produced of this design was chassis number 3135. Assuming that this number is correct, the works records show that four of this model were fitted with Triumph TR2 engines and the remainder with Standard Vanguard engines.

The 'Interim' model was not the happiest of designs, and before long the third and final stage of the change took place when the curved grille/stoneguard was fitted. This, with only minor changes , has remained the hallmark of Morgan models as we know them today.

When one considers that the Plus 4 'Interim' and cowled radiator models were produced for only 4 years, it is amazing just how many variants there were and the changes that occurred to each type. Other than the change to the cowled radiator, the other major body style change was to do away with the twin upright spare wheels which had been a feature of the Morgan since 1936. The single spare wheel was mounted partially within and at the same angle as the sloping rear panel.

The Plus 4 four-seater Drop Head Coupé September 1951- February 1956

The closest we have seen to a family-type four-wheeled Morgan is this variant of the Plus 4, which was introduced in 1954 and at £879 became the most expensive car in the range.

Weighing 84lb more than the Plus 4 four-seater tourer, its performance was obviously restricted by comparison with the two-seaters, but this did not prevent examples being used in competition. One, owned by T Dixon Smith, was sufficiently impressive to be described in the motoring press as 'indecently quick.' Two prototypes of this model were produced in 1951 and 1953, both with the then current flat radiator, although subsequently both cars were rebodied with cowled radiators. The 1953 car became the personal property

of H F S Morgan, who used it until his death in 1959, when it was then sold to the USA.

The production models' bodywork differed from the two-seater Drop Head Coupé mainly at the rear, with a single spare wheel hidden in a small luggage compartment. Other differences were in the greater size of the door windows and in the hood and hood frame. There was an option of either a 4:1 or 3.73:1 ratio rear axle. All examples sold in the UK were fitted with the Vanguard engine, while export cars were fitted with the twin carburettor Triumph TR2 engine. All had Moss gearboxes. Only 51 cars were produced with this bodywork, including the two prototypes, 29

One of the very few advertisements depicting the 'transitional' (Interim) Radiator model. Few of these cars survive today.

Morgan

being destined for the UK market, 18 for the USA, two for Spain and one each for Belgium and Australia.

Plus 4 Triumph TR2 engine September - 1954 June 1958

The first Morgan fitted with the Triumph TR2 engine was a works experimental flat radiator Plus 4, but this car was eventually rebodied as a cowled radiator model. The TR engine was first offered as an option to the public in October 1953, but it was not until April 1954 that it became standard. Production of the Vanguard engine version was phased out as the stock of engines was used up. Initially, the prefix letters 'P' for Vanguard engines, and 'T' for TR engines were used on the chassis numbers, but this was dropped after the former engine ceased to be available.

When the Triumph TR2 engined Morgan Plus 4 was put on sale at approximately £830, it was the cheapest car available on the British market capable of reaching the magic 100mph, although more than one tester found the last two or three mph took a long time to achieve.

Plus 4 Triumph TR3 engine March 1956-February 1964

Introduced in March 1956 at just under £894, this model was outwardly identical to the TR2 engined model it replaced. However, beneath the bonnet the difference became evident; this engine was fitted with larger carburettors and a different inlet manifold. More powerful than the TR2 engine, the car's top speed was 105mph and there was an improvement in the acceleration times, although fuel consumption, which had been a strong

Right: A 'low lamp' interim radiator model. Cars that were partially completed when the lighting regulations changed were mostly left-hand drive for export.

With the curved grille and stoneguard the transition from flat to cowled radiator was complete. Initially the twin upright spare wheels were retained.

point with the TR2 engine, suffered.

Plus 4 Triumph TR4 & TR4A engine October 1962-January 1969

With the introduction of the larger 2.1-litre TR4 engine, Triumph, and therefore Morgan, sports cars were moved into the over 2-litre class.

At the same time performance was considerably enhanced, especially when the increase in capacity was followed by the TR4A engine with a further power increase. This was to be the last variant of the Morgan Plus 4 for many years. It was a model which enjoyed - and still continues to enjoy - a fine competition record.

The TR4 version was introduced in October 1962 and the TR4A engine as an option in 1966. It was priced from £816.

Plus 4 Super Sports February 1961 - May 1968

Ever since its introduction in 1927, the designation Super Sports has traditionally been reserved by Morgan to denote a model which is capable of competition performance. The adoption of this name for a special version of the Plus 4 model came about as a result of the tuning skills and competition successes of one man, Chris Lawrence.

Late in1957 Chris Lawrence decided to try just one more season in racing. For over eight years he had been competing on a shoestring budget with various makes of car and in many forms of racing, but with only limited success. He and his fiancée pooled all their savings to buy a 1956 Morgan Plus 4. Chris was a competent amateur engine tuner who spent many long hours during the winter studying and working on the Plus 4 in order to obtain from the engine the maximum power possible. After three amazingly successful competition seasons in the UK and Europe, he had proved not only his driving ability, but also his tuning skill.

So impressed was Peter Morgan that, by late 1960, the factory had entered into an agreement with Chris Lawrence whereby he would prepare a few engines on a regular basis for the Morgan works, and for customers who sought extra performance.

Originally these cars were not designated Super Sports but received the name retrospectively, although they were not fitted with the 'low line' body. This was a particularly bold undertaking for Peter Morgan. Nowadays, we are used to manufacturers fitting their cars

This Vanguard-engined two-seater coupé was originally built as a flat radiator model and left the works in August 1954. Having languished at a main agent for a year, it was returned to the works to be rebodied with a cowled radiator and re-sold to another agent, leaving the factory in October 1955. Several of the final flat radiator cars were remodelled in this way.

Below & over page: 1951 advertising of the Standard Vanguard-engined Plus 4 two-seater Drop Head Coupé.

Morgan
Plus Four
Drop Head Coupé
We invite you
to visit us on Stand
No 116
INTERNATIONAL MOTOR SHOW
EARLS COURT
One of
Britain's
most
outstanding
post-war
successes
THE MORGAN MOTOR CO. LTD.
MALVERN LINK · WORCESTERSHIRE

Peter Morgan's personal Plus 4 Drop Head Coupé gained a silver award in the 1954 Exeter Trial driven by Peter Garnier and Charles Heywood of The Autocar.

The company's first advertisement for the new Cowled Radiator Plus 4, a body style which has remained virtually unchanged for the past forty years.

Above & right: Cowled radiator Plus 4 Drop Head Coupé with hood and sidescreens erect. This model's tail encloses a single spare wheel with a small luggage compartment. The boot can accommodate a small suitcase, while another could be secured to the open lid.

with engines prepared by specialist companies but, in 1960/61, the Morgan Motor Company was the only one to do this: the Lotus Cortina and its ilk were yet to come.

With this regular order from the Morgan factory, Lawrence was able to open workshop at Westerham Motors, Acton, West London, where he formed the now famous Lawrencetune company. Here he race tuned Triumph TR series engines, not only for the Morgan factory and Morgan owners, but for the occasional Triumph owner as well.

In1961 Lawrence attempted to enter the Le Mans 24 Hour race. The

Only one flat radiator Plus 4 was fitted with a Triumph TR2 engine. Drop Head Coupé chassis number P2750 was eventually rebodied with a cowled radiator. This car is now in America.

Above: The TR3 engine introduced in 1956 provided a useful increase in power over the original TR2. The inlet and exhaust manifolds were separate castings, the inlet manifold in two halves joined by a balance pipe.

Below: Charlie Curtis, Morgan's chief tester until he retired in the early 1970s, prepares Peter Morgan's early TR-engined rally car KUY 387. Note the white tape on the front wing; this was an indicator for stopping astride lines, a feature of rally special tests in those days.

Another example of the Morgan Motor Company's advertising strategy of appealing to both sporting and family orientated members of the public.

Right, top: Classic Plus 4. This is an original US export model from 1967. By then, black-faced instruments were standard. Note that the rear lamps and indicators had yet to change to the stalk fitting used later.

Right, bottom: Immediate recognition point of the Plus 4 Super Sports is the air intake on the side of the bonnet which houses the twin Weber carburettors.

French scrutineers refused the entry as they were convinced it was basically a prewar car. This upset Peter Morgan: apart from hurt pride, the rejection would adversely affect the company's already falling sales figures. Success at Le Mans became a target, necessary to restore the marque's international reputation. The company therefore decided to give Lawrence full works support for an effort in 1962. This included rebodying the car with a 'low line' 4/4 body. This, it was reasoned, would result in a substantial reduction in drag and weight. The lowering was easily achieved thanks to the reduced height of the engine when fitted with side-mounted Weber carburettors, and also a special header tank arrangement. So was born the first Super Sports al-

This 1963 Plus 4 TR4 engine has twin SU carburettors but Stromberg carburettors were available as a factory option. The car belongs to Edgar L Beere Jnr. of Ohio, USA.

A273·704
1967 VIRGINIA

JOHN H. SHEALLY II
VIRGINIA BEACH
MPH
Morgan
AMPS
OIL
RPM

Left, top: The TR4 and 4A engines used in the Plus 4 Super Sports were tuned by Chris Lawrence. The conversion consisted of stripping the engine, balancing all moving parts, polishing and gas flowing the cylinder head and fitting a modified inlet manifold with a pair of Weber carburettors.

Left, bottom: The Super Sports dashboard - this one is very original apart from the personalised speedometer!

John H Sheally's concours winning Plus 4 Super Sports shows the disc front/drum rear brake system that was standard on this model.

though, of course, the actual production model differed in many ways to this specially-prepared race car.

The car's performance at Le Mans in 1962 was such that any doubts anyone may have had were eliminated. Honour was restored and the success was reflected in the sales figures. When the Super Sports model was announced late in 1961 it offered arguably the best value for money in terms of performance. With acceleration figures of 0 60mph in 7.8 seconds and 0 100mph in 22.8 seconds, it was tailor-made for competition. These acceleration figures were comparable with those of the Plus 8 when it was introduced eight years later.

Engines that were to be fitted in the Super Sports were sent from the Morgan factory to Chris Lawrence's workshop, where they received the Lawrencetune treatment. This consisted

One of the body designs submitted for consideration as the Plus 4 Plus was this all-enveloping open two-seater. It was built by EB Plastics, whose alternative coupé proposal was finally adopted instead. The open prototype was found in need of restoration in 1978 and is now owned by a German enthusiast.

of completely stripping the engine and sending the clutch, flywheel, connecting rods and crankshaft to Jack Brabham (Motors) Ltd, for balancing. While this was happening, Lawrence's workshop team polished and gas flowed the cylinder head, the compression ratio was raised to 9 to 1, and a high lift camshaft fitted. In addition, the engine was fitted with a cast alloy inlet manifold and a pair of either 42 or 45D COE Weber carburettors were added, plus a specially designed four branch exhaust manifold.

Initially, the engine developed 115bhp at 5500rpm but later increased to 125bhp. When the engine was returned to the factory it was fitted to a standard Plus 4 chassis which, in turn, was fitted with a light alloy 'low line' body and wings (this style of body was adopted for all Plus 4s from September 1966), and 72 spoke wide rim wheels. Uprated E-type stub axles were fitted after Chris Lawrence suffered breakages of the standard stub axles in competition. Many examples were fitted with individual bucket seats in place of the conventional bench-backed seats. An oil cooler was mounted in front of the radiator, but behind the grille. The total weight of the car was 1736lbs, compared with 1884lbs of the normal Plus 4.

The Super Sports was produced mainly to order until May 1968, and, in all, 104 were made, including five four-seaters and four Drop Head Coupés. Several other Plus 4s were fitted with Lawrencetune engines but, for various reasons, are not recognized by the factory as Super Sports.

Plus 4 Competition model October 1965 April 1967

The Competition Model was introduced

Wooden former for the Plus 4 Plus coupé, shown here arriving at EB Plastics, was panelled in aluminium to make the mould for the glassfibre body shells.

because of public demand for a cheaper version of the Plus 4 Super Sports. It was nearly £100 more expensive than the ordinary two-seater, but considerably cheaper than the Super Sports. At £938 including tax, the Plus 4 Competition Model was dubbed 'The Poor Man's Super Sports'.

Its 'low line' body was almost the same as the Super Sport. A standard 2138cc Triumph TR4 engine was used, extra power being obtained by fitting it with the four-branch, free-flow exhaust system from the Super Sports. Wire wheels were standard but not the special wide 72-spoke variety. Announced in October 1965 at the Earls Court Motor Show, it was produced just in two-seater form, and only to order. Only 42 of this model were made, the last car leaving the factory in April 1967. One car was used by the works as an experimental car; 15 were sold in the UK; 11 to the USA; four to Canada; three to France; two each to Germany and Sweden and one each to Australia, Belgium, Panama and Switzerland.

Plus 4 Plus Fixed Head Coupé November 1964 February 1967

To date the only 'fully enclosed' Morgan to emerge from the Malvern factory has been the Plus 4 Plus, which was introduced at the 1963 Earls Court Motor Show to a very mixed reception. The closed glassfibre body was described by some of the motoring press as having a Jaguar XK150 front and a Lotus Elite rear, while diehard Morgan purists were almost lost for words ... for them, this was the final blow; it had been bad enough when, ten years earlier, their 'true sports car' had been 'cissified' when it was fitted with a cowled radiator.

Peter Morgan had been considering the possibility of introducing a glassfibre-bodied all-enveloping Morgan for some years, and had discussed the idea at great length with his father. HFS had been reluctant to use glassfibre, preferring to wait until the new material had stood the test of time on other vehicles; he reasoned that the whippiness of the Morgan chassis would cause cracking to the body.

After making a very detailed study of a Debonair S1 GT and other similar glassfibre bodies, Peter Morgan contacted E B Plastics Ltd, of Tunstall, Stoke-on-Trent in 1962. This company had been producing two-seater open sports car bodies for several years, and also the Debonair GT saloon, which was introduced at the 1961 Racing Car Show in London. The Debonair received motoring press acclaim, a typical quote being *'the EB Debonair is an outstanding example of the immense strides made in the production of glassfibre bodies, and is very attractive indeed'*. E B Plastics were one of the country's leading glassfibre manufacturers, which also produced cabs for Foden and ERF trucks and Austin Parkinson electric vehicles, plus varied glassfibre mouldings for the motor industry in general.

Peter Morgan laid down a specification: some of the classic lines must be retained (this was interpreted as the Morgan-style frontal appearance, including a somewhat similar radiator grille); there must be no overhang at the

The end result: the Plus 4 Plus is a coupé of distinctive style. Not well accepted at the time of its introduction, this model has since become a much sought-after collector's car.

front and no mechanical changes were to be necessary to the chassis. The founder and managing director of E B Plastics, John N Edwards, produced drawings of several variations on the Debonair theme which met Peter Morgan's criteria. After much consultation and discussion, one of the designs was decided upon, but before the final go-ahead was given, a scale model of the proposal was submitted for approval. A few years ago another model at about 1/5th scale of a glassfibre-bodied Morgan was discovered; it was, apparently, a proposal made by another company asked to submit designs.

So successful were E B Plastics in keeping to Peter Morgan's brief, the only chassis change necessary was to fit a pair of sheet steel extensions bolted on each side of the engine. These linked the front suspension mounting to the bulkhead, thus making the whole struc-

Though there was no mistaking the Plus 4 Plus for anything else, the rear quarter view shows some similarities with the contemporary Lotus Elite.

Left-hand drive Plus 4 Plus shows the separate bucket seats that were a feature of this model. Facia was a glassfibre moulding but of similar layout, and with the same instruments and fittings, as the traditional models.

ture much sturdier and, it was hoped, minimizing the risk of body cracking due to chassis flexing.

The next stage was to construct a full-scale pattern, using wood formers, panelled in light alloy. Many hours were spent producing a high-quality cellulose finish. Then the first mould was made from epoxy resin and, while the first body was being made in this mould, the next was 'laid down'. A Plus 4 chassis from the production line was sent to the Tunstall works.

Peter Morgan was determined the new body should be capable of being fitted on a standard Plus 4 chassis without the need for any special alterations, other than those stated. This ensured the new model could be built without interruption to normal production. The prototype body was mounted, cellulosed, fitted with lights and other necessary equipment and trimmed. It was fitted with a pair of individual seats with a common bench-type back, but all subsequent cars of this model were supplied with bucket seats. The prototype was the only example produced in this way, all future cars being finished, cellulosed and trimmed at the Morgan factory after the body had been fitted at the Tunstall works.

When complete, the prototype was registered 869 KAB, and collected from E B Plastics by Peter Morgan personally. Extensive road testing was carried out, including a tour of Spain and France. These tests revealed a problem associated with the very stiff front end of the car, which meant the front suspension was working harder than it normally did with the standard unbraced chassis, which absorbed some of the worst shocks. In spite of this, the new model still retained the expected Morgan roadholding capabilities. The late John Bolster test drove the prototype for *Autosport* in the summer of 1963: *'In a veritable cloudburst I drove along Ross Spur in sheets of spray, reaching 105mph. The famous controllability of the Morgan was fully in evidence, fast curves being taken at speed in complete confidence. The car rides better than previous examples of the marque because the rigid chassis ensures that the springs do more work.'* Bolster was impressed by the maxi-

The SLR was a very successful exercise in giving the racing Plus 4 a more aerodynamic shape. When it first appeared in marque racing it caused controversy amongst other competitors and organisers, who were reluctant to accept it as a Morgan.

mum speed of around 110mph, and consumption figures of about 29mpg. Once again, the new model was proved in the competition field, both in rallies and trials. The car performed well in the skilful hands of Peter Morgan and the results went a long way toward dispelling the critics' doubts. In spite of this, and the fact that the car had been quite well received by the motoring press, public reaction was less satisfactory and the model was produced only spasmodically over the next four years. In all, only 26, and two spare bodies, of this model were made, including the prototype which was crashed, the remains of which are still retained at the factory.

The SLR Morgan (Sprinzel Lawrencetune Racing)

After his successes in the early 1960s at Le Mans, Spa, Monza and elsewhere, Chris Lawrence realized that if he wanted to remain competitive he needed a faster car than his basically standard-bodied Morgans. Late in 1963 he was offered a new Porsche 904 direct from the Porsche factory but, wanting to remain loyal to Morgan, he decided to use the TR4-engined Plus 4 as the basis for the design and development of a new car for sports car racing.

Chris Lawrence had tuned a couple of rally cars for John Sprinzel. These had proved quite successful, and so, after some discussion, the two drivers teamed up to form Sprinzel Lawrencetune Racing. John Sprinzel succeeded in obtaining some financial support from BP which, together with money put in by each party, plus definite orders for two cars, provided sufficient finance to begin development.

The regulations governing Group 4 GT racing at that time allowed for body modifications, provided the weight of the original car was not cut by more than 15 per cent. The aerodynamic body of the SLR was designed jointly by Sprinzel and Lawrence and was produced in aluminium by Williams and Pritchard. It was fixed to the bare chassis via very light subframes. No other chassis modifications were required, except for a small amount of strengthening to the transverse bulkhead and the need for subframes to be provided for mounting the shock absorbers. Only three cars were built on Morgan chassis and one on Neil Dangerfield's ex-Sid Hurrell Triumph TR4 (SAH 13 7). One of the Morgans was the works prototype of the Plus 4 Super Sports (170 GLP), which was originally built as an exact replica of Lawrence's Le Mans car, for which it had acted as back-up for that event.

The modified TR4 engine had a new crossflow cylinder head with staggered 90 degree angled valves. This was a development financed jointly by Lawrencetune and the Morgan Motor Company, but it was never actually completed. This cylinder head was reported in the motoring press as being a 'Blunstien Head'. Bill Blunstien worked for Lawrencetune and produced the casting drawings, but the design was mainly by Chris Lawrence. Many problems were encountered with the lubrication of the rockers, and another irritation was the oiling-up of the narrow heat band 10mm spark plugs. As Chris explained: *'If you could get on the front row of the grid and could get away at the start you were all right. Otherwise, the plugs nearly always oiled up.'* Another complication of this cylinder head on the 2-litre engine was that the valves were so large it was impossible to obtain the ideal compression ratio of 10 to 1. A modest 6.8 to 1 compression ratio was settled on, but even so the engine developed more torque than the standard engine.

The cars were fitted with larger stub axles and steel hubs, developed as a result of breakages that had occurred in the 1961 and 1962 seasons. It was during this period that Lawrencetune-prepared race cars were developing such high cornering forces that the standard hubs could not take the strain; the manufacturers were reluctant to accept this at first but in due course were persuaded to change the material and increase the size. These stronger hubs were subsequently fully race proven and have since been adopted as standard equipment for the Plus 8.

The SLR Morgans achieved some notable competition successes, and Chris Lawrence used one to lower the class lap record at Goodwood by three seconds. He also took a brilliant third place overall at Spa, splitting six Porsche 904s, which could never have been described as road cars. Once again, Porsche invited him to join its factory race team ... Lawrence was considered a considerable threat to its race domination at that time. One wonders exactly what may have happened, and just how much the SLRs may have achieved, had not Chris Lawrence suffered a serious road accident in 1964, which effectively ended his racing career and prevented any further development of the car.

At least two of the three Morgans and the Triumph still exist. They are very well documented and, from time to time, are still seen in action, giving a very good account of themselves.

V

MAINTENANCE, REPAIR & MODIFICATION

Here, I have concentrated on routine maintenance and common repairs, particularly for those models prior to the 1600 series, but a great deal of the information applies to all models.

I have also included many extracts from technical articles which have appeared in club magazines and other publications. In this way the chapter contains a wealth of information not readily available to the average Morgan owner which, it is hoped, will help you to get the best from your Morgan.

Important. Before undertaking any of the procedures described in this chapter it is essential that you read the safety and working procedure Appendix on page 268. Also read the safety notes on pages 97 and 124, if applicable.

Coventry Climax-engined 4/4

• LUBRICATION

Every 5000 miles remove the sump and clean out with paraffin, allowing it to drain thoroughly before replacing.

The route of the oil drawn from the sump is as follows: the oil is pulled up the suction pipe, the mouth of which is situated about $1/4$in (6.3mm) from the bottom of the sump. It then passes through a gravity-operated ball valve to a highly efficient gear-type pump, from where it is fed under pressure to a gallery case in the cylinder block, and in turn by branch ducts to the main bearings of the crankshaft which, being drilled, permit the oil to be conducted to the connecting-rod big-ends. An overhead rocker feed pipe also takes its supply from the aforementioned gallery. Here it is fed through drilled bosses in the cylinder head to the base of the rocker bracket, which is drilled down both the pedestal and through the rocker pivot arms. Lastly, the rocker itself is drilled to allow oil to get to the pad in contact with the inlet valve and also to the pushrod cup ends. Flow is restricted to the rocker's lower parts by the insertion of a wick in the rocker oilway.

Surplus oil drains through channels cast in the cylinder head, which lead to holes in the head of the cylinder block through which the pushrods operate. Flowing down these the oil feeds its way back through the crankcase to the sump, lubricating the tappets at the same time. To enable the timing chain to be lubricated, surplus oil from the oil release plays directly upon the chain and chain wheels. Cylinder bores are lubricated by crankcase oil vapour thrown from the big ends.

The oil pressure under normal running conditions - with the engine hot - should be around 40psi at 30mph. The pressure is higher, of course, when the engine is cold and oil viscosity greater. If difficulty is experienced in maintaining oil pressure, it is possible that foreign matter has lodged under the ball valve of the oil pressure release, which is located at the front of the engine on the nearside. After removing the aluminium cover the ball and spring should be examined; if the ball is pitted or ridged it should be renewed.

• CYLINDER HEAD REMOVAL & VALVE SEAT GRINDING

To remove the cylinder head, drain all the water from the radiator, take off the dynamo belt and disconnect the top water hose. Remove the ignition wires from the spark plugs, carefully numbering each one. Remove the two overhead inlet valve rocker covers and disconnect the brass unions which convey the oil via copper pipes to the

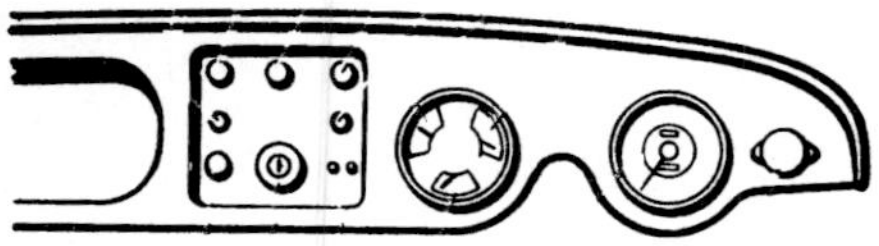

overhead rocker oil system. Disconnect the pipes from the carburettor and induction pipe (in one unit) from the cylinder head and exhaust manifold. Following this, unscrew the large hexagonal nuts which hold down the overhead rocker mountings, then the rocker mountings, each with its pair of rockers, may be lifted. Withdraw the pushrods upwards and remove the cylinder head holding-down nuts. The cylinder head may now be lifted from the cylinder block. If the head is tight it may come free easier if it is tapped under the lugs which will be found at the side. Do not try to drive a wedge between head and block. Remove carbon from the combustion chambers and clean all traces of the gasket/jointing compound from head and cylinder block faces.

Compress the valve springs to enable the cotters to be removed, followed by the springs and valves themselves. Valves and springs should be refitted in the location from which they were removed. If fitting new guides note that the inlet guides are shouldered but exhaust guides aren't; press in with the chamfer inwards, leaving 0.56in projecting downwards into the valve chest. Inlet valaves have a head diameter of 1.43in and exhaust valves a diameter of 1.22in. In each case the valve face angle is 45 degrees. It's helpful when grinding in valves to place a light spring under the valve head to lift the valve off its seat when the valve is turned from time to time during the grinding process to avoid formation of ridges on the seating.The aim is to obtain a thin, unbroken ring of bright metal around both valve and seat. Clean the valves, ports and combustion head of all traces of grinding paste. Replace valves and springs and adjust tappet clearances as described later.

When replacing the cylinder head it is advisable to ensure that the new gasket is perfectly flat and in good condition and to apply a small quantity of good quality jointing compound to each side of the gasket. It is important that the cylinder head nuts should be tightened evenly; start at the centre of the head and work outwards and diagonally. It is essential to go over the nuts again with a spanner after the engine has been warmed up. The torque to be applied should be 35lbf.ft.

The inlet valve tappet clearances cannot be checked until the pushrods, rockers and rocker mountings have been replaced. Also the head holding-down nuts should be finally tightened when the rocker mountings and induction pipe have been bolted home. Do not replace the rocker covers until the engine has been run for at least 10 minutes and the holding-down nuts been retightened and the tappets finally adjusted.

• TAPPET/VALVE CLEARANCE ADJUSTMENT

Tappets should be adjusted to the following clearances:

Inlet 0.006in
Exhaust 0.008in.

They should each be set with the relevant piston at the top of the compression stroke.

• IGNITION TIMING & FIRING ORDER

The ignition should be set with No.1 piston (nearest front) at Top Dead Centre, the contact breaker points just separating and the micrometer adjustment in the midway position. The gap between the contact breaker points when fully separated should be 0.012-0.015in. The firing order of the engine is 1, 3, 4, 2.

• DYNAMO BELT ADJUSTMENT

Slacken the dynamo securing bolts and swing outwards and upwards (this movement being governed by the pivoting bosses on the water outlet pipe and the adjustment link plate). When the correct belt tension has been obtained - 0.5in deflection on longest run - retighten the securing bolts.

• TIMING CHAIN ADJUSTMENT

To adjust the timing chain the three bolts holding the distributor drive housing should be slackened off a little, then the aluminium flange, together with the drive housing, can be pivoted about the bottom securing bolt, thus by swinging the assembly outwards the chain is automatically tightened. The chain, however, must not be dead tight but should have a slackness of about 0.25in. To check this, remove the plug from the timing cover and insert a finger through the hole to feel the chain. When the correct adjustment has been obtained, the housing securing bolts should be retightened, care being taken not to alter the adjustment while doing so.

• COOLING SYSTEM

Thermo syphon; no fan or thermostat fitted. Normal running temperature 90 degrees C.

• ENGINE & GEARBOX REMOVAL

Except when undertaking major work on the engine or clutch it will not be necessary to remove the engine from the chassis. The easiest way to remove the engine is to remove the gearbox, clutch housing and engine as one unit. Although not essential to do so, I have found this far easier if first the radiator

is removed. It should not be necessary to jack the car.

1 Remove the bonnet and disconnect the battery.

2 Remove the headlamp bar or swing it to one side, disconnecting the wires at the snap connectors.

3 Remove the radiator (two bolts at bottom), having remembered to disconnect the top and bottom hoses.

4 Disconnect all pipes, control cables, wires, etc. from the engine.

5 Disconnect the exhaust pipe at the manifold, swing clear and tie back (it is not necessary for the pipe to be removed from the silencer).

6 Remove the pedal board.

7 Unbolt and unscrew the board carrying the coil, cutout, *etc*. Do not remove (the board can be lifted slightly to allow the gearbox to clear it).

8 If car is a four-seater, remove the battery.

9 Remove the propshaft tunnel.

10 Disconnect the speedometer drive, handbrake and front end of the propshaft.

11 Remove the engine and gearbox mounting bolts.

12 Lift the engine, preferably with a loop attached to the nearside front and offside rear cylinder head bolts, tilting to clear the front upper tube.

Replace the engine in the reverse order of the above. Care should be taken that all mounting bolts drop in easily; this will ensure the correct alignment within the chassis.

• PISTON REMOVAL

It is generally advised that removal of the pistons on the Coventry Climax engine is not possible without first removing the cylinder head. However, in an emergency, it *is* possible and here's how.

Place the crankshaft at Bottom Dead Centre for the cylinder line to be stripped with the camshaft in position for inlet valve closing on that cylinder. When the connecting rod is free, draw it downwards on the camshaft side of the engine. There is just enough room to remove the piston.

• PISTON REPLACEMENT

Once again, place the crankshaft and camshaft in the positions referred to in the removal operation. Enter the piston into the bore, again from the camshaft side of the engine, compressing each ring with your fingers (use extreme care) as it slips into the bore. Make sure that the stop peg in the bottom ring groove is in the ring gap.

• PISTON & RING CLEARANCES

At bottom of skirt on thrust side, zero-0.0005in.

At top of skirt on thrust side, 0.002-0.003in.

Gudgeon pin, push-fit in piston when hot.

Two compression rings and one scraper ring per piston, gap 0.015in, free without play in the grooves.

• CRANKSHAFT & FLYWHEEL

Three main bearings with bronze-backed white metal-lined shells dowelled in the housing; running clearance 0.0015-0.002in. The end float of the crankshaft is controlled by the front main, which has a thrust washer behind the timing gear; float distance 0.003-0.007in.

The flywheel with integral starter gear has 88 teeth and is bolted to the crankshaft flange and locked by tab washers. If removed, renew the tab washers on reassembly. Round felt oil seals are fitted in grooves in the sides of the rear bearing cap. Bearing diameter is 1.75in for the two outer mains, 2in for the centre main and 1.5in for the crankpins.

• CAMSHAFT

The engine must be removed to enable the camshaft to be extracted. No timing marks are provided so wheels should be marked before dismantling. Inlet and exhaust closes the same angle either side of Top Dead Centre.

• CLUTCH

Borg & Beck 8in plate, adjusted by nut and lock nut on the front end of the pushrod to give slight free movement at pedal. Engine and gearbox must be removed as one unit for work on the clutch.

• CARBURETTOR

Solex horizontal 'self start' carburettor, type 3011 BFG; SU electric fuel pump. Carburettor choke 30-23mm. Jet sizes: main 115mm; pilot 55mm; compensating 210mm.

Standard Special-engined 4/4

• LUBRICATION

Every 8000 miles remove the sump and clean out with paraffin, allowing it to drain thoroughly and dry out well before replacing (Note: do not remove the oil gauze which surrounds the pump as the set screws also hold the pump cover in place. Do not attempt to clean out the sump with paraffin without removing it from the engine as any paraffin will cause dilution of the new oil.

The flow of oil from the sump is as follows: the oil is drawn through a filter by a gear-type pump and delivered under pressure to the crankshaft journals, crankpins, camshaft bearings and pushrod guides, the spray from the crankshaft bearings being ample to provide efficient lubrication of the piston little-ends and timing chain.

The overhead valve gear is lubricated by direct feed through a passage formed in the cylinder block and connected to the cylinder head via an external copper pipe, the oil returning to the sump by gravity through passages formed in the head and cylinder block. Oil return scrolls are embodied at the front and rear ends of the crankshaft, which effectively prevent oil leaking along the shaft.

A by-pass oil filter is incorporated in the oil circuit and the only attention this requires is for the filtering element to be removed and cleaned by rinsing in petrol every 2000 miles and a new element fitted every 10,000 miles. It is essential that these simple operations be carried out at the specified periods to ensure proper filtration of the oil.

When cleaning or fitting new elements, the following procedure should be adopted:

1 Disconnect the inlet and outlet pipes from the filter.

2 Loosen mounting straps and remove complete filter.

3 Ensure that both inlet and outlet pipes are clear by blowing through them after removing them from the engine.

4 Ensure that the calibrated orifice in the inlet union is quite clean and clear.

5 If fitting a new filter, transfer all removable fittings from old to new filter, using new sealing washers supplied.

6 Install the new filter, making sure the pipe connections are undamaged and are properly tightened. Run the engine for a few minutes to check for oil leaks, after which re-check tightness of connections (it is advisable to check them again after a few days' running to ensure absolute freedom from oil leaks).

The oil pressure should be 40-60psi under normal conditions. This will be higher, of course, when the engine is cold and oil viscosity greater.

• VALVE CLEARANCE ADJUSTMENT

The correct running clearance of both inlet and exhaust valves is 0.022in with the engine cold. To adjust the clearances, remove the rocker cover and turn the engine with the starting handle for half a revolution after the valve to be adjusted has fully closed. Slacken the locknut on the rocker and adjust the ball-pin until the feeler gauge is a close sliding fit between the stem and the rocker. Now tighten the locknut and re-check the clearance (tightening the locknut will sometimes alter the setting). Repeat this operation on each valve and replace the rocker cover.

• VALVE TIMING

When any part affecting the valve timing is removed, take particular note of the markings on the chainwheel teeth; if these are indistinct, scribe marks on the face of the wheels before removing the chain. This will save work resetting the timing when the parts are reassembled, as they can simply be set to the marks.

If the marking operation has been neglected, the camshaft can be set as follows by observing the cam lobes for No.1 cylinder (closest to front):

Inlet opens 10 degrees before Top Dead Centre.

Inlet closes 50 degrees after Bottom Dead Centre.

Exhaust opens 50 degrees before Bottom Dead Centre

Exhaust closes 10 degrees after Top Dead Centre.

• IGNITION TIMING

The ignition timing can be reset as detailed should it have become disturbed. The ignition advance is automatic and should be set to fire at Top Dead Centre and will automatically be at full retard when the engine is at rest.

It should be noted that the cylinders are numbered in sequence from the rear of the engine forwards, with No.1 cylinder being furthest from the radiator.

Turn the engine until No.1 valve closes and continue turning until No.1 piston (nearest back) reaches Top Dead Centre. This position is indicated by a mark on the flywheel, but if the clutch housing is in position, this mark will not be visible, in which case the Top Dead Centre position for No.1 cylinder is mid-way between the point of opening of the inlet valve and the point of closing of the exhaust valve on No 4 cylinder when the engine is turned in a clockwise direction.

If No.1 sparkplug is removed and a bent, small-diameter rod is inserted through the plug hole, it will be possible to feel when the piston is at the Top Dead Centre position.

To obtain the correct firing point, turn the crankshaft in a clockwise direction, viewed from the front of the engine.

If the engine is in the chassis frame the crankshaft can be turned with the starting handle, but if the engine has been removed from the chassis, the turning may be done either by pulling on the fan belt or by way of a spanner applied to the starting handle nut on the front end of the crankshaft.

Slacken the clamp bolt and turn the distributor body until the contact breaker points are just separating on

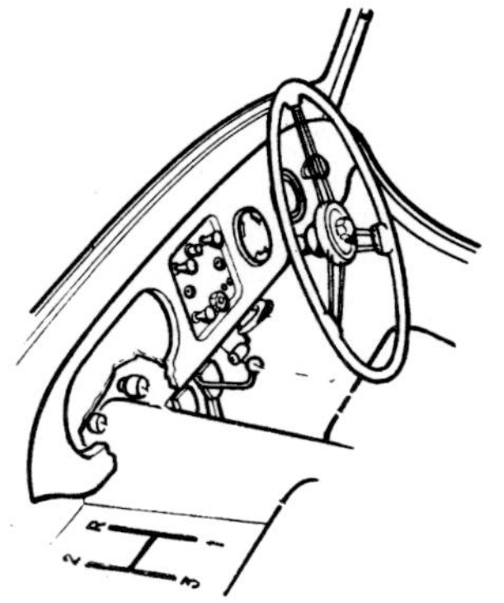

the flank of the cam, when the distributor arm is opposite the No. 1 segment in the cover, then re-tighten the bolt.

Maximum power is obtained by giving the greatest possible advance without pinking. The setting recommended above should be regarded as a starting point, as individual engines may require a greater or lesser degree of advance than this. Do not, however, alter the setting by more than one degree at a time, or two degrees if measured on the flywheel. The firing order is 1, 3, 4, 2.

• CYLINDER HEAD REMOVAL

Drain the water from the cooling system and uncouple the radiator hose from the cylinder head. Next, remove the rocker cover and, by undoing the two nuts at the base of the distributor tower, remove the distributor and coil together, complete with plug leads. Do not slacken the clamp bolt on the distributor body as this would upset the timing.

Now remove the inlet and exhaust manifolds by uncoupling the exhaust pipe and removing the manifold nuts, but first remove the petrol pipe connections to the carburettor and the control connections. Next, remove the water outlet pipe from the top of the cylinder head. Remove the rockers complete with the shaft and towers by undoing the nuts at the top of the towers. The pushrods cannot be removed as they are held in position by return springs, acting at the bottom of the pushrods.

The oil pipe from the oil gallery to the head, which is the feed to the rockers, should now be undone together with the bolt which is screwed into the head to assist in fixing the dynamo.

Do not at first remove the cylinder head nuts completely, but slacken them individually, commencing in the centre and working outwards and diagonally, thereby avoiding any tendency to distort the head. Finally, remove the cylinder head. (Note: do not insert any instrument, such as a screwdriver, between the cylinder head and the cylinder face in an attempt to free the head, as this will damage the gasket and may also damage the cylinder head and block faces. If the gasket joint does not separate freely when lifting the head, a blunt knife may be very carefully inserted to free the joint. Should any water have found its way into the cylinder bores, wipe it away immediately.

• VALVE REMOVAL

Before attempting to remove the valves, a small wooden block slightly smaller than the combustion space but a little thicker, should be prepared. Place the cylinder head on the bench with the wood block in one of the combustion spaces. The valve springs can now be dismantled, the valves removed and the operation repeated on the remaining combustion spaces. Care must be taken not to mix the valves and collars: to ensure they are replaced in the correct order, appropriately number both valves and the cylinder head.

Before starting to clean off the carbon from the piston crowns, turn the engine by hand until two of the pistons are at Top Dead Centre, and fill the remaining bores with clean rag to prevent any chips of carbon falling into the cylinders. It is recommended that the ring of carbon formed inside the top edge of the cylinder be left intact; its removal could adversely affect oil consumption.

Now proceed, removing all traces of carbon from the piston crowns and cylinder faces, using an old screwdriver or similar blunt instrument in a chiselling manner. Do not exert too much pressure on the piston crowns, otherwise the soft aluminium heads may be deeply scratched or scored, and it will be impossible to remove the carbon which will form in these scores during subsequent decarbonizing.

Having completed the cleaning of the first two pistons, give the crankshaft a half-turn, transferring the rags to the two cylinders which have been dealt with, and treat the remaining pistons in the same manner. (Note: if the engine is turned in a direction opposite to normal rotation, the vertical distributor driving shaft will tend to lift and may cause the helical gears to disengage. To prevent this, slight pressure should be applied to the top of the shaft whilst the crank is being turned).

• DECARBONIZING

This operation should be carried out with the head removed at approximately 10,000-mile intervals. First, remove the sparking plugs and scrape clear the combustion spaces and valve ports, taking care not to scratch the valve seats. When completed, wipe clean with a rag dampened with paraffin.

Do not attempt to polish the ports with emery cloth or other abrasive because, unless proper facilities are available, it is extremely difficult to ensure that all traces of the abrasive have been removed; should any such particles find their way into the cylinder bores, serious harm may be caused.

• VALVE GRINDING

The object of this is to ensure that the valves are a gastight fit with the bevelled seat of the valve and that they make perfect contact when fitted together. This is achieved by grinding the two surfaces together, but care must be taken that each valve is ground to its correct seat in the head, as indicated by the numbers stamped on the valves.

Use first coarse, and then fine, grinding paste. A light coil spring may be fitted under the head of the valve for the purpose of lifting it from its occasionally seat during the grinding operation. Continue until both surfaces have an even, matt appearance, then wipe away all traces of the paste to avoid any chance of it finding its way to the bores and causing serious damage.

Having completed the valve grinding operation, the valve may now be reassembled. Follow the previous dismantling instructions in reverse order. The cylinder head should be tightened, starting at the centre and working outwards and diagonally. After the engine has been run for a short time, tighten the nuts again.

• REPLACING THE DISTRIBUTOR

When the cylinder head is removed, the distributor vertical driving shaft may be accidentally lifted out of engagement with the oil pump at the bottom end. Before attempting to replace the distributor, make certain that the lower end of the shaft engages with the pump by pressing on the top end while an assistant turns the engine with the starting handle.

To engage the distributor spindle with the top half of the driving shaft, remove the distributor cover and turn the spindle until it engages with the shaft. There is only one position in which the parts can be engaged, and thus the timing is not disturbed. (Note: do not use force in re-fitting the vertical bracket and distributor. When the surfaces are correctly engaged, the bracket will touch the cylinder head flange, and the nuts may then be tightened).

• ENGINE REMOVAL

Should it be necessary to remove the engine from the chassis, the best method to adopt is as follows:

1 Drain the radiator.
2 Remove the bonnet.
3 Disconnect all pipes, control cables and wires.
4 Disconnect the distributor leads.
5 Disconnect the exhaust system.
6 Remove the headlamps, *etc.*
7 Remove the radiator.
8 Remove the pedal board.
9 Disconnect the clutch operating lever.
10 Disconnect the gearbox and re move.
11 Remove the mounting bolts at the front of the engine.
12 Remove the clutch housing mounting bolts.

The engine is now free and can be removed. Replacing the engine is a reversal of the above procedure.

• RENEWING MAIN BEARINGS

To carry out this work the engine must first be removed from the chassis, as described previously. Then continue as follows:

1 Remove the cover from the timing chest.
2 Withdraw the sprockets with the timing chain.
3 Remove the front and rear engine bearer plates.
4 Remove the clutch.
5 Remove the flywheel.
6 Remove the sump.
7 Remove the big-end caps (note the numbers and keep in order).
8 Remove the main-bearing caps.

The crankshaft should now be free and easily removed.

When reassembling, reverse the above sequence. There are, however, two points that must be remembered:

1 Replace the bearing caps from whence they came. They are all numbered and providing you noted the numbers as you removed them there should be no difficulty replacing them.
2 Take care when replacing the timing chain sprockets to avoid disturbing the valve timing.

• RENEWING THE BIG-END BEARINGS

It is possible to replace these without removing the engine, using the following method:

1 Drain the sump.
2 Remove the sump.
3 Undo the big-end caps.
4 Push out the old shells (you can use the new shells to do this).
5 Insert the new shells.
6 Replace the caps.
7 Tighten the nuts.

If the crankshaft is scored and requires re-grinding the engine must be removed. If the pistons have to be removed they can be drawn through the bottom of the block.

• CABLE BRAKE (GIRLING) MAINTENANCE & ADJUSTMENT

The majority of owners of cars fitted with this type of braking system seem to believe it is not as efficient as hydraulic brakes. In my opinion, this not true provided the cable brakes are correctly maintained and adjusted. Therefore, I feel I cannot do better than produce the relevant parts of an article written by 'DICKON' (a leading authority on Morgans) on this subject, and recommend that his directions be followed.

'First, try the adjustment of the brakes. Screw in the adjuster on each brake back plate until it will go no further. Go on, really tight! Make sure

that no reasonable amount of effort will persuade the adjuster to move another click. Use a spanner, which fits the flats really well, not an adjustable spanner, please, unless you really must. A set spanner about 4in long is right. Get all the weight on you can with that. Now slack back the adjuster to the last click and then one click more.

'When you have done all your brakes in this way, sit in the driving seat and give the brake pedal three or four hearty jabs. Go on, put some weight on! It won't bend! Then go round the adjusters and try them again. See if you can get that other click, not forgetting to slack back as before. Another couple of jabs at the pedal and you should be ready for a trial on the road.

'If the braking system is in good order you will be able to stop in the most convincing fashion. Watch the chap behind, his brakes are probably no good at all. "Neither are mine", you say?

'Better check the balance of the braking system.

'If the brakes are not adjusted to this setting, adjust the Bowden cable adjuster on the rear axle until the correct setting is obtained. The brake pedal can be wedged at the rubbing point of the front brakes whilst adjustment is made. Do not forget, front brakes just rubbing, rear brakes just free.

'The front brake rod should be left alone unless the pedal travel is excessive. In this case, take up the front brake rod to give reasonable travel, and adjust the rear brake cable as described previously.

'Normally, the rod or cables from compensators to brakes will not need adjustment, but if, in a moment of misplaced enthusiasm, someone has upset the adjustment, the rods or cables can be adjusted so that the compensator crossheads are about 1.5in before mid-position (at right-angles to the rods) when the brakes are hard on. When this adjustment has been made, leave it alone, and do not carry out any adjustment whatever to brake cables or rods, either for compensation or balance, unless the individual brakes have been correctly adjusted as first described.

'If the brakes are still no good, have them down and relined, checking the drums are unscored and free of grease. Run for a time and adjust the brakes as described, when they should delight you.

'A point to watch is that the nuts on each side of the expander plunger on the back of the brakes have not been tightened up. They are fitted with spring washers and should be free enough for the expander units to follow the brake shoes when the brakes are applied.

'After relining the brakes, the nuts on each side of the adjusters can be eased off until adjustments have been made, and then they should be tightened up really well. But again, not those on the expanders.

'Finally, don't expect to stop without putting some pressure on the brake pedal. It has been stated that 4-4 brakes are potent, but it is not suggested that the operation is very light; nor is the operation of some hydraulics. There is a 1951 car of famous make, fitted with hydraulics of equally famous make, which required 160lb pedal pressure for 80% stopping. You will get 80% on a 4-4 with less than 160lb if the brakes are in correct adjustment. But even 150lb is a good pressure (try standing on one foot to get the feel of it), and you must be prepared to press as much as that at times for really fierce stopping.

'Having done all this, look after them; the adjustment does not last for ever.'

Standard Vanguard and Triumph TR-engined Plus 4s

Most of what follows applies to the above engines, but it may be necessary to make minor alterations depending on the engine being worked on.

• ENGINE/LUBRICATION

Every 3000 miles it is desirable to drain the engine oil and refill with fresh oil. The oil cleaner cartridge should be replaced at the same time; ensure absolute cleanliness both inside and out when changing the cartridge. Remember not to tighten the securing bolt more than is necessary to obtain an oil-tight joint. Approximately one pint of oil will be lost due to the removal of the cartridge and the sump should be topped up before using the car.

If, at any time, the cleaner body is removed from the crankcase, great care must be taken to fit the joint washer correctly, otherwise damage will be caused through the 'blanking off' of the oil passages when the engine is next started. It is advisable to fit the washer to the crankcase and ensure that the holes in the washer match those in the crankcase before attaching the body. The filter on early engines was Fram type FHMS 800, element no. C800. Later engines used Purolator type MF5201, element no. MF52.

The flow of oil from the sump is as follows: drawn through a filter by the Holbourn-Eaton gear-type pump, it is delivered under pressure to the crankshaft journals, crankpins, connecting-rod little-ends, camshaft bearings and rocker shaft. Oil returning from the rocker gear lubricates the tappets and cams. The jets of oil from the connect-

ing-rod bearings lubricate the pistons and the timing chain is fed with oil from the camshaft bearing. Suitable oil seals are embodied at the front and rear ends of the crankshaft and effectively prevent oil leaking along the shaft.

• CYLINDER HEAD

Remove all parts that would obstruct removal of the cylinder head. Next, undo the cylinder head nuts, starting from the centre and working diagonally outwards towards each end. Remember to loosen all the nuts a little at a time to avoid any distortion of the head. (Note: the procedure of turning the crankshaft in order that the compression will 'break' the seal of the cylinder head should not be practised with this type of engine).

Once the cylinder head has been removed it is very important that the crankshaft is not rotated unless the cylinder sleeves are firmly clamped down against their seatings. This can be accomplished by using two tubes and washers fitted over the cylinder head studs, each secured with a cylinder head nut. They should be fitted on the studs on opposite sides of the block which are situated in the centre of each pair of cylinder liners. Ensure that the washers are large enough to cover the lips of the liners without overlapping the bore. This done, the crankshaft can be safely rotated to bring the pistons to Top Dead Centre to remove the carbon deposit. If the washers and nuts are not fitted, the sleeves may rise and, in consequence, water will leak into the crankcase after assembly.

On removal of the head and withdrawal of the pushrods, the passageways in the block leading to the camshaft - and thence to the crankcase and sump - must be sealed off with a clean rag or masking tape. The same applies to the rocker shaft oilway, both in the head and the block, to prevent the entry of carbon particles which could do serious harm to the engine.

A stick of soft solder or a very blunt knife are the most suitable tools for removing the carbon. On no account should emery cloth, sandpaper or the like be used.

The tappets should be removed and the inside cleaned just prior to refitting of the cylinder head to ensure that any chips of carbon which may have fallen into the chamber are removed from the pushrod seatings.

A new cylinder head gasket should be fitted each time the cylinder head is removed. The plain side of the gasket should be placed downwards against the cylinder block and coated on both sides with non-setting sealing compound.

When replacing the cylinder head nuts, tighten them gradually, starting in the centre and working diagonally outwards towards each end. This will produce an even pressure on the gaskets and prevent undue strain on the cylinder head casting. Finally, tighten to a pressure of 100lbf/ft. It will be necessary to tighten them again after the engine has thoroughly warmed through and also after 500-1000 miles.

• TAPPETS/VALVE CLEARANCE ADJUSTMENT (VANGUARD ENGINES)

Tappets should be adjusted to the following clearances:

Inlet 0.010in (cold)
Exhaust 0.012in (cold)

The method of adjustment is as follows: remove the rocker cover and turn the engine crank for half a revolution after the valve to be adjusted has closed (turning the engine is easier with the sparking plugs removed). Slacken the locknut and adjust the rocker screw with a screwdriver until the gauge is a sliding fit between the top of the valve stem and the rocker face. Now tighten the lock-nut and check that the clearance has not altered.

The tappets are plain cylindrical, sliding directly in the crankcase. Lift out upwards after removal of the cylinder head. The pushrods, with the exception of the end ones, can be extracted by pushing the rocker aside, but it is far easier to lift the rocker gear.

The rockers are bushed and are of two types; offset left-hand and right-hand. Nos 1, 2 and 3 are identical and No. 4 is drilled for oil feed and has a set screw to locate the shaft. The rockers are arranged on either side of the pillars and separated from each other by springs (the middle spring being longer). End rockers are retained by double coil spring washers and end caps retained by taper pins.

• VALVES

The inlet and exhaust valves are not interchangeable. Double springs are fitted, but the inner springs do most of the work. The inner springs are retained by a figure-of-eight slotted collar, locating on tapered registers on the stems.

To remove the inner valve spring, hold the valve and press the collar down clear of the tappet and move sideways. The top of the stem passes through the larger part of the hole.

Outer springs are retained by caps which are spigoted on the valve stems and held in place by the rockers. The caps are slotted to allow a feeler gauge to be inserted between cap and valve stem for clearance adjustment. When the rocker gear has been removed, the outer springs and caps can be lifted off.

On all except the earliest engines, the exhaust valves have auxiliary inner springs. All springs should be fitted with their close coil ends to the head.

• VALVE GUIDES

Valve guides are plain, have no shoulder and are interchangeable. Fit by pressing in until the guide projects 0.78in from the face of the head. Inlet valve diameter is 1.5in, exhaust diameter 1. 28in and face angle in each case is 45 degrees.

• VALVE TIMING

If, for any reason, this has been disturbed, it should be reset as follows:

Inlet opens 10 degrees before Top Dead Centre.

Inlet closes 50 degrees after Bottom Dead Centre.

Exhaust opens 50 degrees before Bottom Dead Centre.

Exhaust closes 10 degrees after Top Dead Centre.

• IGNITION TIMING

Ignition contact breaker points should be set to be just opening at 4 degrees before Top Dead Centre on No. 1 cylinder (closest to front). To obtain Top Dead Centre turn the crankshaft until the small hole on the belt pulley is in line with the pointer attached to the timing cover. The vernier scale should then be centralized and adjusted by means of the knurled screw. Each division on the vernier scale represents 2 degrees on the distributor, or 4 degrees on the crankshaft. The gap between the contact breaker points when fully separated should be 0.012-0.015in. The firing order is 1, 3, 4, 2.

• ENGINE REMOVAL

The Vanguard engine can be removed from the chassis with or without the gearbox; the gearbox can also be removed separately. (Note: the unit should never be split at the bellhousing until the gearbox has been removed. This is the removal procedure:

1 Remove the headlamp bar and brackets.

2 Remove the bonnet.

3 Disconnect the radiator stays at the front end.

4 Drain the radiator.

5 Slacken the hoses.

6 Undo the two radiator securing bolts at the bottom.

7 Remove the radiator.

8 Disconnect all the wires and controls to the engine.

9 Disconnect the four pipes connected to the front suspension lubricator on the dashboard.

10 Remove the centre dash panel over the bellhousing.

11 Remove the gearbox cowl.

12 Remove the propeller shaft cover.

13 Disconnect the handbrake cable at the front end.

14 Disconnect the speedo drive.

15 Disconnect the clutch linkage.

16 If the gearbox is to be removed separately, remove the rear mounting bolts leaving the Silentbloc bushes *in situ*. Support the centre tube with the aid of a jack.

17 Remove the rear brake cable at the hand lever. Disconnect at the rear end.

18 Remove the setscrews holding the gearbox to the clutch shaft housing.

19 Draw the gearbox back, bringing with it the clutch shaft and muff coupling.

20 If the engine and gearbox are to be removed as one unit, remove the gearbox cover complete with the lever and selector forks.

21 Remove the front mounting setscrews.

22 Fit slings round the front and rear of the sump.

23 Lift the engine, tilting to clear the front suspension.

To reassemble, the reverse procedure to the above should be adopted. However, remember to replace the sponge rubber ring which fits round the clutch shaft housing behind the bellhousing and seals against the dash panel.

• CRANKSHAFT

The crankshaft has three main bearings, the Thinwall, steel-backed, white metal-lined shells being located by tabs. The end float is controlled by split thrust washers recessed in each side of the centre main bearing and located by tabs in the cap. Main bearings and thrust washers can be changed in an emergency without the removal of the crankshaft by using the new bearings to push the old ones out.

The flywheel, with shrunk-on starter ring gear, is spigoted on the rear flange of the crankshaft and retained by four setscrews and located by one dowel. The flywheel has two dowel holes at 90 degrees and the shaft has two holes at 180 degrees so that the flywheel can be fitted in any one of four positions. The clutch spigot bearing bush floats on the end of the shaft.

The front and rear main bearing caps fit in square recesses in the crankcase. The rear cap has felt seals at the side; insert the felt after fitting the cap, using a long punch. A split aluminium oil retainer is spigoted and bolted to the back of the main bearing; it has an internal oil return thread and must have clearance to the shaft. The front clearing cap has an aluminium bridge-piece containing the sump flange and is

screwed to the cap with cheese-head screws. Cork seals are fitted at the ends. Main bearing diameter is 2.48in, crankpin diameter 2.09in. Running clearances are 0.001-0.002in for mains, 0.006-0.0025in for big-ends and end float should be 0.004-0.006in for mains and 0.0075-0.105in for big-ends.

• PISTONS

These are Aeroflex split-skirt in aluminium alloy. The gudgeon pins are located by spring rings; fit pistons and rods with the split in the skirt and the big end cap towards the near (camshaft) side. The top compression ring is chromium-plated.

The big ends will pass through the cylinders so remove and insert pistons and rods through the top. The liners must not be disturbed or the bottom joint may be broken. Piston clearance should be 0.0015-0.002in.

• CAMSHAFT

This has a Duplex roller endless chain drive. The camshaft sprocket is keyed to the camshaft and retained by two setscrews. The sprocket has four bolt holes, offset to give half-tooth variations in timing. It is also reversible to give quarter- and three-quarter-tooth variations.

The camshaft runs in four bearings, the second, third and fourth bearings being integral with the crankcase, and the front bearing a cast iron bush bolted to the front of the crankcase; endfloat is controlled by a bush between the sprocket and flange on the shaft.

The camshaft cannot be removed with the engine in place, as the front suspension crossbar is in the way. With the engine out there is no need to remove the sump.

Production timing marks are punched dots on the sprocket and camshaft and radial lines scribed on the faces of both sprockets in line with centres, with No.1 piston (nearest front) at Top Dead Centre on the firing stroke. To re-time the engine in the absence of the marks, remove the valve caps and outer springs from both valves on Nos. 1 and 4 cylinders, and set tappets at 0.014in clearance on the valve stems. At Top Dead Centre with both No.1 valves shut, both No 4 valves should be just open and should shut equally either side of Top Dead Centre (0.035in down piston stroke). Camshaft bearing clearance should be 0.003-0.005in and endfloat 0.004-0.012in.

• THE COOLING SYSTEM (VANGUARD ENGINES)

The engine is cooled by liquid (water). A bellows thermostat in a housing is bolted to the front of the cylinder head.

To remove the water pump, which has a carbon and rubber seal unit, first remove the radiator. Take out two short setscrews holding the body to the cylinder block, then undo the long setscrews passing through the flange of the bearing housing behind the pulley flange and take out the dynamo stay setscrew.

The fan belt should be adjusted by swinging the dynamo until there is about 0.75in of deflection on the vertical run of the belt.

Carburettor (all Fat Radiator 4/4 models)

Solex carburettors were a standard fitment on all flat-rad Morgans, and although the carburettor type varied according to the model of car (Coventry Climax engine, sidedraught; Coventry Climax-engined Le Mans and TT Replicas, downdraught; Standard Special engine, downdraught; Standard Vanguard engine, downdraught), their similarity of design enables me to quote methods of adjustment relevant to them all.

• SLOW RUNNING

This adjustment should only be made when the engine is at normal running temperature. The satisfactory setting of the slow running is a combination adjustment of (1) the brass volume control screw and (2) the steel spring-loaded slow running adjustment screw. Screw in the brass screw until it is lightly seated, then unscrew one complete turn. Next, start the engine and set the slow running adjustment screw until the engine is running at about 500-600rpm. Then unscrew the brass screw until the engine begins to run unevenly or 'hunt'. When this happens, screw the brass screw back in until the hunting just stops.

If this increases the engine speed again, drop the speed by unscrewing the steel screw to 500-600rpm. If this causes the engine to hunt again, re-adjust as previously described except that is only necessary to rotate the brass screw slowly in a clockwise direction until the engine is idling smoothly.

• PILOT JET

If the engine has had a fair amount of wear and tear, but does not yet require a rebore, it may not idle satisfactorily, no matter how carefully you carry out the adjustments listed above. If so, a larger pilot jet will frequently restore stable idling.

• MAIN JET

An engine may occasionally take a size

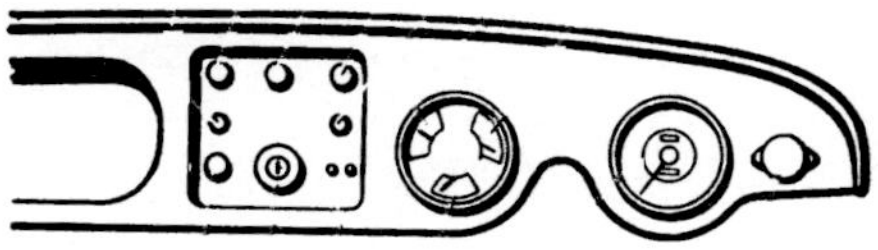

smaller main jet in the carburettor without any appreciable loss of performance, particularly in the summer months, although it may work efficiently in winter as well.

If ample power is being obtained and there are no symptoms of a weak mixture, such as spitting back excessively in the carburettor in cold weather when opening the throttle, then continue to use the smaller size jet, which will result in lower petrol consumption.

• FLOODING/LEAKAGE

If the carburettor leaks (but not at the joints), first examine the float. A shake will tell if there is petrol inside it. If so, a replacement float is the proper remedy because, if the leak is soldered, the weight of the float may be altered and cause trouble in other ways, besides increasing the fuel overflow due to the float tilting and sticking in the chamber.

If there is no choice but to solder the float, be very careful to first remove all petrol from inside it. This can best be done by placing the float in an saucepan of simmering water which will cause the petrol to boil out of the leak. When it stops, carefully remove the float, taking care not to squeeze it. Then solder the hole using the minimum amount of solder necessary to avoid changing the float's weight.

If the float is sound, dirt in the needle valve may be the cause of the flooding. Remove it bodily from the float chamber and swill in a container of petrol or paraffin. If the leak persists then the seating has probably been scored. If it is not scored too deeply, a new seat can be made by tapping the needle onto its seat with a light, double-ended spanner or similar. Care should be taken to rotate the needle each two or three taps. If this is to no avail, the only alternative is to fit a new needle valve.

Steering & suspension (all Flat Radiator models)

For the first few months of production, 4/4s were fitted with a simple reduction gear mounted halfway down the steering column, which was a very slightly modified version of the steering fitted to the three-wheeler. However, this was soon changed to the now familiar Burman worm-and-nut box.

The Burman steering system is not adjustable for wear, though end-play in the column can be taken up by the two large lock-nuts under the steering wheel. Special flat spanners must be used due to the limited thickness of the nuts. Care must be taken to avoid overtightening the nuts, as this will cause undue tightness in the ball-race and result in stiffness in the steering.

• TESTING FOR WEAR

First, ensure there is no movement between the drop-arm and the splined shaft by checking the tightness of the clamp nut. With the aid of another person, hold the drop-arm and gently rock the steering wheel back and forth: there should not be more than an inch of freeplay measured around the circumference of the steering wheel before the drop-arm is moved from one direction to the other.

To check the play in the steering box, jack the front offside of the car and remove the wheel. Some books recommend undoing the thin diagonal tension rod by releasing the back nut at the bottom and swinging it clear, but I have found this unnecessary. Remove the steering box cover end-plate, which is secured by three 1/4in BSF bolts; with the cover removed the end of the column and the nut can be seen and you can check for any perceptible backlash on the thread. The nut should begin to move immediately the column is turned, and if it fails to do so the trouble must be attended to. However, before fitting a new column and nut, test for play in the rest of the box. Hold the drop-arm firmly with both hands and try to push and pull it; if movement is possible, wear exists on either the rocker shaft or the bronze rocker shaft bushes, or both. The final test is for wear between the peg and the bronze nut. This is done by removing the cover from the top of the steering box (secured by 1.25in BSF bolts). Ask a helper to turn the steering wheel, then place your fingers on the rocker arm with the finger tips touching the nut. You should be able to detect if there is a slight time-lag between movement of the nut and movement of the rocker arm. If so, there is play between the sleeved hole in the nut and the peg.

The only method of rectifying these faults is by fitting new parts. If play exists between the nut and the column, a new 0.005in undersize nut should be fitted. (Note: there are two types of steering nut, with 5 or 6 start threads, and it is advisable to send a pattern when ordering). If, however, the play is excessive, a new column and nut are necessary. To remove the old column, first undo the large nut in the centre of the steering wheel; on postwar cars the wheel may be removed with the aid of a puller, but on prewar cars it is easier to split the wheel by undoing the four 1.25in nuts on the wheel boss. This leaves just the small rear part of the boss to be removed. The two large lock-nuts and cone are then unscrewed. Undo and remove the drop-arm and tap up the rocker shaft, when the column and nut may be withdrawn through the end of the box. (Note: care must be taken to prevent any of the loose ball bearings at the top of the column falling inside).

The bronze bushes may be drifted out with the aid of a long bolt; a fairly heavy hammer should be used as movement is restricted. The new bushes can be driven in using a piece of wood between the bush and the hammer or mallet. The top bush can be easily identified as it is bevelled off to leave room for the nut to turn. Both bushes are made with a slight lead on them to assist with easy insertion.

Before reassembly absolute cleanliness of all parts is essential. Assemble the column first and adjust the two locknuts until all endplay has been removed but there is no stiffness. Screw on the steering nut (if it is an undersize nut, a slight lead may have to be scraped or filed on the column) and assemble to rocker shaft. Ensure that the peg is directly over the sleeved hole before exerting any pressure, thus avoiding damage to the nut. Thin paper gaskets and jointing compound should be used on the steering box cover plates to ensure they are oil-tight. Only an SAE 140 oil should be used in the box; grease should never be used as this will not penetrate to the rocker shaft bearings and will therefore cause excessive wear.

• THE DRAG-LINK

The drag-link is designed so that wear can be taken up at its screwed ends. Remove the split-pin and, by inserting a screwdriver into the end of the rod, remove the inner screw. Clean and grease the screw and the ball, reassemble and tighten the inner screw right home, then replace the split-pin. This should be done at both ends of the drag-link.

• THE TRACK ROD

To test for wear in the track-rod ends, get a helper to turn the steering wheel just enough to move the track-rod and wheels. Any wear will be immediately noticeable. The only remedy is to fit new parts; two new bolts and conical bearings are all that will be required.

• SWIVEL-PINS, BUSHES & HUB BEARINGS

Wear can easily be checked by pulling and pushing the top of each front wheel. If there is much perceptible movement they require attention.

To refit swivel-pin bushes, jack the car as high as possible and remove the front wheels and shock absorbers. Unscrew the grease nut, remove the split pin and nut and extract the hub by the use of a puller. Strip off the brake drums and shoes and the brake cable, unscrew the nut at the top of the swivel-pin and, if the car is a postwar model, unscrew the two bolts which retain the rebound spring at the bottom tube. The swivel-pin can then be removed downwards. The springs, dust cover (if fitted), brake backplate and large bronze washer will come away. (Note: in the case of Plus 4s, remember to remove lubrication pipe and bolt).

Unless there is a fair amount of wear on the swivel-pins, it will only be necessary to renew the bushes. The old bushes may be drifted out and the new bushes inserted and driven in, using a wooden block to avoid damage to the bronze. The swivel-pins and stub axle assemblies require the bushes to be reamed out to fit the swivel-pins. If you are unable to do this yourself, any good motor repairer will handle the job for you at a moderate charge. Reassembly is the reverse of the dismantling instructions. (Note: absolute care must be taken to keep everything clean during reassemby). Check the ball-races before refitting, and if there is any roughness after they have been cleaned, or if they can be rocked, however slightly, they must be renewed. Do not forget to repack the hubs with grease and adjust the brakes.

Now set the tracking. The toe-in should be 0.12in maximum and on no account should the wheels be allowed to toe-out. If the steering is erratic over rough ground, it is most likely due to trouble with the shock absorbers. Also check the tyre pressures.

• STAY RODS

I am often asked the correct method of fitting and tensioning the two diagonal stay rods on the Morgan. The best method is to screw the lower end right in and lock with the nut, then tighten the upper nuts evenly so that the rod is neither in tension nor in compression.

• FRONT SHOCK ABSORBERS

The upper bushes of the stub-axles are fed with oil through drilled kingpins and pipes from a foot-operated valve connected to the engine gauge line. The foot plunger should be operated by pushing down about once every 50 miles in normal driving. The plunger operates a mushroom valve, which admits pressured oil to the pipe lines leading to the front suspension.

If too much oil reaches the suspension it could indicate that the driver is keeping his foot on the plunger too long, or that the valve is leaking. To rectify a leaking valve, disconnect all four unions and detach the valve from the dash panel. Insert a screwdriver in

the front hole to hold the valve head while the cap is unscrewed from the valve stem. This will free the releasing spring and the conical packing gland. Grind in the valve with fine grinding paste, and after ensuring that everything is absolutely clean, reassemble, then secure the cap by light riveting.

• FRONT SHOCK ABSORBERS
Newton hydraulic dampers of the 11,000 series with a 2.75in (70mm) stroke were fitted to the 4-4s. These require very little attention and, if they are giving a satisfactory ride and are not leaking, the best thing to do is leave them alone. The early series 11,000 units are of a sealed construction and cannot be dismantled; this type can be recognized by the end of the cylinder, which is rolled over on to the gland.

The later type is adjustable and, to adjust, disconnect the lower end, push it up as far as it will go and turn until a click is felt as an adjuster screw engages with a slot. Turn clockwise to increase damping and *vice versa*, pull out and refit. It is also possible to top-up this type, but from experience I have found that, should this be required, the unit is almost certainly faulty and rather than delay the inevitable it is best to change it.

The Plus 4s were introduced with Girling dampers No. DAS 4/131, of the usual telescopic type. They are a sealed unit with no provision for topping-up. The bearing bushes should be inspected every 6000 miles or so and when necessary they should be replaced. Later, the Girling equipment was replaced by Armstrong shock absorbers and, again, no provision is made for topping-up.

• REAR SHOCK ABSORBERS
Andre Multiplex shock absorbers were fitted to the 4/4. These require no maintenance but can be adjusted as follows: should the spring action seem too free, increase the frictional resistance of each shock absorber by turning the centre adjusting nut clockwise, but by not more than one graduation at a time. If the spring action feels too retarded and stiff, reduce the frictional resistance by turning the adjusting nut anti-clockwise. Careful adjustment as described should produce the ideal condition.

It should be noted that the optimum setting of the shock absorbers can only be tested at comparatively high touring speeds, and adjustments should be made to suit these conditions. Testing and adjusting over bad roads or at low speeds will only produce poor results. (Note: the frictional resistance required to effectively control the action of the springs is comparatively small, and care should be taken not to increase the pressure more than is absolutely necessary when adjusting to obtain the desired results. Inevitably, a considerable increase in pressure may be required for competition use).

Plus 4 models were fitted with Girling piston-type hydraulic dampers, model no. PV5: 1951-52 - right hand SA 237/13, left hand SA 237/12; 1952-55 - right hand SA 237/17, left hand SA 237/16. Periodically inspect the units and check the rubber bearing bushes for wear or signs of perishing.

Provision is made for topping-up with the dampers in place: thoroughly clean the unit before removing the filler plug and place a shield over the unit before the plug is removed. Next, check that the filler can is absolutely clean inside and out. Use Girling piston-type damper oil (thin); it is impossible to overfill. This applies to those dampers fitted up to 1958; after which they were sealed for life.

• FRONT SPRINGS
Should it become necessary to change the coil-type front top springs, the same procedure as outlined for the changing of swivel-pins and bushes should be followed. When changing the rebound springs a similar procedure should be adopted, but in the case of the 4/4 it should not be necessary to disturb the hub, brake drum back plates, *etc*. If the car is a Plus 4 the following procedure is recommended:

1 Jack the car and remove wheel and lubrication pipe.
2 Remove the track rod from the steering arm.
3 Remove the lubrication bolt at the top of assembly.
4 Remove the shock absorber from the hanger pin.
5 Remove the two bolts from the flange on the bottom of the lug.
6 Place jack beneath kingpin and raise.
7 Remove lubrication bolt from the kingpin bush.
8 Remove jack and the kingpin and rebound pin should fall away.

To replace the spring, carry out the above operations in reverse order.

• REAR SPRINGS
The semi-elliptic rear springs have Silentbloc bushes at their front anchorages and their rear ends work in a tubular crossmember which is flange-bolted to the chassis frame side-members. If the trunnion block is refitted, see that the solid half is at the top. Each spring is located against the bearing surface of the trunnion by a bolt passing through the block.

To remove the springs, jack the rear of the car and drive out the anchorage bolt. Jack the axle, remove the 'U' bolts and slide the spring forward out of the rear trunnion.

Coil-type springs are an integral part of the sliding-pillar independent from suspension fitted to all 4/4 and Plus 4 models.

• STUB AXLE BUSHINGS & WHEEL SHAKE

High wear points on the Morgan - and, believe it or not, there are a few - are the brass bushings that are pressed into the front wheel stub axles and ride on the vertical kingpin (centre pin). To check for wear, jack the front end and have a helper jiggle the front wheel about the horizontal (hold the top and bottom) and look for movement of the lower end of the spindle with respect to the lower rebound spring. Bushing wear results in a certain vagueness in steering and a feeling of front end looseness, but otherwise is not serious except in the extreme. Wheel shake is not cured by rebushing. Although the shake may disappear at first, when the bushings wear in the shake will return. Wheel shake is usually caused by loose damper blades or unbalanced wheels.

In the event of steering shake, check for missing parts. Make sure shock absorbers are secure at both ends, damper blades are intact, chassis stays and the centre pin bolt and nut are tight. Your helper should rock the steering wheel about centre while you check for vertical motion of the steering drop arm (as previously described) or motion of the damper blade components. Make sure the damper blade is sandwiched between chassis and metal strip but can move freely. Older (drum brake) Morgans do not have the aluminium spacer. The spacer is mounted on the chassis section allows the damper blade to move freely.

Areas that may require attention:

1 *Shock absorber replacement.*
2 *Upper shock bracket replacement.*
3 *Damper blade replacement.*
4 *Front wheel bearings.*
5 *Centre pin and bushing replacement.*
6 *Cross-axle stay adjustment.*

• SHOCK ABSORBER REPLACEMENT

1 Use two 9/16in open-end spanners to remove the two nuts at the top of the shock absorber (one nut is used as a locknut).

2 Push strut into the body of the shock absorber.

3 Remove the 5/8in nut on the bottom mounting stud of the shock absorber.

4 Work off the shock absorber.

Access for shock absorber replacement is achieved by jacking the front end and removing the wheels. I suggest putting a light coat of chassis lubricant grease on the rubber fittings at the bottom of the shock absorber so that it moves easily on the mounting stud.

• UPPER SHOCK ABSORBER BRACKET REPLACEMENT

1 Remove the shock absorber from the top attachment (bracket) as above.

2 Remove bolt that holds the wing brace and bracket to upper cross-axle (this may be hard due to limited access at lower end). Spanner size is 1/4in open-end.

3 Use 3/16in open-end spanner to remove one-shot lubricant fitting from the centre pin bolt.

4 Use 7/16in spanner to remove centre pin bolt and free the shock absorber mounting bracket.

Access is as per shock absorber removal, but cross-axle must be resting on support, front wheels off the ground, to prevent the centre pin from separating from the upper cross-axle. (Note: the car must not be moved with the centre pin bolt removed).

• DAMPER BLADES

Unless the brass portion that fits between the spindle and upper rebound spring has worn out or is missing, repairs will not require disassembly of the front suspension.

The damper blades are designed to prevent the upper rebound spring from influencing the steering due to spring 'wind-up'. In fact, they dampen minor front wheel shimmy, hence the name. Unless you have an early Series I 4/4, your car should have damper blades.

The damper blade assembly consists of a bronze plate sandwiched between the upper rebound spring and the stub-axle casting. A wide strip of spring steel is bolted to the bronze plate and runs to the frame, where it is either bolted down solid (Series I 4-4) or sandwiched between a plate and the frame. There is a triangular metal shim at each end of this sandwich and the damper blades lie between them. Any fore and aft movement of the damper blade can, therefore, be adjusted with the shims by merely slackening one of the bolts holding the assembly and tapping on the shim. This is an easy but important adjustment: if the slack in the damper blades is not taken up, wheel wobble (shimmy) often results.

If the damper blades show excessive wear they should be replaced. Simply unbolt from the brass ring and slip them out of the clamp on the frame. Slip in the new blade and bolt to the brass ring. Do not forget to adjust the shims after replacement of the damper blades.

Whilst your assistant rocks the steering wheel, note the fore and aft movement of the damper blade, which should not exceed 0.03in. If it does, adjust the shims.

• CROSS-AXLE STAY ADJUSTMENT

The purpose of the cross-axle stays is to triangulate the cross-axle beams into a rigid structure, thus preventing flexing during bump and rebound. The cross-axle stays are adjusted so that they are always in tension. Too little tension and the front will sag, too much may snap the stay. Tension on the cross-axle stays should be adjusted before having the front end aligned and checked after alignment.

1 Place the crossframe on four axle stands; two inside the 'cross' and the other two between the outer edges of the 'cross' and the stub axles.

2 With the front wheels removed, place a jack under a hub and jack slowly upward. The cross-axle should lift simultaneously from both the inner and outer stands.

3 If the cross-axle lifts from the outer stand first, the cross-axle stay is too loose. Care should be taken not to jack the car too far off the stands - 0.75 to 1in should be sufficient. If the outer end is more than 0.25in off the stand and the inner end has not lifted yet, the stay is much too loose.

4 Tighten the cross-axle stay until the cross-axle lifts off both stands simultaneously. The cross-axle is now properly tensioned. There is no advantage in increasing the tension further.

5 Repeat the above steps for the other side.

Steering/suspension work safety notes (all models)

1 If front hubs have been removed, when they are refitted ensure that the hubs have not been reversed. Look at each knock-off and make sure it threads in the same direction as the rear wheel on the same side. If you have removed the hubs and installed them in reverse direction (threads going the wrong way), you will loosen a wheel.

2 Check every nut and bolt again after you have driven about 100 miles. Also check for wheel bearing play and take up if required.

3 Have wheel alignment checked after bushing replacement.

4 When working on the front suspension, the cross-axle must be seated on wheel stands. Do not work on the car with it raised on a jack.

5 Make sure the caliper securing bolts have been safety wired in such a way that the wire will tend to tighten the bolts.

6 If the pipes have been disconnected, be sure to bleed hydraulic brakes.

Decambering front wheels (Plus 4s)

I will not attempt to go into the pros and cons of decambering your Morgan; suffice to say that all of the racers do it and it's recommended it if you use radial ply tyres.

To decamber your Morgan, you must move the bottom of the kingpin outboard in respect to the top. This can be accomplished in several ways. On the Le Mans cars, Chris Lawrence had longer bottom tubes made for the cross-axle. The local slalom crowd has been using 'decamber plates'. These are replacement baseplates with the hole for the kingpin moved 0.75in outboard. This is a very simple modification to make and the only drawback seems to be that a greater strain is put on the two mounting bolts for the baseplate (because of the greater offset). Grade 5 or better bolts should be used here.

This is the procedure:

1 Disassemble the front suspension as previously described. (Note: if you are only installing the decamber plate and not replacing the kingpins and bushings, disregard steps 1 to 4).

2 The end of the cross-axle will have to be modified for the larger offset. If it has not cracked already over the years, cut away the outer edge of the flange on the cross-axle.

3 Mount the new baseplate to the kingpin and reassemble the suspension. Be sure to use 5/16in grade 5 or better bolts to hold the baseplate to the cross-axle.

4 The last step is to adjust the toe-in. As you have effectively increased the distance between the two front wheels, toe-in will be excessive unless the track rod is also lengthened.

Toe-in measurement (Plus 4s)

The easiest method (aside from taking the car to your friendly local alignment shop) for adjusting toe-in is to use a plumb-bob. Any weight, like a nut on the end of a string, will do.

1 First, jack the front of the car, spin the front wheels and mark the approximate centre-line of the tread with a piece of chalk.

2 With the wheels pointed straight ahead, drop the plumb-bob from both the front and rear centre-lines of each front wheel. Mark the positions on the garage floor with a piece of chalk.

3 Now, measure the distance, rear to rear, front to front.

4 The toe-in is the difference between the front and rear measurements. If the rear distance is greater, you have toe-in; if the front measurement is larger, you have toe-out. You should

Wheel bearing part numbers (Plus 4s)

• PART NUMBERS FOR BALL BEARING CARS		
Inboard	Federal	LS11
Outboard	Federal	MS7
Seal	Aeroquip	962492A
• PART NUMBERS FOR ROLLER BEARING CARS PLUS 4S AND 4/4 (0.62IN TAPER)		
Inboard		
	Inner race (rollers/cage)	Timken 1988
	Outer race	Timken 1922
Outboard		
	Inner race (rollers/cage)	Timken 03062
	Outer race	Timken 03162
Seal		
	Aeroquip	962487 A
• RACING SUSPENSION PLUS 4 (0.75IN TAPER)		
Inboard		
	Inner race (rollers/cage)	Timken 14125 A
	Outer race	Timken 14274
Outboard		
	Inner race (rollers/cage	Timken 09067
	Outer race	Timken 09195
Seal		
	Aeroquip	900295 A

have 0.12-0.19in of toe-in, *ie* shorter in front.

Toe-in adjustment (Plus 4s)

1 Loosen the locknut(s) at the end(s) of the track rods between the front wheels and also the clamp on the rod from the steering drop-arm to the track rod. Note the position of the clamp on the track rod.

2 Rotate the track rod to lengthen or shorten it, as required.

3 When the toe-in is correct (0.12-0.19in), tighten the locknut(s) on the ends of the track rod.

4 Tighten the clamp on the tie-rod in the same position it was in before. (Note: if you forget to mark the position of the clamp, or just want to check the centring of the steering, do it now before you tighten the clamp on the tie-rod).

Front wheel bearings (Plus 4s)

Early (drum brake) cars up to about SN 4203 have ball bearing front hubs with a distance place between the two races. The outer race of the inside bearing is held by a screwed ring, which should be drilled and cotter-pinned in place. Failure to do this may cause loss of the wheel. Later disc brake cars have roller bearings.

Hub removal (Plus 4s)

On disc brake cars, the caliper - attached with two bolts (18mm), usually safety wired together - will have to be removed. Do not remove the brake hose from the caliper. Hang the caliper in some out of the way place, until required for refitting.

1 Remove the cotter key on the axle nut. On wire wheel hubs, rotate the hub until the two holes are aligned so that the pin can be worked out.

2 On drum brakes, back off the adjustment cams.

3 A 1in socket is required to loosen the spindle nut. Use 26mm if the nut has been damaged.

4 Slip the hub off. Ball bearing hubs will require a puller. Screw the spindle nut partway back on when breaking loose with the puller as this will prevent spreading of the spindle end from puller pressure.

Clean the wheel bearings thoroughly and inspect for signs of pitting on either balls/rollers or races. Ball bearings can be spun (dry) while holding by the inner race at about 45 degrees from the horizontal. You can usually feel wear (pitting) as a 'grabbing' feeling. Pack the bearings with clean grease and install in the reverse order of disassembly. Be sure to cotter pin the retaining rings on inboard ball bearings.

On wire wheel hubs, be sure that the right and left sides are not mixed. Mark them on removal. Tension tapered roller bearings by tightening them to 20lbf/ft of torque, and them back off one 'flat' on the nut.

Front suspension strip/rebuild & bush renewal (Plus 4s)

First, degrease the front suspension. The front end becomes unbelievably oily from the one-shot lubricater and road dirt. Use paraffin as a solvent.

1 Remove shock absorber (see earlier section). (Note: steps 2 through 4 may be disregarded if the front suspension is to be disassembled to replace the brass damper blade ring only).

2 Separate the spindle from the track rod. Remove the cotter key and castle nut (5/8in open end). Use a snap-on puller (P/N CJ89A/CJ89 3) to separate. Slip the puller 'C' disc between the rubber seal and spindle arm and install the puller. Screw bolt end of puller on to the rod and bolt. Things will separate with a loud snap.

3 Remove disc brake calipers. Hang the caliper out of the way, as in hub removal. On drum brake cars the fluid line must be disconnected.

4 Remove the hub (see hub removal).

5 Remove bolts (two) at the bottom of the suspension, one also holds the stay that attaches to the chassis (1/4in-W).

6 Place a jack, extended about 8 inches, under the lower centre pin retaining nut (between the bolts just removed). The car must be at least 12 inches off the ground to enable removal of the centre pin.

7 Remove the one-shot lubricator fitting connection to the centre pin

(1/4in-W).

8 Remove the centre pin bolt (7/16in-W). As the bolt is removed, the centre pin will move down due to the spring force. However, the jack will permit control of this force. After the bolt is removed, let the jack down slowly. Next, carefully jerk the upper rebound spring out at the top. Be very careful as this spring will still be slightly compressed.

9 The upper rebound spring and the cover tube can be removed along with the damper blade at this point.

When you have all the parts disassembled you will probably find major wear to the lower bush, primarily on the inboard side (away from the hub). In extreme cases the spindle holder may also be worn.

Removal and replacement of bushes consists of three major steps:

1 Remove the old bushes.
2 Install the new bushes.
3 Ream the bushes to fit the centre pin.

Bush removal. Removal requires a bushing tool that will push out the bushes and clear the internal diameter of the spindle. Use a jack stand to hold the spindle and a drift to drive out the bushes. A brass knock-off hammer is about the right weight. Both bushes are driven out from the same side.

Bush installation. Installation is best done using a hydraulic press (most garages can do this) or, alternatively, use a 0.5in drive socket reversed and a hammer, or a specially made drifting tool.

Bush reaming. The post-installation internal diameter of the bushes is smaller than prior to installation. Therefore, even though the bushes fit the centre pin before installation, they will require reaming after installation. A reamer is adjusted progressively to remove more metal until the centre pin will slide through the spindle with low resistance and very little play.

A modification. A useful way of extending bushing life, and keeping the front suspension clean, is to install an 'O' ring to keep the oil in and the dirt out. Drive the bottom bush in about 0.25in deeper during installation. Cut off about 0.19in from your old bush. After reaming, insert an 'O' ring in the bottom and hold it in place by driving in the 0.19in piece cut from your old bush. Be sure that the grease hole at the top end of the bush is not obscured. If it is, drill it through, taking care not to damage the threads.

• FRONT SUSPENSION REBUILD

1 The lower centre pin retaining plate must be removed from the old centre pin and installed on the new one.

2 Force the upper rebound spring and cover into position at the upper suspension member (do not forget the damper blade!) and drop a screwdriver through the centre pin bolt to hold the spring in position.

3 Lubricate the spindle bushing area and centre pin with chassis grease prior to assembly and perform a routine chassis lubrication when the job is completed.

4 Slide the centre pin up through the lower rebound spring, spindle upper rebound spring and cover. Place the jack under the centre pin nut at the bottom.

5 Work the centre pin into the suspension using the jack. Watch out for the spindle hanging up on the cover (inside the upper rebound spring).

6 You will probably find that the jack will not force the centre pin completely in as the jack will lift the whole car! The retaining plate can usually be drawn home by using long bolts. These will have to be replaced after the top is secure. Remember this when placing the jack so a bolt can be inserted into the holes. (Note: by installing an eye bolt in the garage floor, the cross-axle can be chained down so that it will not lift. The assembly can now be jacked right home. This is a 'must' if the engine is out of the car while doing suspension work).

7 Install the centre pin bolt from the top into the centre pin. Use a small mirror and torch to centre the pin under the hole. This can be difficult! When the centre pin is under the hole the bolt can be threaded home, which completes the assembly process. (Note: do not try to thread the centre pin bolt into the centre pin by 'feel'. If you miss and crossthread the bolt, you will ruin both the center pin and the bolt. For the same reason *do not* use the centre pin bolt to jack home the assembly).

8 The hub, if removed, can now be installed. If your car has tapered roller bearings, adjust for minimum play (tighten the hub nut for minimum play, 20lbf/ft, then back-off one flat). Fit a new split pin.

9 Install the steering track rod if removed.

10 Install the caliper when the hub is on (disc brakes only). Be sure to safety wire the bolts properly.

11 Install the one-shot oil pipes.

12 Adjust drum brakes.

Trunnions (Fat & early Cowled Radiator models)

Owners of flat radiator or early cowled radiator models will find that

their cars are fitted with trunnions in the rear frame crossmember, through which the main leaf of the rear spring slides.

If you were to unbolt each side of the tubular crossmember (the one that holds the twin spares) and remove it from the frame, you would see that there is a steel cylinder fitted in each end of the crossmember. In this cylinder is a round steel pin on which the main leaf of the rear spring rolls. It is often found that the pin has become flattened because the pins are rusted in place. A local machine shop can easily make some new ones for you (and not at great expense, either) from 0.37in diameter tool steel.

After cleaning out the cylinder's two holes in which the pin rotates and greasing them and inside the crossmembers where the cylinders fit, everything will work like new and you will be surprised at the difference.

Steering adjustments (Plus 4s)

These steering adjustments are settings that, once made properly, do not change very much. They are easy to check and make, so if you have done any suspension work or you do not know when steering set-up was last checked, it would be wise to check the following.

• CHECKING TIE ROD ENDS

Jack and securely support the front of the car. The area of the steering box, its mounts and the tie rod ends, drag link, kingpins and springs should be clean and properly lubricated with chassis grease. The kingpins and bushings should have no more than 0.12in play, and the spindles should be perfectly free to turn on the kingpins.

Check the tie rod and drag link ends using a large pair of locking pliers, squeezing the pin into its socket. There will be no appreciable movement in a good tie rod end. Approximately 0.06in movement is the maximum acceptable.

• CHECKING STEERING BOX CENTRING

Check the centring of the steering box in relation to the position of the roadwheels by winding the steering wheel from full right to full left lock, counting the number of turns, which should be 2.66 revolutions. Now, count back 1.33 turns from each full lock position. This should centre the steering worm, with the road wheels pointing straight ahead and the steering wheel spokes positioned where they are most comfortable for you in straight line driving.

If the road wheels are not pointing straight ahead, the drag link clamp needs adjustment. This adjustment procedure assumes that the Pitman arm (drop arm) is in its proper position on the rocker arm shaft.

The Pitman arm. There are usually marks stamped into the end of the rocker arm shaft and the Pitman arm during manufacture which must align. Do not try to pull the Pitman arm, which is on a tapered spline and is a very tight fit, off the rocker shaft without a very strong puller made especially for the purpose. It is very easy to damage the steering box by trying to remove the Pitman arm with the wrong tools.

The drag link. To adjust the drag link clamp position, loosen the pinch bolt that tightens the aluminium clamp around the tie rod (track rod). Clean any rust or paint from the tie rod for about six inches on either side of the clamp, and make sure that the clamp will slide freely on the tie rod.

The procedure. The steering box must be centred on its worm and, at the same time, the road wheels positioned so that they are straight ahead, and the clamp tightened in this position. The best way to do this is to centre the steering box and adjust its play with the drag link loose, then position the road wheels and tighten everything.

• ADJUSTING THE WORM

Centre the steering worm by winding the wheel from lock-to-lock as described previously. At the centre position of the worm there should be about 0.5in of free play at the rim of the steering wheel, which will become greater toward either lock. There is a 'high spot' manufactured in the centre of the worm, which is felt as a slight stiffness for about 1-2 inches of travel as you wind through the centre position of the steering box. In a good, unworn box, it should take about 12inf/lb to move the steering through this 'high spot'.

To adjust the box itself, make sure that the end and top cover bolts are tight, loosen the locknut on top of the box and back it up the screw about one or two full turns. Then, gently but firmly, tighten the adjusting screw down into the box until you feel it contact the rocker shaft and bottom (be sure to distinguish between tightness caused by damaged adjusting screw threads and that caused by a bottomed screw). Now back up the adjusting screw about 1/8th of a turn and temporarily tighten the locknut to prevent the screw from turning.

Wind the steering wheel back and forth across the dead centre position, moving at least a half revolution to either side. Note how it feels as you cross dead centre. Do you feel that 'high spot'? When the adjustment is correct, the 'high spot' should require the aforementioned 12inf/lb to cross. This is barely perceptible when turning the wheel with your finger at the rim. There should be no real binding.

It may take two or three attempts to get the correct adjustment: a very small movement on the adjusting screw has a substantial effect. Try moving the screw about the width of its slot at a time, but re-lock the nut after each adjustment or the movement of the rocker arm inside the box may alter the position of the adjusting screw.

Any roughness felt at the steering wheel while making this adjustment indicates damage to the steering worm which may be bad enough to require worm replacement.

When the box is adjusted to your satisfaction, permanently tighten the locknut and make a final adjustment check, as the setting may be changed by the upward pulling action of the tightened locknut.

• DRAG LINK ADJUSTMENT

Centre the road wheels by either sighting across the sidewalls of the front tyres to a reference point on the rear tyre or fender, or by rolling the car along a straight line. When the wheels are centred, re-centre the steering box and tighten the drag link clamp pinch bolt where it positions itself along the tie rod. (Note: the pinch bolt should only be tight enough to securely lock the clamp onto the tie rod. Over-tightening can break the aluminium clamp. Should this occur, you will lose control of the steering).

• CENTRING THE STEERING WHEEL

With the wheels straight ahead, note the position of the steering wheel spokes. If you wish to change the position, loosen the steering wheel retaining nut (use a rag over the nut to avoid damaging the chrome finish), remove the steering wheel and replace it in the position of your preference for straight ahead driving. Tighten the retaining nut.

• STEERING BOX LUBRICATION

Fill the steering box with 90 EP gear oil through the filler plug in the top cover. Then remove the rubber plug on the steering column jacket (about 18in up the column from the box itself), fill the column with the same oil and replace the plug. If you need a new plug, VW Beetle brake adjuster hole plugs are exactly the right size.

The column full of oil serves as a reservoir that will keep the box lubricated if it has any leaks. It also maintains a full supply of oil to the upper worm bearings. It is beneficial to add a molybdenum disulfide gear oil additive to the steering box oil as this prevents wear and gives slightly lighter steering.

• WHEEL ALIGNMENT

The best handling and steering characteristics are achieved with the following alignment specifications:

Toe-in 0.06 to 0.12in

Camber 0 to 0.5 degree negative

(Note: the factory specification for camber is 2 degrees positive).

Caster 4 degrees

Adjustments and alignment settings should be made in the following order:

1 Caster
2 Camber
3 Toe in
4 Steering box centring
5 Drag link pinch bolt centring
6 Steering wheel positioning.

Adjusting headlamps (all models)

Adjusting the headlamps on your Morgan is a relatively simple task. All you need is a screwdriver and a wall to shine the light beams on.

Park the car about three feet from your garage door (or other wall) with the lights on low beam. You may need to do this at dusk or even in the dark to tell exactly where the light is shining on the wall.

Mark the outline of the beam shining on the wall with a pencil and then move the car back 25ft from the wall. The top of the low beam should be no higher than the top of the circle on the door, and no lower than the centre of the circle.

If the beam shines outside the circle, use the screwdriver to turn the adjustment screws on either side and above the headlamp housing. Turning the screws will cause the beam to move left to right, or up and down on the wall. A little experimenting will allow you to align the lights with the marks on the wall. Once the low beams are adjusted, the high beams should be in the right spot also.

If you install halogen headlamps, you will find that the highly concentrated white beam does an excellent job of illuminating the road at night. However, halogen lights must be adjusted more precisely, because the beam is more concentrated and there is less margin for error.

Clutch lining condition (all models)

A lining in good condition is a mid-brown colour and is polished very highly, with the grain showing through the polish - this is assuming that no oil has reached the lining. If small quantities of oil reach the lining it will burn off in normal use, which has the effect of

gradually darkening the lining. However, provided the grain still shows clearly through the polished surface, the efficiency of the clutch is very little affected. If increased quantities of oil or grease reach the clutch lining, one or more of the following conditions may exist:

1 The oil may burn off and leave on the surfaces a carbon deposit which assumes a high glaze and causes clutch slip. This is a very thin but quite definite deposit which, in general, hides the grain of the lining material.

2 The oil may partially burn off and leave a resinous deposit on the faces, which frequently produces a fierce clutch and, possibly because of the tendency of the linings to adhere to either the flywheel or pressure plate face, a spinning clutch.

3 A combination of conditions (1) or (2) can cause clutch judder during engagement.

4 Still greater quantities of oil on the clutch produce a black, soaked condition of the lining resulting in either clutch slip, fierceness or judder in engagement, according to conditions. If conditions (3) or (4) are experienced, the only remedy is a new clutch lining and prevention of oil reaching the clutch.

Cooling system maintenance (all models)

All cars sooner or later suffer from overheating, frequently after an overhaul which, instead of rejuvenation, results in boiling, knocking and other ills which make the unfortunate owner wonder if it was all worthwhile.

In an ideal world the radiator should always be filled with soft rainwater, but this is rarely done. So, after a few years of using ordinary tap water in the radiator, it will often be found that deposits of scale have formed on all surfaces exposed to the water. These cause resistance to the flow of heat, as a result of which, the engine's internal temperature rises. It is possible to experience a seriously overheated engine without boiling the coolant; in such a case, scale is probably the culprit.

Many good methods have been devised for removal of scale and one of the author's used to great effect on a number of badly scaled engines calls for small quantities of sodium metaphosphate in the cooling water, combined with regular draining and flushing of the cooling system. The effect is gradual, but very definite.

When scale forms in a cooling system the principal offenders are compounds of calcium, which form a crystalline cement with other non-crystalline compounds. The sodium metaphosphate has the property of exchanging bases with the calcium compounds, so destroying the bond of the scale and reducing it to a sludge which can be washed out. The treatment, therefore, is to put a small quantity of sodium compound, say 1/4oz, into the cooling water which, after a few hours of normal use of the car, should be drained, then the system washed out and refilled with clean water and a fresh dose of phosphate. The treatment is continued with the car in normal use until inspection shows the cooling system is free of scale.

With some engines, particularly the Coventry Climax fitted to prewar 4-4s, various odd corners, especially round the exhaust ports, form sludge traps; these deposits must be washed out otherwise the circulation in some parts of the cooling system can be blanked off, causing overheating. Climax engines which show severe signs of running-on can often be cured by attention to this point and by a general cleaning of the water system.

When the cooling system is free of scale it is worthwhile examining hoses to ensure that there is no obstruction to the flow of water and to clean the outside of the radiator core of the dust, oil, dead flies and paint which so frequently impedes the free transfer of heat from the radiator to the surrounding air. One final point: when the cooling system is really clean, why not keep it so by using really soft water? Filtered rain water is ideal.

Another useful method of cleaning scale is to remove the radiator and allow it and the engine block to dry out completely. Then, with the aid of a pressure water hose, flush the radiator and engine separately in both directions, first top to bottom and then bottom to top. Carry on until the water flows cleanly in both directions. This removes a lot of loose scale and rust, but remember it does not remove deposits from the water channel walls; for this you must use the previously described method. However, you will be amazed at the amount of scale and dirt that pressure flushing will get out of the system.

Problems sometimes occur with the electric water temperature gauge. The most common faults are the gauge reading hot when the ignition key is turned on, or the gauge showing a broad fluctuation of temperature while driving. The remedy is generally two-part -

First, the sender unit in the radiator may not be electrically earthed. The cure is to clean the paint off the radiator brace rod on the right side of the car where it bolts to the radiator. Next, loosen the nut on the rod where it bolts

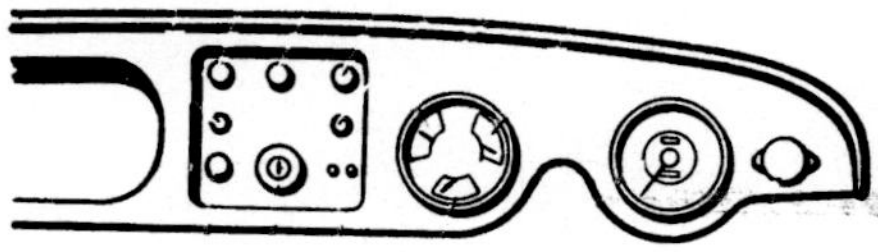

to the tool box. This point is insulated by two rubber washers - one inside the tool box and one on the outside. Run a piece of wire or solder from under the clamping nut, loop over the washer and clamp it between the washer and the tool box.

The second problem involves the wire leading from the radiator sender unit which is taped to the right-hand radiator bar. This wire tends to break inside its insulation at two points - where the wire leaves the sending unit and where it leaves the rod and goes into the wiring loom.

Another fault is when the temperature gauge reads hotter than the engine actually is. Tests have revealed that the temperature gauge is often a little off in its markings; between the 'N' and 'H' mark is about normal temperature, or about 175 degrees F.

One last point - if you use a rust inhibitor in the coolant, do not use one that contains chromates which will soften the gum rubber hoses that the Morgan has.

Compression testing and valve clearance adjustment (Plus 4s - TR2 &TR3 engines)

If you have a compression gauge available, it is a good idea to check each of the four cylinders prior to setting the valves. A normal/good engine, with all four plugs removed and cranked with the starter motor should produce from 165 to 185psi on the gauge.

All the manuals recommend two sets of cold clearances: 0.010in inlet; 0.012in exhaust for 'normal' driving; and 0.013in inlet and exhaust for 'fast' motoring or competition. You will have to be the judge of this. A 0.013in setting would appear to be the most satisfactory for all types of driving, although it does tend to clatter somewhat. Incidentally, the clearance you use on the valves will affect the micrometer setting on your ignition.

To make the actual adjustment, use the system of 9, which ensures that the valve train is properly clear. This means that you check clearance on No.1 (nearest front) valve when No. 8 (1+8=9) is fully compressed, No 2 when No 7 is compressed, and so on, turning the engine as required for each setting. After you have set all the valves, crank the engine with the starter motor for a few revolutions and re-check all your settings. You may find have apparently altered. This can be caused by worn or sticky lifters or improper seating of the pushrods. Continue to check the valves this way until all 8 settings check out two or three times in a row.

At this point it is desirable to test the compression. If all four cylinders are within 5 or 10psi on the gauge, your car's engine is in good shape, unless they are all below 150psi. In this case the engine needs at least a valve grind and possibly new rings. If one or two are quite low and the others above 165 to 170psi, there may be a burned or warped valve (usually exhaust) and you may be able to cure the problem with a valve grind.

Please bear in mind that if your compression gauge has turned up a problem you cannot possibly achieve a really satisfactory tune until it is corrected. Tuning is entirely a matter of co-ordination and balance of all the factors involved.

Carburettor adjustment (TR & Vanguard engines)

• TR-ENGINED PLUS 4S WITH TWIN SUs

First, run the engine to normal operating temperature and, when it is warm, run the engine up to 4000rpm a couple of times in the lower gears. This will keep the plugs from oiling up. Loosen the clamping bolts on the throttle spindle connection and disconnect the control link by removing one of the clevis pins. Be sure the throttle control levers are as far forward as they will go.

Set the idling speed to about 750-800rpm by adjusting the throttle screws on both carburettors, maintaining balance on the two carburettors by listening for an equal hiss at each air intake: they must sound alike. A stethoscope is fine for this, or if you do not trust your ear you can invest in a special airflow gauge which is available from most motor accessory shops and not too expensive. This gives a fairly accurate measurement of the rate of airflow through each intake and can be used on any carburettor.

When the throttles are synchronized, set the mixtures. This is done by holding the throttle levers forward and adjusting the jet adjusting nuts (on the bottom of the carburettor) until the engine runs smoothly. The use of your stethoscope or gauge here is also recommended to maintain balance. If the idling speed increases more than you want during this procedure, reduce it by resetting the throttle screws, but be careful to still maintain balance.

When the mixture is correct on both carburettors, lifting the piston on one of them should cause the engine to run rough and misfire due to excessive leanness of the mixture. This should be checked on both carburettors and the engine should behave the same for both, regardless of which piston is lifted. If the engine speeds up when you raise the piston of one carburettor, the other is set too rich. If it stalls - too lean - and you must adjust accordingly.

When the carburettor mixtures and throttles are set, tighten the clamps on the joint throttle spindle, replace the clevis pin and start final tuning. This last is a matter of the final balance between ignition and carburation. It is a simple procedure about which you can be as fussy as you please, or ignore it completely. All that is necessary is go back to the micrometer adjustment on your distributor and adjust it for smoothest running. Then you go back to the carburettors and see if you can adjust the mixtures still further, and then back to the ignition and so on, until you are either convinced it is as sharp as you can get it, or you are bored with the whole silly business.

As you go through this procedure, check for two things: 1) smoothness of operation and, 2) speed of response when you push down hard on the throttle to rev up to 3500-4000rpm. When the engine really barks back at you and seems to jump eagerly at a touch of the throttle, you have it made. The fun of driving a well-tuned car is well worth the effort of setting it properly.

• VANGUARD-ENGINED CARS

There is very little to adjust on the earlier cars equipped with the Vanguard engines and Solex carburettors. About all you can do is vary the main and air correction jets to change the performance of your car and, in tuning, you can only set the 'slow running' adjustment to give the idling speed you want. After this adjustment is made, final tuning is the same as it is for the TR2 and 3 engines.

High performance tuning & race preparation (TR2, 3, 3A and 4 engined cars)

This engine was designed to produce very good low speed torque with high power output at medium revolutions, and although the maximum revolutions can be raised by 2000 over the safe recommended peak of 5000rpm, it is most inadvisable to use 7000rpm. Continuous high-speed testing has proved that a crankshaft can break after 20 minutes at 6250rpm, so therefore the revolutions must be kept below this figure. On standard (*ie*, untuned) engines, maximum torque is produced at 4400rpm so there is little or nothing to be gained from exceeding this figure when accelerating. On a tuned engine, maximum torque will be produced at higher revolutions, but it is still advisable to restrict these to below 6000rpm.

For racing, the crankshaft, flywheel and clutch pressure plate should be balanced to relieve the internal stress imposed by the out-of-balance forces and give a smoother unit at high revolutions. Much stronger valve springs are not necessary because of the restrictions upon revolutions, but slightly stronger springs are available which will lengthen the effective life of the valve seats by making them more gas-tight. These springs have not been tested over 6500rpm, but no signs of valve float or bounce have been apparent up to these revolutions.

Even if nothing breaks, bearings are liable to fail due to overloading and high oil temperature. Every tuned engine should have an oil radiator fitted which, under racing conditions, reduces the working temperature of the oil by 20-25 degrees C, giving a safe margin.

Apart from the fitting of twin-choke Weber carburettors, the component which gives the greatest increase of power without imposing any additional loading on the bearings is an extractor exhaust manifold. This is proved by dynamometer test to develop 7-8bhp more at 5000rpm and weighs approximately one-third the weight of the cast-iron manifold.

The high-speed inlet manifold for SU carburettors provides the nearest approach to individual feeding of each inlet port, with minimum interference from the adjoining inlet, only improved upon by separate carburettors on each port or twin-choke carburettors. It also gives perfectly even distribution to all four cylinders, whereas the standard inlet is inclined to deliver a richer mixture to numbers 2 and 3, as it feeds through the balance pipe. Weber twin-choke instruments give an increase of 18bhp over two single carburettors and are supplied with cast light alloy manifolds. Webers should only be used with the extractor exhaust manifold, otherwise the additional performance potential will be reduced by the restriction imposed by the standard manifold. It is no use pushing in more air, if it cannot be effectively disposed of.

Modified camshafts must be correctly designed to produce good low-speed torque, owing to the limitations on engine revolutions. It is quite easy to produce much higher revolutions and high-speed power, but these designs have no power below around 3000rpm, so on a Triumph engine are of little value. A good modified cam has been carefully developed for good low-speed torque, which is present from 1000rpm upwards in addition to giving approximately 6bhp more at 5000rpm.

Greatly increased torque and power are produced by conversion to 2.2-litres, by fitting 37mm bore liners and pistons which also increase the compression ratio to 9.24:1 - with the standard combustion chamber. Even with a

modified higher compression ratio cylinder head, the engine is smooth and perfectly tractable, due to modern scientific methods of modifying the head for better gas flow and combustion.

Ram pipes for SU carburettors provide a few per cent more power, and a cold air intake allows a great weight of mixture to be sucked in, with more power being developed.

The standard pushrods are of light tubular steel so there would be nothing gained by changing these for light alloy. These days, carbon-fibre pushrods are used for racing.

The standard inlet valves are of optimum diameter and must not be enlarged, better cylinder filling being obtained by gas flowing ports and seatings as effected on the modified cylinder head. Bhp tests prove that any alteration to valve size will result in a drop in power due to lower gas speeds.

A modified distributor can provide an ignition advance curve more appropriate to a tuned engine.

Roadholding and high-speed cornering are vastly improved by the fitting of an anti-roll bar and adjustable shock absorbers, used in conjunction with competition springs. Under racing conditions, cornering speeds have been increased by 15-20mph over standard by these modifications.

• RACING PREPARATION

The standard SM needles should be changed for RV Racing but these will mean heavy fuel consumption so are not recommended for road use.

It is better to use a synthetic oil, like Mobil 1, with the larger clearances and to warm up the engine before high speeds. Piston clearances can be increased to 0.008in at the expense of heavier oil consumption; pistons should be buff-polished to minimize carbon deposits.

Do not remove the thermostat, otherwise overheating will result.

When installing new contact breaker points, check point gaps after use, owing to 'bedding in', correct gap 0.015in.

A high-pressure electric fuel pump should be fitted to provide a more constant head of fuel.

For serious racing, the crankshaft should be checked for flaws or cracks by X-ray or similar method. It can be strengthened by shot blasting the webs and hardened by nitriding. The main and big end journals can be micro-polished or hard chrome-plated 0.001in undersize for the high pressures and temperatures encountered with racing. The connecting rods should be checked for straightness and correct fit on both big and little ends.

• HOMOLOGATION

Modified inlet and exhaust manifolds, the oil cooler kit, modified camshaft and alternative Weber carburettors have been homologated by the FIA for competition.

Engine - standard & competition differences (4/4 Series V)

This section deals with the differences between the competition and standard engine versions of the Morgan 4/4 Series V.

The 4/4 competition engine is identical to standard except that the cast iron intake manifold is replaced by an aluminium alloy casting with separate tracts for each port. A water-heated hot spot is incorporated, connected to an adapter on the front of the head. The hot spot surrounds the riser from the dual-choke carburettor and a drain tube for surplus liquid fuel is fitted at the lowest point in the riser. A short strut is required to locate the manifold and attaches to the block. Appropriate small fittings and hoses are necessary to connect between the water passages in the block and the intake manifold. The carburettor is changed to a Weber, 28/36 DCD 16/18 two barrel compound, from the single-throat Zenith. The carburettor has progressive linkage to the second throat which is activated near the end of the accelerator travel. The manifold and carburettor are interchangeable between the Cortina 1500 engine (standard) and the Cortina 1500 GT (competition).

The simple cast iron exhaust manifold is replaced by the sub-divided type, optimising pressure-wave action in the gases to augment scavenging and charging. It is welded from steel tubing, with long branches for each exhaust port, joined to combine the flows from the inner and outer pairs of cylinders and finally merging into a single downpipe. The junction with the cylinder ports is formed by a separate flange at each branch-end. The usual combined gasket of reinforced asbestos is used. The Ford Cortina GT manifold is not usable on the 4/4.

In order to make use of the superior breathing allowed by the above modifications, the engine is changed internally. The compression ratio is increased, the valve size is changed, valve lift is increased, ignition timing is altered and valve timing is changed, resulting in increased overlap; all this means a significant horsepower increase of 31 per cent between standard and competition units.

The standard cylinder head casting is employed but the combustion chambers are modified and dimen-

4/4 Series V engine - standard & competition version specifications

	bhp at rpm	lb/ft at rpm	CR	valve diameter Inlet	valve diameter Exh	valve lift Inlet	valve lift Exh
Standard	59.5@4600	81.5@2300	8.3:1	1.437in	1.188in	0.315in	0.343in
Comp	78.0@5200	91.0@3600	9.0:1	1.410in	1.250in	0.319in	0.335in

	static timing	spark plug	valve timing	open	close	valve setting hot	valve setting cold
Standard	8° BTDC	Champ N5	Inlet	17° BTDC	51° ABDC	0.010in	0.008in
			Exh	51° BBDC	17° ATDC	0.017in	0.018in
Comp	10° BTDC	Champ N4	Inlet	27° BTDC	65° ABDC	0.014in	0.012in
			Exh	65° BBDC	27° ATDC	0.021in	0.022in

sioned for a compression ratio of 9:1. A special high-lift cam is used which has the number 116E stamped on its rear end. The pistons are graded selectively for the cylinder bores in the usual manner, but have extra clearance for sustained high engine speeds.

If you decide to upgrade your standard unit to competition specifications, you will have to purchase a replacement carburettor, intake manifold, exhaust manifold and camshaft, as well as have work done on the combustion chambers and new valves fitted.

The basic 1500 Cortina engine (122E) is produced with the 8.3:1 compression ratio and 7.0:1 is available as an option. The block is identified by the cast marking 120E-6015-AB on the lower left-hand side. The compression ratio is denoted on the machining location pads above the valve ports, 'H' for 8.3:1 and 'L' for 7.0:1. Factory compression figures (cranking with starter, all plugs removed, and carburettor full open) are 173psi for 8.3:1 and 144psi for 7.0:1 at sea level. It is unlikely you will have a low compression model unless the car was purchased in England. The 120E differs from the earlier block in that it has a thicker clutch housing mounting flange with reinforced webs. The camshaft is marked 109E - 6250.

The 1500 Cortina GT has the 9:1 compression ratio, with compression of 179psi at sea level. Other markings are as for the 1500 Cortina.

The 4/4 Competition model is delivered with a tachometer as standard equipment and a temperature gauge can be ordered as an extra on both models. The standard rear end ratio is 4.56:1 with 4.1:1 available to special order. With the standard ratio at an indicated 65mph, the 4/4 is turning 4100rpm.

Engine tuning (4/4 Series IV and V)

The following deals with the 1500cc Ford-powered Morgans (116E through 122E). Specifications and data are for the competition model (Ford denotes the power plant as the 1500GT) with the standard specifications shown where they differ. Most Morgan owners know little about the Ford engine and it took some research to discover exactly the difference between the competition and standard versions.

This article covers a complete tune which can be done at home with normal tools and should bring the car up to optimum performance as per factory specifications. Further improvement would require dynamometer time and detail adjustments to suit the individual engine. To do a comprehensive job I recommend the following steps. It may seem a lot of work but when you are finished your car should be in excellent tune.

Tuning steps

a) Road test.
b) Adjust valves.
c) Compression test.
d) Adjust timing and ignition.
e) Change spark plugs.
f) Adjust carburettor.
g) Road test.

Parts and tools required: set of points and condensor; set of plugs; valve cover gasket; distributor cap and rotor; compression gauge; timing light (optional); torque wrench (optional).

a) In order to see if performance has been improved, undertake a simple road test. On a given stretch of flat road, record the time required to accelerate from 10-30mph in second gear, and then from 20-60mph in third gear. Comparison of your times before and after tuning should show improvement. The recorded time for each test should be an average of three runs on identical roads.

b) Valve clearance adjustment is very important and must be done to assure a valid compression test. If you have access to a torque wrench I recommend re-torquing the head prior to valve adjustment. I do not know if the Ford is prone to head warpage - I do not think so - but one preventative measure is to keep the head torqued properly. Unless the proper extension for your torque wrench is available, you will have to remove the valve train and thermostat housing for access to the head bolts. Slacken the bolts slightly and then torque-tighten to 65-70lbf/ft, using a sequence which spirals outward from the most central bolts. Reinstall the rocker shaft assembly (torque to 17-22lbf/ft) and proceed with valve adjustment.

Competition model valve clearances -
Inlet: 0.010in (0.012in hot)
Exhaust: 0.023in (0.022in hot)

Standard model valve clearances -
Inlet: 0.009in (0.010in hot)
Exhaust: 0.019in (0.017in hot)

Use the fan belt to turn over the engine with the spark plugs removed. The following chart will reduce the turning of the crank:

with 1 & 6 full open, adjust 3 & 8
with 2 & 4 full open, adjust 5 & 7
with 3 & 8 full open, adjust 1 & 6
with 5 & 7 full open, adjust 2 & 4
(valve numbers are counted from the front of the engine).

The valves can be adjusted with the engine running (hot) but it is a messy job (oil) and it helps to have two people (one to adjust, one to measure). The cold setting should be used when the engine has been sitting overnight.

c) The compression check is performed while cranking the engine with the starter and recording the compression in one cylinder with all other plugs removed. Use the button atop the starter solenoid but make sure the car is not in gear. Normal compression pressure at cranking speed (400rpm) is 179psi at sea level (for the standard model, 173psi.) Actual pressure will vary according to many factors but it should be above 130psi and each cylinder's readings within about 10psi of the others. The test should be performed with the engine warm and the throttle fully open. Emphasis is placed on the evenness of the readings with the actual value of less concern; as long as the compression is even, the engine can be given a good tune. If you are out as much as 25psi between readings you are wasting your time - though economics of the situation may dictate going ahead. To determine if the rings or valves are at fault, squirt some oil into the plug holes and re-check the compression; if the readings are about the same, then the valves are leaking, if the readings rise a significant amount then the piston rings are not sealing. If low compression is found between two adjacent cylinders it is possible that the head gasket is blown. It would not be difficult to perform a valve or ring change on the 4/4 due to the excellent accessibility and small size of the components involved.

d) Set the points next. Remove the distributor cap and check it and the rotor for excessive pitting or erosion of the contacts; check the carbon high tension brush for spring tension by pushing on it. If replacement is not required, clean with a dry rag and proceed to the points. If they are badly pitted they should be replaced and then, with the breaker arm resting on one of the four cam lobes, set the gap to 0.015in with a feeler gauge. If a replacement is not available the points can be removed and dressed with emery paper or a point file until all signs of pitting or burning are removed. The condensor will outlast many sets of points and generally does not require regular replacement. When it fails the points will arc and wear rapidly. Oil the distributor by smearing grease on the distributor cam, place a drop of engine oil on the breaker base plate through which the distributor shaft emerges, and two drops of oil on the head of the screw in the distributor shaft exposed by pulling off the rotor. Be sure not to get any on the points.

The timing is set statically; do this every time the points are set. Ignition timing is the relation between the crank rotation (combustion cycle) and the spark ignition, as opposed to valve timing - the relation between crank rotation and valve actuation. Turn the engine over until cylinder No.1 (nearest front of engine) is starting its compression stroke (you can tell by feeling the compression build up, also, with the rotor installed the points should be starting to open with the rotor pointed at 1.30 to 2 o'clock). Continue turning (using the fan belt) until the notch (rear flange of pulley) on the crankshaft pulley is lined up with the timing mark on the upper left portion of the timing chain cover. This gives the static timing of 10 degrees BTDC (8 degrees BTDC for the standard model).

Now at the distributor check that the fifth (fourth on the standard model) line of the ignition timing scale counted from the vacuum diaphragm housing is in line with the edge of the distributor housing. At the fifth line the distributor is set at 10 degrees static advance (fourth line equals 8 degrees).

Next, undo the distributor body clamp bolt and rotate the distributor a little way anti-clockwise, then clockwise until the points just start to open. The exact location of the point opening can be accurately determined by using a 12-volt light bulb connected between the earth and low tension terminal on the upper side of the distributor. When the ignition is on, the light will illuminate only when the points are open (rotate clockwise until the light comes on). The engine is now statically timed to either 10 or 8 degrees before Top Dead Centre. The clamp bolt should be tightened.

e) A new set of plugs is installed next. Spark plug types -

Competition

Autolite AG22
Champion N4
KLG B7E

Normal

Autolite AG32/AGR32
Champion N5
KLG B6E

The specified spark plug gap is 0.023-0.028in. Plugs should be installed by hand, although the cast iron head is not prone to cross-threading. Check the old plugs for condition and note the colour as an indication of mixture. If the plug has an oily deposit you may have a broken ring or loose valve guides allowing excessive oil into the combustion chamber. If the plug has a black sooty deposit the mixture is very likely over-rich; if the plug tip is

whitish, the mixture is weak. The plug should be a dark grey colour and all should look about the same; if they do not there is a fuel distribution problem - this is more common on multi-carburettor engines. The plug should be torqued just enough to flatten the crush washer.

Part d) can now be completed. A timing light should be used to check the automatic advance mechanism (controlled by engine intake manifold vacuum and rpm - through centrifugal weights in the distributor). Attach the timing light at the number one plug lead and start the engine. Illumination of the crank pulley should reveal the notch opposite the upper timing mark. If the notch is above the timing mark, unloosen the distributor and move it counter-clockwise to retard the ignition; if the notch is below, rotate the distributor clockwise to advance timing. The advance operation can be checked by slowly increasing the throttle and observing that the notch moves upward, away from the timing mark; as the throttle is backed off the notch should return to the timing mark. Any erratic motion, such as the notch jumping suddenly away from the timing mark, indicates that the springs are weak in the centrifugal advance system and replacement is in order.

f) Carburettor tuning is the last step of the tune-up. Before detail adjustments are made, routine cleaning should be done as follows:

Competition model (Weber dual-throat carburettor): the top should be removed (six screws) and the float bowl inspected and cleaned. The filter at the fuel inlet connection should be cleaned. Remove each of the jets - primary main, secondary main, primary idling, secondary idling and accelerator pump - and clean. Be sure not to interchange any jets during re-installation.

Standard model (Zenith VN2): remove the float bowl (four screws) and disassemble by removing the float and the emulsion block. Clean the bowl and check the accelerator pump by pushing on the piston (jet will squirt petrol). Remove and clean the main and compensating jets in the emulsion block. Carefully reassemble.

Start the engine and allow it to reach normal operating temperature. A vacuum gauge is very handy and can be attached through a blanking plug on the intake manifold. The idle and mixture should be adjusted for maximum vacuum (minimum gauge reading) while maintaining correct idle rpm (around 800). In the absence of the gauge, tune for the smoothest engine operation.

Compression ratio (general)

We all know that this term is one of those used to help indicate the power of the engine, but how many of us really know what is means? Hopefully, the following will help clarify the expression.

When the piston is as close to the top of the cylinder as possible, that is, at Top Dead Centre, the small volume (which is the combustion space) is called the clearance volume. The volume of cylinder that the piston moves through is called the displacement. It depends on the bore (that is, the diameter of the cylinder) and the stroke (or the distance through which the piston moves). The compression ratio is defined as the piston displacement plus the clearance volume (which is the total volume when the piston is at Bottom Dead Centre) divided by the clearance volume. The compression ratio of an engine is very important because, when changed, other factors affecting the performance and efficiency of the engine also change. An increase in compression ratio:

1 *Increases the compression pressure.*

2 *Increases the compression temperature of the charge.*

3 *Decreases dilution by exhaust products.*

If the compression ratio is too high (for a particular fuel) in a gasoline engine, a phenomenon called detonation, or knock, will occur because the charge ignites too soon.

Detonation (general)

The automotive and petroleum industries spend millions of pounds a year on the study of detonation, and still there is not complete agreement among engineers as to just what it is and what to do about it. Briefly, it may be explained as follows: when the mixture is ignited, combustion starts at the spark and spreads through the combustion chamber behind what is called the flame front. As the temperature of the burned gases rises, they expand and compress the unburned gases. If the compression ratio of the engine is too high the unburnt gases are compressed to the point where they get so hot that spontaneous combustion begins at some point remote from the spark. There is a rapid increase in pressure at this point, which results in violent, vibrating pressure waves, the audible evidence of which is a metallic click or knock when these waves strike the wall of the combustion chambers. Excessive detonation is accompanied by severe pressures, which may eventually damage the engine. In addition, there is a small loss of power when detonation occurs. However, since the engine is operating at close to maximum thermal efficiency when detonation is incipient, maximum efficiency is often achieved when there is a slight detonation or "ping".

It should be noted that compression ratio is not the sole determining factor affecting detonation. For this reason, especially in recent years, a number of developments have been made that have permitted compression ratios to be increased without detonation. These include:

1 Special substances (called anti-knock agents), such as tetraethyl lead and triptane, that can be added to a fuel to help it resist detonation.

2 Optimum shaping of the combustion chamber to reduce detonation.

3 Increasing the rate at which heat is carried away from the combustion space by the use of thinner sections in the head, the use of materials of higher heat conductivity, such as aluminium alloy and bronze for the head, and by improving the conductance of heat from the top of the piston.

4 Reducing the number of local hot spots in the combustion space by better finishing, which removes small projections that may become red hot and cause spontaneous ignition, and by more efficient cooling of various parts near or in the combustion chamber, such as the spark plug and the valve seats.

Increasing compression ratio (general)

Milling of the cylinder head is the least costly method of raising compression ratio (this will not work well for an engine which uses a recess in the piston top as all or part of the combustion chamber), but the range is limited to about a one point increase in compression ratio. One of the arguments against milling is that OHV heads are one of the engine's weakest links which can ill-afford any loss of material.

Increasing bore or stroke (or both) automatically cause a higher compression ratio because the swept volume of the cylinder is increased while the combustion chamber volume is hardly changed. If bore, stroke and head are left as standard, then the only way to increase compression ratio is by means of high pistons. Where cost is not a factor, this method is always preferable to head milling. "High dome" pistons are available for Triumph engines. However, because of the high installation costs, they should not be considered until other major engine work is required.

How high can you go on compression ratio with today's fuels? For all-round best results with a normal unblown engine - for best street performance, gas mileage, power and spark plug life - one school of experts say 10.5 to 1. The absolute maximum is 11 to 1, they say, and when you start closing in on 11.25 to 11.5, big problems crop up in a hurry. With fuel injection and the best in head work and ignition you can get on the ragged edge of 11.5 to 1. Go higher and you defeat your purpose.

Periodic brake maintenance (all models)

Start by putting the car on stands and removing the wheels. Wire wheels should come off from time to time anyway to keep them from rusting onto the splines. Check the condition of the brake linings. On disc brake cars, this is quite a simple matter as the pads are quite visible at the rear of the calipers when the wheels are off. Front brakes wear at about twice the rate of rear brakes because of the weight transfer during braking. If the lining material on the pads is less than 1/4in thick, it is time to replace them. This is also a simple task on disc brake cars; just remove the keeper pins and withdraw the pads by pulling on the tab with a pair of pliers. While the pads are out, clean any oil or grease off the disc and check for fluid seepage from the pistons. The pistons must be pushed back into the calipers using a tyre lever or other suitable tool. The new pads (Mintex 20064) are then slipped in and the keeper installed. Before the pistons are pushed back, remove the master cylinder reservoir cap and wrap absorbent rag around the reservoir as brake fluid may overflow.

Rear brakes and front drum brakes require removal of the drums to inspect the linings. Early cars have the front drum and hub as one piece and require a puller to get it off - good time to inspect the front wheel bearings also. The linings should be at least 1/8in thick and 1/16in above the rivets if that type. If in doubt, replace - linings are much cheaper than drums, which will be destroyed if rivets or shoes grind into them. If the linings are thick enough and dry, then they are okay. Clean away the accumulated brake dust with a soft rag but try not to breathe the dust as most brake linings contain asbestos.

Check the backing plates and wheel cylinders for signs of fluid seepage. Carefully lift the edge of the rubber dust covers on the wheel cylinders and look for signs of fluid leakage. Check around the axle seals for signs of grease leaking onto the backing plates. If the shoes are wet, there is fluid or grease leaking from somewhere. Leaky wheel cylinders should be attended to immediately: the cheapest thing to do is to rebuild the offending cylinder. If, after dissassembly, you find that the bore is so pitted it will not clean up with a hone, the cylinder must be replaced. An alternative to replacing is to have the old cylinder honed and sleeved; several companies offer this service. After sleeving, a rebuild seal kit is installed, which will probably cost about the same as replacement would. Someday, as parts become scarcer, this may be the only alternative. Don't throw away your old wheel cylinders!

If the brake shoes are wet or worn they should be replaced. Most auto supply stores with a workshop can reline either bonded or riveted brake linings and some may even have exchange parts.

After the wheels are replaced, check the flexible brake hoses. Morgans have three flexible hoses: one at each front wheel and one over the rear axle. Each hose should be carefully inspected visually for signs of degradation - cracking, weathering, abrasion, rotting or nicks are reasons to immediately replace these hoses. If all the hoses pass visual inspection, then a "hands-on" test should be carried out. Firmly grasp the hose in your fist while an assistant applies maximum pedal pressure to the brakes. Any "feel" of hose expansion is a sign of weak walls and replacement is necessary. Be sure you don't confuse a stiffening of the hose with expansion, as the 700-900psi line pressure will make the hose feel like steel pipe. Finally, be sure the hose has a free operating area and will not come into contact with body, suspension or exhaust parts during turning and spring travel. If in doubt about the condition of any flexible brake hose, replace it.

Check the steel lines for cracks, pitting or rust. Some restorers favour copper brake lines; if you find any my advice is to replace them with steel because I have known vibrating copper pipes to work harder and fracture.

Drum/hydraulic brake spares (Plus 4s)

Most current catalogues only list spares for cars from about 1960, so on the opposite page is a reference list starting with the Plus Four. It comprises brands and the names of mass-production car models which shared the same components (some Morris, Riley and MG models contemporary with the Austins mentioned shared components, *too).*

Silicone brake fluid (general)

If you use regular brake fluid in your car, change it once a year and use a good quality product such as Castrol LMA (low moisture activity). Brake fluid is hydroscopic (it absorbs water from the atmosphere) and the absorbed water is what causes pitting of the cylinders and breakdown of the cups. Consider changing to silicone brake fluid.

There are many advantages to using silicone brake fluid and almost no disadvantages. Some of the advantages are:

1 *It repels moisture, ie, no corroded or sticking brake or clutch cylinders.*

2 *It promotes seal life and helps stop small seepages.*

3 *It operates excellently to extremely high temperatures for hard driving.*

4 *It does not require periodic flushing and changing.*

5 *It will not attack painted surfaces if spilled upon them.*

With the silicone fluid currently available, it is not necessary to flush or dismantle the brake system to get rid of the old brake fluid, as has been reported in some journals although, of course, you should follow the fluid manufacturer's advice. Silicone brake fluid is usually compatible with all automotive brake fluids and all brake system parts. In fact, silicone brake fluid can usually - be used to top-up a system containing regular brake fluid (or *vice-versa*). Convertion of your Morgan to silicone brake fluid can be achieved by following these steps:

1 Place car on jack stands and remove wheels.

2 Attach a piece of clear plastic tubing to each bleed valve and run it into a clear container.

3 Open a bleed valve and have a helper pump the pedal to produce a a flow of fluid. This is done to make sure you blow the bleed ball off its seat, as it may be rusted in place. Close the bleed valve. Repeat this procedure at each wheel.

4 Take a suction bulb and suck all the fluid from the reservoirs under the bonnet.

5 Open all the bleed valves at the wheels and pump the pedal slowly at least 25 times to expel as much fluid as possible.

6 Remove the bleed valves and balls from each wheel cylinder, leave the caps off the reservoirs and let the car sit with the system open for 48 hours. Regular brake fluid evaporates, and letting air flow into the system through openings will help evaporate some of the remaining fluid.

7 Silicone fluid has a tendency to absorb air molecules, so do not agitate it prior to use. It is best if the container of fluid sits at least 24 hours prior to use in the car.

8 Install new, or cleaned and inspected old, bleed valves and balls in the wheel cylinders. Fill the reservoirs with silicone fluid, being careful not to

Brake component part numbers (Plus 4s)

Part numbers are Girling except where otherwise stated.

FRONT SHOES	Plus Four 1950-55	GB 41144 and GB 41145 also Austin A40 1951/55 4/4 Series II	Springs GB 40454
	Plus Four 1955-57 from P 3333	GB 41332 and GB 4133 also Austin A40, A50, 1955-56 Consul, Zephyr, 1951/55	Springs GB 40454
	Plus Four 1958-59	GB 45792 or Ferodo F 210 G Also Austin A50/A55 1956-59	Springs GB 47618
FRONT WHEEL CYLINDERS	Plus Four 1950/51 4/4 Series II	390320 and 390321 Also Austin A40/A50 1951/56 Consul, Zephyr 1951/55	Rubbers SP 2049
	Plus Four 1958/59 from 4096	64673308 and 64673309 or Unipart GWC114 & 115. Also Austin A50/A55 1956/59 Consul/Zephyr 1956/62	Rubbers SP 2034
FRONT HOSES	Drum Brake Plus Four 4/4 Series II	3700628 W superseded by 74047437	
REAR SHOES	Plus 4 1950/55 4/4 Series II	GB 41114 and GB 41115 also Austin A40 1951/55	Springs GB 40655 GB 40897
	Plus Four 1956/57 from P 3550	GB 14705 also Austin A40/A50 1955/56	Springs GB40655 GB 40897
	Plus Four 1953/59 from 4096also later Morgans	GB 45796 and GB 47989 also Austin A50/A55 1956/58 & many other cars	Springs GB 44061 GB 46747
REAR WHEEL CYLINDERS	Plus Four 1950/55	390120 and 390121 Also Austin A40, Consul/Zephyr 1951/55	Rubbers SP 2050 SP 2057?
	Plus Four from P 3550	390430 also A 40/ A 50 1955/56	Rubbers SP 2049?
	Plus 4 from 4096 and later Morgans	64673804 or Unipart GWC 1104 also A 50/ A 55 1956/59	Rubbers SP 2032
REAR HOSE	Plus Four & 4/4	3700625 W	
MASTER CYLINDER	Plus Four 1950/55	64067536 (very scarce)	SP 1975
	Plus Four 1955/59 4/4 Series II	3112344 W SP 1967/4 769	Rubbers

(Note: the Lockheed KB 1044 is apparently a replacement for the Girling GB 45792, but the lining seems to be a very hard grade and is not recommended unless your car has vacuum servo assistance.

splash or unduly agitate the fluid.

9 Begin bleeding at the right rear wheel. You will need to leave the bleed valve open at first while your helper pumps the pedal slowly to get the master cylinder to pull the new fluid from the reservoirs. As soon as you observe the fluid level begin to drop in the reservoir as the pedal is pumped, you can begin bleeding in the following manner.

10 Open the bleed valve, have your helper push the pedal slowly to the floor and hold it there. Close the bleed valve whilst your helper slowly lets the pedal come back up to rest. Repeat this action over and over to bleed the brakes. The secret to bleeding the silicone fluid is to pump the pedal slowly as the agitation caused by rapid pumping dissolves air in the system. You should take 2-3 seconds to depress the pedal and at least 4-5 seconds to let it return.

11 You can tell when the silicone fluid starts to appear at the wheel bleed valve by a slight change in fluid colour and the fact that the remaining regular fluid will appear as small balls held in suspension within the silicone fluid. Continue to pump fluid through until you have predominantly silicone fluid emerging. Do not concern yourself with the amount of air bubbles at this time.

12 At the left rear, right front and, finally, left front wheels, repeat steps 9 through 11. Do not forget to top-up the reservoirs from time to time as you proceed.

13 Repeat the operation at each wheel. Bleed until the fluid is relatively free of bubbles.

14 After all the wheels have been bled, the pedal will still be somewhat spongy - don't worry about it. Most of the small air bubbles dissolved in the system will bleed back into the reservoirs.

15 After a few days, perform another bleeding operation. The first pump of the pedal should result in fluid with a few air bubbles. By the second or third pumps the fluid should have no bubbles of air present. Small amounts of the regular fluid still present in the silicone fluid will have no effect on the performance of the system.

16 At this point you should have a solid pedal, no leaks, and a brake system that will not need attention for many years to come. If you are careful and reasonably conservative with the fluid, a quart of silicone fluid will be enough to convert the clutch and brake systems.

17 When converting clutch systems, you may have to manually pull the pedal back up.

After conversion to silicone brake fluid, the need for an annual fluid change is eliminated. If the damage to cylinders from the years of using regular fluid has not been too great, you should never have anything to do except check the level in the reservoirs until the brakes need relining: unfortunately, even silicone brake fluid won't stop lining wear!

Definition of fit (general)

We've all heard about the type of 'fit' necessary on certain parts of the car, but do we really know what the term means?

The following definitions of the various types of fit are given to assist the restorer in arriving at the correct amount of clearance between moving parts, as well as to enable a better appreciation of why the various tolerances must be adhered to. Generally speaking, all bores are made to a standard size (so standard reamers, plugs, gauges, *etc*, may be used) with a plus tolerance. The maximum size of the male parts is usually a standard size less the minimum clearance required for the fit desired. The minimum size for the male parts is this maximum size minus the tolerance.

1 Running fit. A running fit is one providing enough clearance for there to be a continuous film of oil between the two parts. A running fit usually requires 0.001in for an oil film plus 0.001in for each inch of diameter.

2 Slip fit. A slip fit exists when the male part is slightly smaller than the female part and involves less clearance than a running fit. An example of the minimum allowable clearance for a slip fit would be a piston pin that, from its own weight, would pass slowly through the connecting rod bushing. In most cases (except where a limited movement of the parts is involved) slip fits are specified where, due to anticipated expansion of the female part, enough additional clearance will result to change this type of fit to a running fit and provide adequate clearance for a film of oil.

3 Wring fit. A wring fit is the type of fit required between a bore and a plug gauge to determine the inside diameter of a bore. With a wring fit, it is necessary to turn or wring the plug gauge or part to force it through the bore. This type of fit does not provide for a film of oil.

4 Press fit. A press fit is one that requires force to enter the male part into the bore. Accepted practice for press fit is to have the male part larger by 0.001in for each inch of diameter of the bore into which it is to be pressed. An example of a press fit is the bushing into the spindle on Morgan front suspension.

5 Shrink fit. Generally speaking, a

Engine component fit/clearance data (Plus 4s)

Location	*Original fit*	*Wear limit*	*Type of fit*
Cylinder bore taper	-	0.006in	-
Camshaft bearing clearance	0.0025in to 0.0045in	0.006in	Running
Clearance between lifter and lifter bore	0.0005in to 0.0015in	0.005in	-
Connecting rod side clearance	0.0006in	0.020in	-
Piston pin clearance in connecting rod	0.0002in to 0.0005in	0.0015in	Slip
Piston pin clearance in piston	0.0001in to 0.0002in	0.0015in	Slip
Piston and cylinder	0.0047in to 0.0054in	0.010in	Running
Top piston ring to groove side clearance	0.0015in to 0.003in	0.004in	-
Rest of piston ring to groove side clearance	0.001in to 0.0025in	0.0035in	-
Piston ring end cap	0.003in to 0.010in	0.035in	-
Crankshaft end play	0.002in to 0.006in	0.008in	-
Main bearing clearance	0.0005in to 0.0025in	0.005in	Running
Rod bearing clearance	0.0015in to 0.0035in	0.005in	Running
Stem to valve guide clearance (Exhaust)	0.003in to 0.005in	0.006in	Running
Stem to valve guide clearance (Intake)	0.0015in to 0.0035in	0.005in	Running

shrink fit is tighter than a press fit. The amount of shrink will range from 0.001-0.002in for each inch of diameter and in some cases even more. Parts having a press fit may be assembled by shrinking rather than force if desired. There are two methods of shrinking two parts together, either or both of which may be used. One method involves expansion of the female member by heating and the other involves contracting the male member by chilling with dry ice.

• EFFECT OF EXPANSION ON FIT
Allowances are made in establishing fit on parts that are exposed to higher temperature to take into account the anticipated expansion of the part during operation and still provide for the type of fit required. Allowances must also be made for unequal expansion of dissimilar materials. Absolute minimum allowance for expansion of parts exposed to flame or exhaust gases (pistons, piston rings and valves) is 0.001in for each inch of diameter of length. In anticipating the expansion of a piston to make allowance for the additional clearance required in the cylinder, 0.001in for each inch of diameter is added. In anticipating the expansion of a piston ring to make allowance for the additional gap required between the ends of the ring, 0.001in for each inch of length (0.003in for each inch of diameter) is added.

The use of a torque wrench (general)

The importance and proper use of a torque wrench are sometimes not appreciated - under-tightening of fasteners (bolts, nuts, *etc*) causes metal fatigue in moving parts and leaks where gaskets are used; over-tightening causes metal fatigue in the fastener itself (stretching), damage to parts (warping, pull-up and stripped threads) and leaks caused by damaged gaskets.

The real reason for tightening threaded fasteners is to create the proper compression between the parts being joined. Because compression cannot easily be measured between the parts, the desired compression force is converted to the measurement of torque. Nobody needs to be told that a rusty or dirty bolt is harder to turn than a clean oiled one, but some people forget that the required torque can be obtained on a rusty bolt long before the proper compression is. Torque requirements are all set for ideal conditions of clean, lubricated male and female threads. Unfortunately, most of us think that cleaning bolts with a high-speed wire brush takes care of this, but actually the resultant distortion of the threads can produce enough jamming to give faulty torque readings. The best way to clean threads is with a tap and die.

It is also very important to use the proper lubricant to attain correct torque; ordinary chassis grease and engine oils are not ideal. If they are used, the torque reading should be increased by 20 per cent. Anti-seize compound is the best lubricant and should be used not only on the threads but also under nuts, washers and bolt heads. The use of anti-seize is doubly important when bolts are threaded into aluminium or magnesium alloys. These light metals tend to 'grow' and when a dry bolt is removed it's inclined to damage threads.

If a bolt has been over-tightened, 'pull-up' often occurs around the bolt holes. This can easily be determined by running a file over the holes which will reveal any high spots. When this condition exists, required torque can be obtained without the proper compression and gaskets may leak. The best way to correct this is to counterbore or chamfer the holes.

The torque wrench is one of the most important tools in your garage, but if it is not used properly you might just as well go back to an open-ended spanner and guess work.

Clutch replacement (early Plus 4)

Replacing the clutch on a Plus 4 is not difficult if you have a reasonable amount of time and a decent place to work. Before starting this project, purchase a new clutch, disc and linkage. Rebuilt clutches break so don't make yourself do the job twice. This is an ideal time to replace anything else in the area you have been putting off, *eg*, the brake cable.
Other essentials are a good set of tools, jack, axle stands, shop manual, hand soap (about a bucketful), rags, Band-Aids, coffee and a friend willing to help. Remember to use containers for the parts removed and if you have a bad

memory, make a note where each goes.

1 Remove drive tunnel.

2 Remove bonnet by loosening the front mount and sliding bonnet off rear mount.

3 Remove seats, squab, floor mats and clean all the grease from the transmission, bellhousing, engine and floorboards, and if you have time clean the battery compartment, differential housing and associated area.

4 Dismount the floor-operated lube valve. Do not disconnect plumbing. Pull up and out of the way as necessary.

5 Disconnect throttle cable from cabs.

6 Remove centre panel from firewall. There are 8 small bolts.

7 Remove steering wheel, seats and squab, this helps the operation.

8 Disconnect front universal joint.

9 Remove rear transmission mount bolts.

10 With floor jack under oil sump and as far back as possible, lift engine and transmission so the transmission will clear the crossmember. (Note: if you don't have a flexible pipe exhaust, disconnect the manifold from the headpipe).

11 Disconnect the clutch linkage by pulling the pins and cotter pins.

12 On the side opposite the pedals, remove two bolts which hold the firewall to the frame.

13 Remove starter.

14 Disconnect emergency brake cable at the rear of the car.

15 Remove emergency brake assembly from the bellhousing.

16 Remove special bolt through which the brake cable passes. Move cable sheath down and slide cable wire out of slot in transmission housing.

17 Unbolt the rest of the transmission from bellhousing.

18 Remove transmission, remembering that there is a 2ft extension shaft on the end of the transmission. (Note: if the extension shaft remains in the bellhousing, remove it by hand).

19 There is a small fibre plug in the end of the transmission that might fall out of the front end of the transmission. If it does, put it back after applying a bearing fit/lock fluid.

20 Unbolt bellhousing and pull back off engine. The top three bolts are studs - do not mix with other bolts as the threads are different.

21 Wiggle out bellhousing using the freedom of loose firewall. The starter hole goes on top.

22 Unbolt clutch evenly and progressively - watch your fingers!

23 There are two dowel pins - found in the bolt ring - that align the clutch with the flywheel. Do not damage these pins! If they are worn, replace them.

24 Check the throw-out bearing by checking for full travel and cracks in carbon block.

25 Clean bellhousing, being careful not to get solvent in throw-out bearing. Paint bellhousing if you wish. The bellhousing is magnesium and can be polished to a mirror shine but must be lacquered to keep its lustre. Sanding and polishing can be accomplished with a reducing machine.

26 Check pressure plate by setting it on a flat surface and seeing if the disc the throw-out bearing presses against is level.

27 Check disc for wobble.

28 You are ready to reassemble.

29 Take extension shaft off transmission. Use mallet if necessary.

30 Use extension shaft to align the disc.

31 Be sure new clutch cover is aligned on the dowel pins. Tighten bolts opposite each other as on a disc wheel.

32 Replace bellhousing. Magnesium is brittle, do not over-stress as it is a tight fit.

33 Go to instruction 20 and retrace your steps back to and including 18. Then put in your new clutch linkage.

34 Retrace 17 through 12 and set transmission back on frame.

35 Adjust clutch linkage for 1/2in of pedal play.

36 Using the starter button, crank engine over several times. Listen for unusual noises like loud banging - if this happens get expert assistance.

37 Retrace steps 9 through 4.

38 Take a short drive, then perform any minor adjustments.

39 Now complete steps 3 through 1.

40 De-grease yourself, tools, driveway and workbench.

Wiring colour codes (all models)

The colours of the wiring of British cars have not been arrived at on the whim of the manufacturer, there is an overall scheme of things, and this scheme was adopted many years ago and is common for all British cars and motorcycles.

Seven basic colours are used for six functional areas. Black is reserved for earth, always.

1) BROWN	-	*The battery circuit*
2) YELLOW	-	*The generator circuit*
3) WHITE	-	*The ignition circuit*
4) GREEN	-	*The auxiliary circuits*
5) BLUE	-	*The headlamp circuit*
6) RED	-	*The auxiliary lighting circuits*
7) BLACK	-	*All earths.*

The solid colour was always used for the main feed (hot) wire, with traces of another colour added to each individual circuit after a switch. Black is

Diagram 1

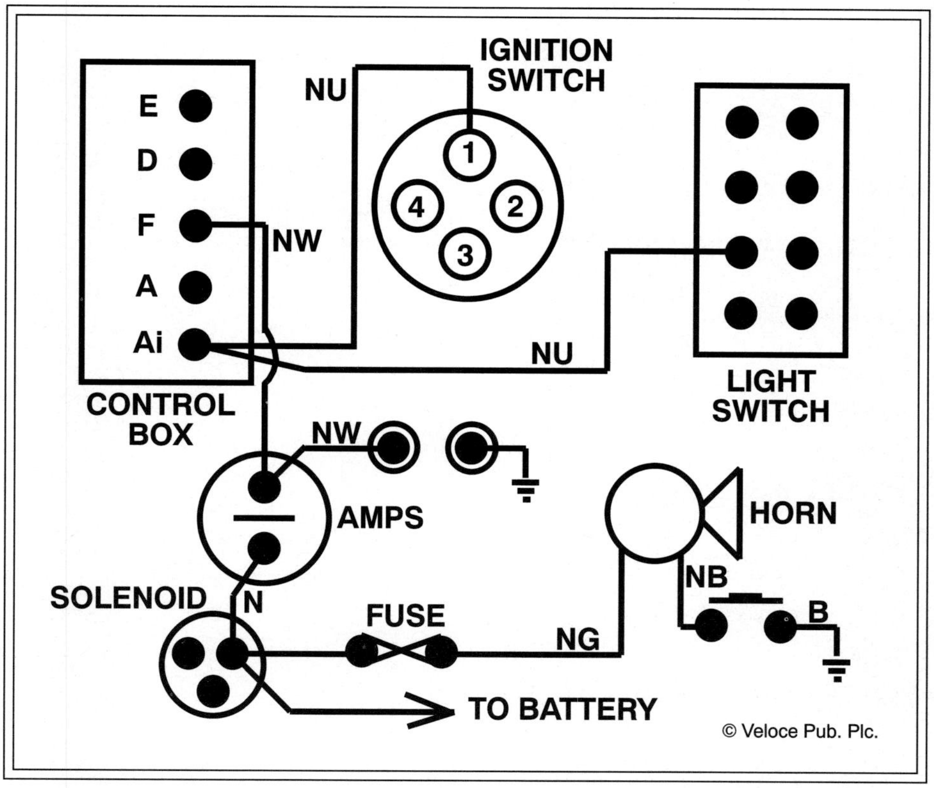

used for the earth wire if the item is not earthed internally. If the switch was in the earth side of the circuit (as with later Plus 4 windscreen wipers), black with a tracer is used to the switch.

Wiring diagrams (all models)

1) BROWN - The battery circuit. From the battery or starter switch to the ammeter and/or control box (voltage regulator) terminal 'A', and feeding the light and ignition switches, and radio from the control box terminal 'A1'.

In addition, brown was the basic colour used for leads from the starter switch to the electric clock, inspection lamp sockets and horn fuse.

In diagram 1 we see that the 'main feed' from the tie point on the solenoid to the ammeter is BROWN. From the ammeter to the 'A' (Ammeter) terminal on the control box the wire is BROWN with a white tracer, as is the wire to the inspection socket (red socket). The wire from the 'A1' terminal on the control box to the ignition switch and to the lighting switch is BROWN with a blue (lights) tracer.

The wire from the horn fuse to the horn(s) is BROWN with green, and the wire from the horn to the horn button (earth side) is - you've guessed it - BROWN and BLACK.

2) YELLOW - The generator circuit, from the generator armature to the 'D' terminal on the control box, from the generator field to the 'F' terminal on the control box, and to the ignition warning light (see diagram 2).

The generator circuit is quite simple. The wire from the generator armature to the 'D' (Dynamo) terminal of the control box is YELLOW, as is the wire from the 'D' terminal of the control box to the ignition light. The wire from the field terminal of the generator to the 'F' (field) terminal of the control box is YELLOW with a green tracer.

3) WHITE - The ignition circuit. This colour is used for all (usually unfused) items that are powered when the ignition switch is turned on, such as coil, fuel pump, starter solenoid switch, overdrive, *etc.* (see diagram 3).

There are not too many WHITE wires in the Morgan system. WHITE from the ignition to the coil, WHITE and red to the solenoid and WHITE with light green

Diagram 2

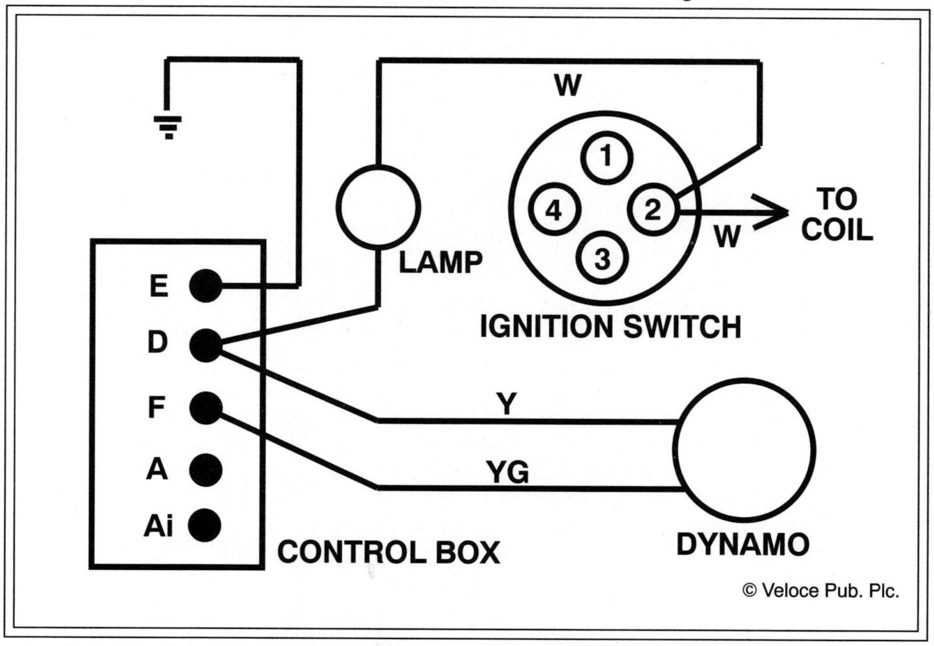

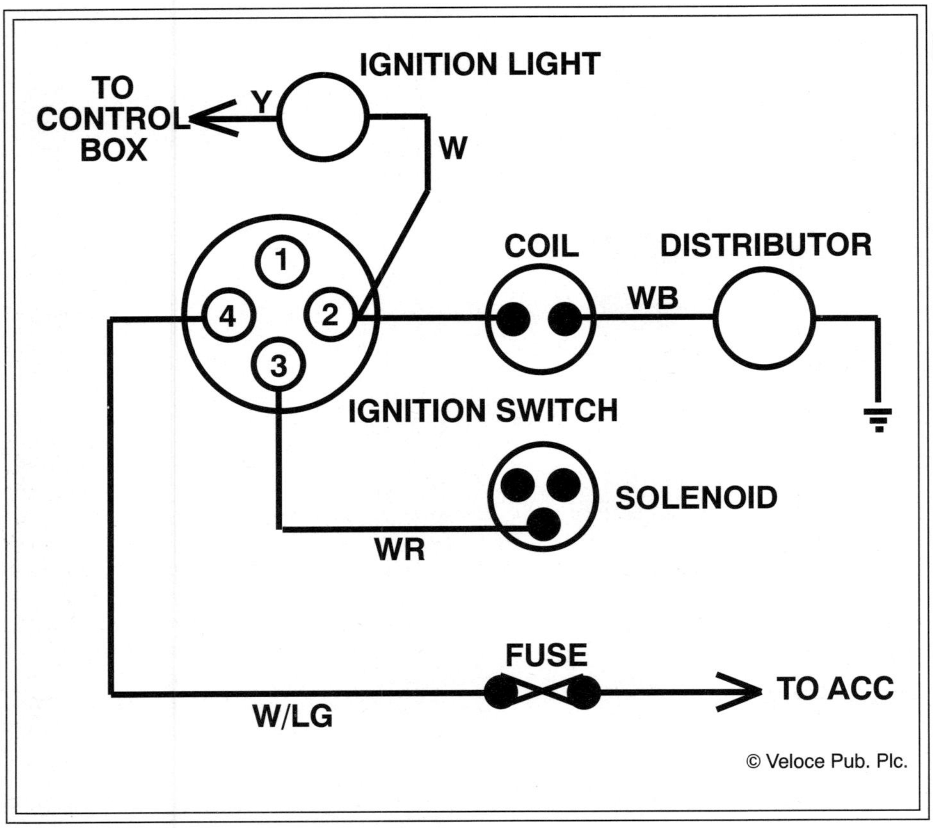

Diagram 3

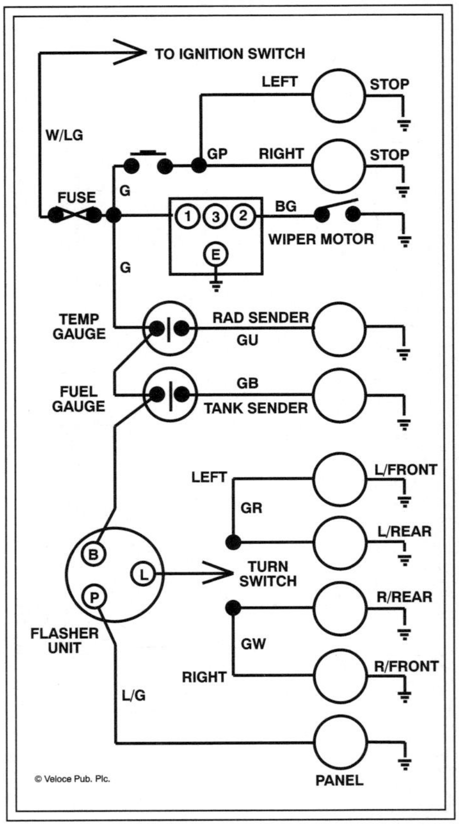

Diagram 4

(accessories) to the accessory fuse (it was a WHITE wire on Morgans with the combined ignition/lighting switch).

4) GREEN - The auxiliary circuits. All circuits through the ignition switch and protected with fuses such as stop lamps, turn signals, windscreen wipers, gauges, *etc*. (see diagram 4).

The primary wiring for the Morgan auxiliary circuits is from the accessory fuse and is GREEN to the stop lamp switch, the windscreen wiper motor, the gauges and then to the turn signal flasher.

The stoplight wires are GREEN and purple, the fuel tank wire is GREEN and black, the wire to the temp sender is GREEN and blue, while the wire to the wiper motor switch (earth side, remember) is BLACK AND GREEN.

The turn signals use lots of colours - light green to the warning lamp and GREEN/brown to the switch from the flasher. GREEN/red for the port side and GREEN/white for the right side.

5) BLUE - The headlamp circuit. This is fed from terminal 'S2' (or 'H') on the headlamp switch (see diagram 5).

The Morgan circuit is BLUE from the headlamp switch to the dip switch. High beams are BLUE/white, while low beams are BLUE/red.

6) RED - The auxiliary lamp circuits. These are fed from terminal 'S1' (or 'T' or 'S') on the lighting switch. These are the sidelamps, dash lights, tail lamps and other lighting circuits, such as fog lamps, which are supplied with power only when the lighting switch is on (see diagram 6).

Side lamps and tail lamps are both RED wires, as is the input to the dash

Diagram 5

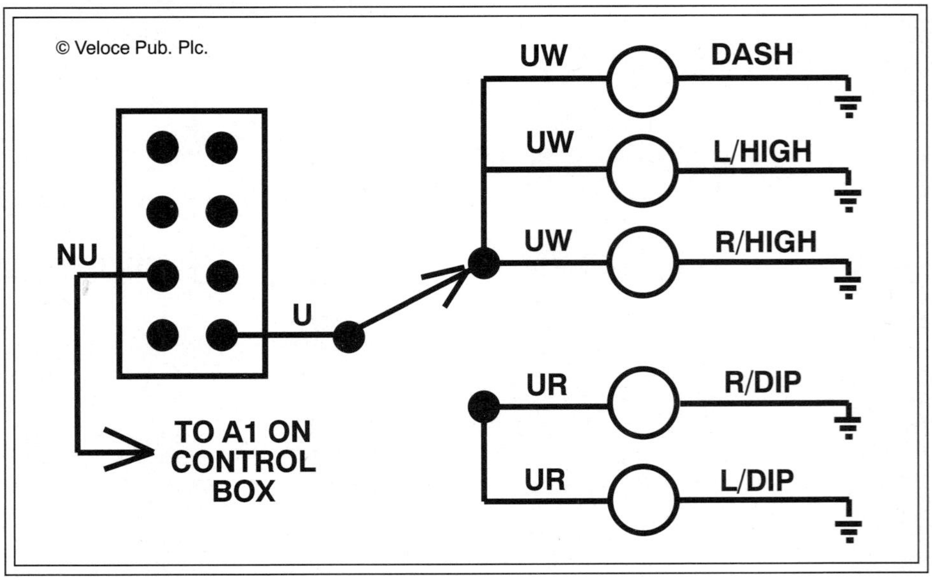

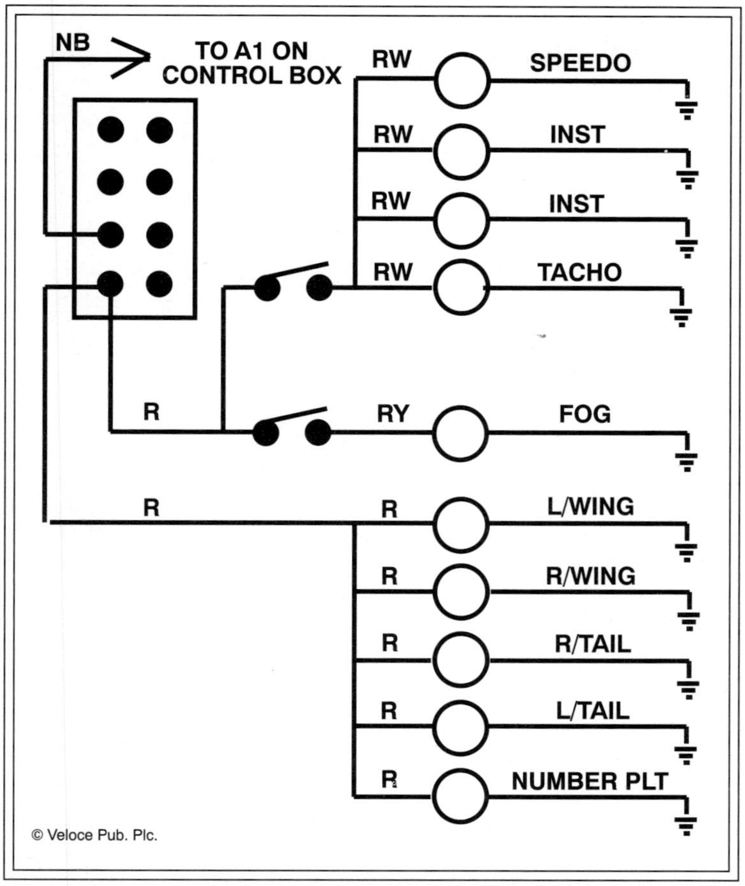

Diagram 6

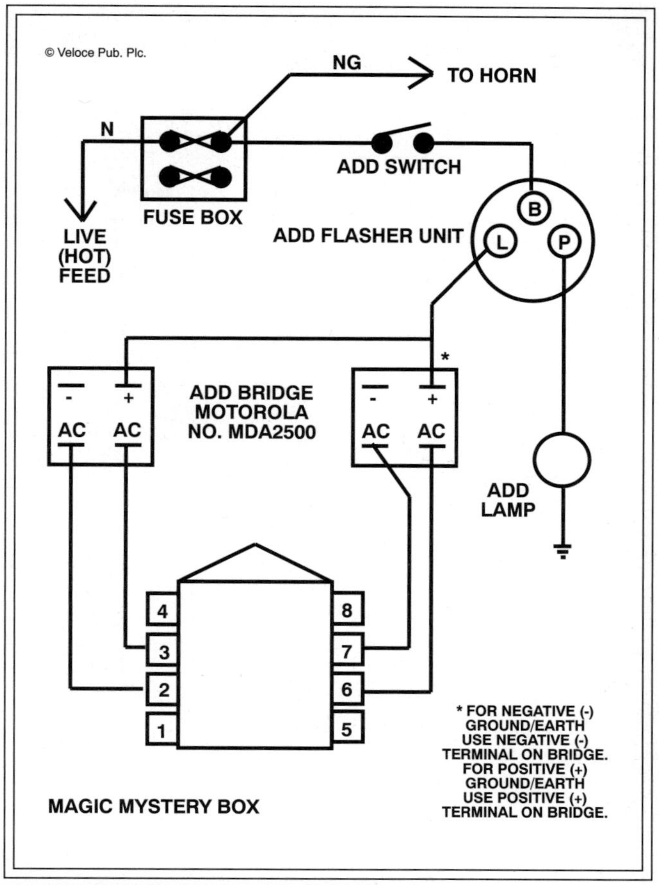

Diagram 7

light and the fog lamp switches. The dash light wires are RED/white and the fog lamp wire is RED/yellow.

Fitting hazard lights (all models)

The fitting of hazard warning light flashers on an older Morgan, not originally so equipped but having directional signals, is a very simple job using a common electronic diode assembly called a bridge rectifier. These devices contain a group of diodes, and certain models have a very convenient package using automotive push-on wiring terminals. The bridge rectifier assemblies are available through Radio Shack (Tandy) or from a semi-conductor parts house.

The unit to use is a 25 ampere, 50-volt bridge rectifier - Radio Shack 276-1185 or Motorola MDA 2500. Under normal operating conditions these diodes stop the normal turn signal voltage from feeding to the unenergized lamps.

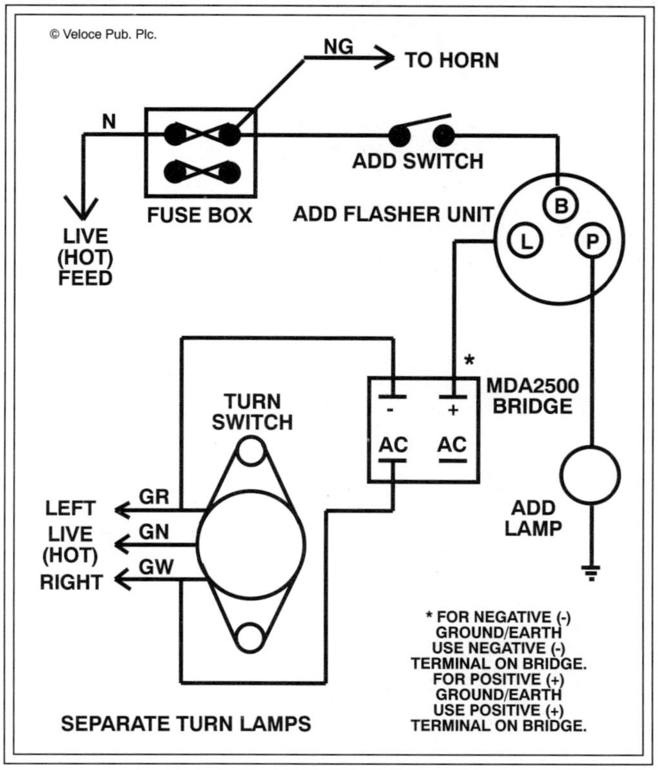

Diagram 8

Colours used in Morgan wiring

1 BATTERY CIRCUIT:			
N	-	BROWN	- Battery (main feed)
NG	-	BROWN/GREEN	- Horn (live)
NB	-	BROWN/BLACK	- Horn (to button)
NW	-	BROWN/WHITE	- Metered
NU	-	BROWN/BLUE	- Live (to ignition, lamp switch)
B	-	BLACK	- Earth
2 GENERATOR CIRCUIT:			
Y	-	YELLOW	- Generator output
YS	-	YELLOW/GREEN	- Field input
W	-	WHITE	- Ignition (on)
3 IGNITION CIRCUIT:			
W	-	WHITE	- Ignition on (to coil)
WR	-	WHITE/RED	- Start
W/LG	-	WHITE/LIGHT GREEN	- Live (to accessory fuse)
WB	-	WHITE/BLACK	- Points
4 AUXILIARY CIRCUITS:			
G	-	GREEN	- Live (fused)
GP	-	GREEN/PURPLE	- Stop lamps
GB	-	GREEN/BLACK	- Tank sender
GU	-	GREEN/BLUE	- Temp sender
GR	-	GREEN/RED	- Left turn
GW	-	GREEN/WHITE	- Right turn
GN	-	GREEN/BROWN	- Turn switch
L/G	-	LIGHT GREEN	- Turn indicator
BG	-	BLACK/GREEN	- Wiper switch
5 HEADLAMP CIRCUIT:			
U	-	BLUE	- Live (switched)
NU	-	BROWN/BLUE	- Live (to switch)
UW	-	BLUE/WHITE	- High beams
UR	-	BLUE/RED	- Low beams
6 AUXILIARY CIRCUITS:			
R	-	RED	- Live (running lights)
RW	-	RED/WHITE	- Dash lamps
RY	-	RED/YELLOW	- Fog lamp

• MAGIC MYSTERY BOX (EARLY PLUS 4S)
Early Plus 4s used a common rear lamp for both stop and directional signals, with a 'magic mystery box' to keep things straight. Two rectifier assemblies are used to feed 12-volt battery power to the front and rear directional lamps. When the hazard lights are turned on, power is fed to all four lamps, bypassing the regular turn switch and flasher.

The markings on the bridge rectifier relate to its use as a recitifier. In this application, they are connected as shown in diagrams 7 and 8.

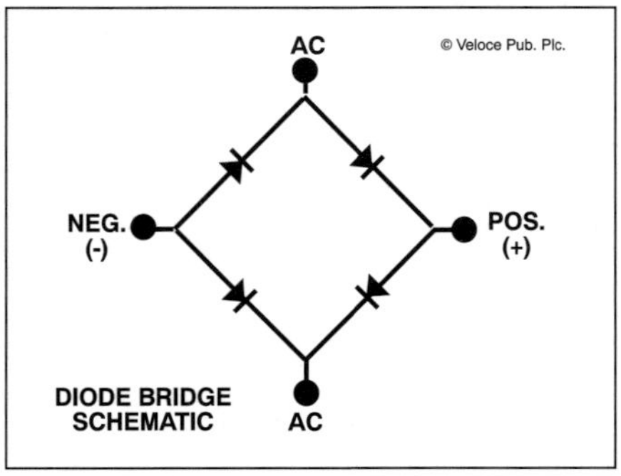

Diagram 9

• INDEPENDENT TURN LAMPS
The circuit shown in diagram 9 is for those cars with independent rear brake and turn signal lamps. As the front and rear turn signal lamps on the left and right sides of the car are connected together, only one diode assembly is necessary. One (AC) terminal of the bridge rectifier is connected to the left lamp feed of the directional switch and the other (AC) terminal is connected to the right lamp side of the switch.

Starting problems (4/4)
One other diode modification can be made to some 4/4 models. The coil often used on the 4/4 (Cortina) cars uses an external ballast resistor to lower the voltage to the coil, which is usually located on the right valance panel (see diagram 10). Sometimes these cars are hard to start when cold and a little hotter spark will help. Very late cars have an extra set of contacts in the ignition switch to short out this resistor when starting (se diagram 11).

What to do with older cars? The author has sometimes just used a clip-lead to jump the resistor when starting. A more permanent modification can be made using a diode circuit.

When starting, the current feeding the solenoid is also sent to the coil through the diode. When the starter is released, the diode stops current from the coil reaching the solenoid (see diagram 12).

Rewiring your Morgan (all models)
A good, reliable wiring system is absolutely essential to the safe, reliable and pleasurable operation of your Morgan, yet this important aspect of restoration is often given only grudging attention. This fact is surprising since it is well known that wiring and ignition problems are the most common cause of vehicle malfunctions. Consider also that faulty wiring constitutes a definite hazard in that it can lead to disastrous fires. The restorer is well advised, I think, to pay considerable attention to his electrical network early in the restoration project. Not only will the attention pay handsome dividends later in trouble-free operation, but a little extra effort and money will add considerable class when it comes to

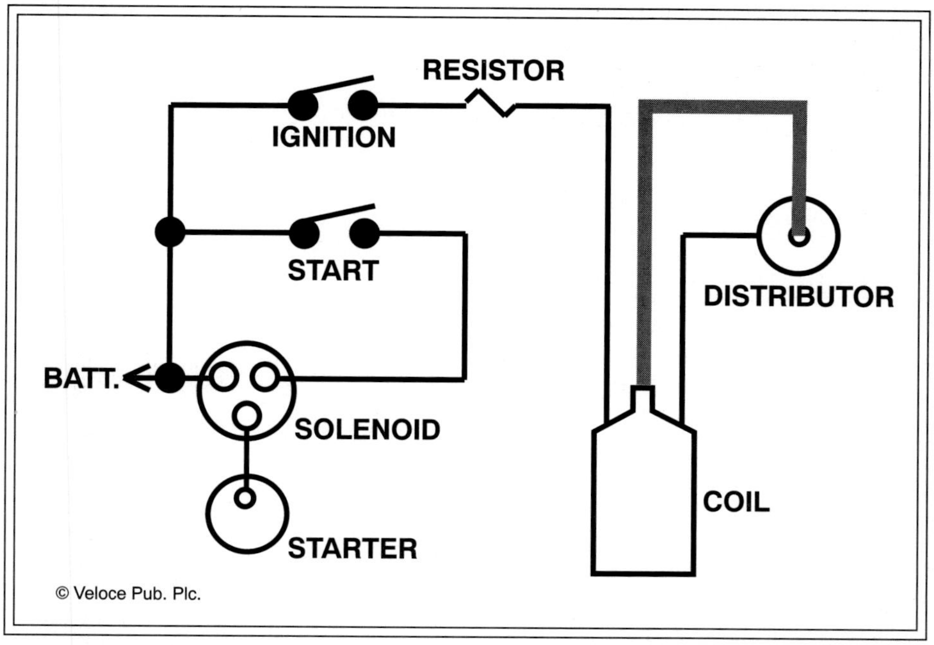

Diagram 10

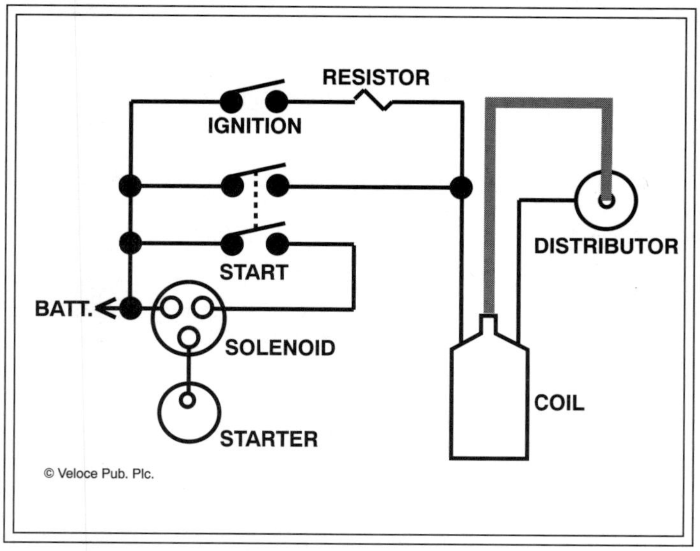

Diagram 11

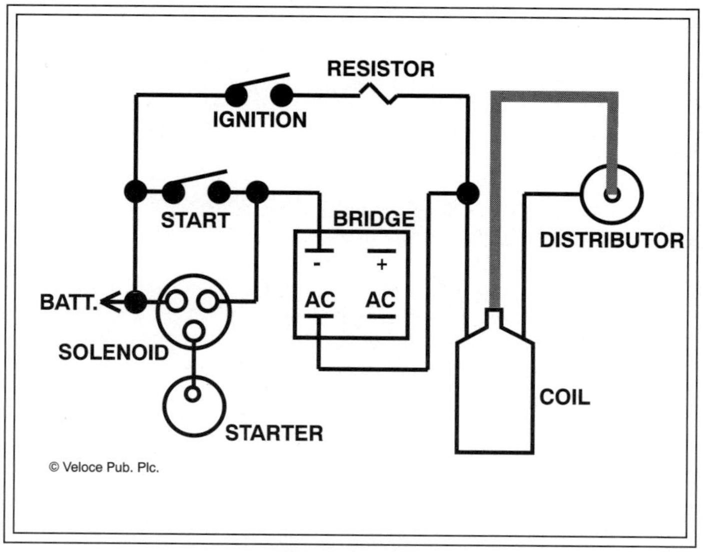

Diagram 12

concours judging.

A number of companies are now offering complete wiring harnesses and most of the products I have seen appear to be very creditable reproductions. A new harness should be a worthwhile investment, in spite of the cost. If not already available from stock, presumably, any one of the firms can duplicate your old wiring harness if you can remove it intact. The wiring harness is the heart of the project but there is still much to do after it is installed, of course.

Remember to use at least 14 gauge for headlamp and horn circuits. 16 gauge is OK for tail lamps, parking lights and turn signals. The connection between the battery and the ammeter carries the entire current for all circuits except starting and maybe the horn, so it should be at least 12 gauge. The wire from the generator carries the entire charging current and should also be 12 gauge. Note that the smaller the gauge number, the larger the wire and hence the current capacity.

Most problems in getting lighting circuits to work properly stem from lack of an adequate earth. Paint or corrosion films may prevent the return currents from the headlights, winglamps or tail lamps from getting back to the battery through the frame. Maybe you did a terrific job of priming on your valances and wings, and then covered every square inch with a heavy coat of paint. Now these parts are perfect insulators. Result? No lights! It may now be necessary to scrape off some of that paint in order to provide a low-resistance electrical path. A pocket ohmmeter is a good piece of equipment to have when looking for high-resistance earths.

• POINTS TO REMEMBER

Make sure that the battery terminal clamps are clean and in first-class condition. If there is any doubt, replace them now!

Check the condition of the engine earthing strap, the wire from the battery to the solenoid and the solenoid to

Electrical wire size/capacity

CURRENT RATING AMPS	UK SIZE	IMPERIAL SIZE OBSOLETE	METRIC SQ MM	AMERICAN WIRE GAUGE
6	14/.25	14/.010	.7	18-19
8.7	14/.30	14/.012	1.0	18
13.5	21/.30	21/.012	1.5	15-16
17.5	28/.30	28/.012	2.0	14
21.75	35/.30	35/.012	2.5	13
27.5	42/.30	42/.012	3.0	12-13
35	65/.30	65/.012	4.5	10-11
42	84/.30	84/.012	6.0	9-10
50	97/.30	97/.012	7.0	8-9
60	120/.30	120/.012	8.5	7-8
99	37/.74	37/.029	16.0	5-6
155	37/.76	37/.030	25.0	3-4
248	61/.90	61/.036	140.0	1-2

Cable produced in the United Kingdom was formerly rated by the number of strands and the diameter (in inches) of each strand, eg 14 strands of 0.012in was shown as 14/.012. Since Britain now uses metric units, this becomes 14/.30 since 0.012in = 0.30mm, which is equivalent to a cross-section Of 1.0sq mm on the European system and 18 gauge American. Note the European and American systems do not tell you how many strands. The stranding of several small wires (instead of using one big wire) makes it more flexible and less likely to break due to vibration. All automotive cable is formed by using such separate strands of wire in a common insulator.

the starter. Replace if worn or if the strands have begun to separate.

Scrape, file or grind the earthing point on the frame and engine earth strap so that a good earth contact is obtained. Check the wiring diagram to determine the proper connection of the battery: is it positive or negative earth?

When working on the wiring, disconnect the earth terminal of the battery to prevent disastrous arcing if you accidentally short or earth wires.

Replace all wires that are frayed, badly worn or oil-soaked. Use stranded wire of the same gauge or larger then the original wires (see chart).

Spark plugs (general)

Spark plugs consist of a threaded metal shell, a piece of porcelain and an electrode and all they do is sit in the combustion chamber and spark at the appropriate time to start the flame that runs your car. That's all. Basically, a plug must fire under very extreme conditions time after time after time. Plus there must be different heat ranges for different types of engines.

For instance, a plug type recommended for a compression ratio of 7:1 would probably burn up rather quickly at 10:1. A plug for 10:1 would probably foul in an engine of 7:1 ratio. The reason being in an engine of 7:1 the cylinder pressures are not too high and the plug can conduct enough heat away to keep from burning and hold just enough heat to keep the electrode end clear of the oil vapour present in the combustion chamber. Now, if we take this plug out and install it in a similar engine with a compression of 10:1 the cylinder pressures are much higher, hence more heat. The plug cannot conduct the additional heat so the electrode starts burning; we would have to install a 'colder' plug or one that will conduct the heat at a higher rate. A cold plug in a low-compression engine would conduct too much heat from itself and not hold enough heat to burn off oil. This causes the spark to earth itself without firing the engine, a condition known as fouling. An over-rich fuel mixture can cause the same condition.

The next question is, how does a plug control its heat conductivity? This is related to how much of the porcelain is in contact with the shell or metal section. If the porcelain is in contact the full length of the shell, much heat is conducted away from the electrode, making it a cold running plug. If the porcelain is in partial contact with the shell, less heat is given up to the head cooling so the plug runs hotter.

The way to tell if a plug is running hot is to install a new set of plugs and run them for a couple of days. Then remove them and arrange them in order. Now see if the centre electrode porcelain insulator is white or has a blistered look. If it is burned, the plug is too hot or is being made to run hot by too lean a mixture, advanced timing or, in the case of late TR3 engines, frayed carburettor grommets. If the plug is sooty or black in colour, the plug is too cold, the mixture too rich or the timing may be retarded. A lot of problems can be identified by reading the plugs.

When the correct plug is used in a correctly tuned engine the centre electrode porcelain insulator will be chocolate brown or dark grey in colour, depending on the brand of petrol used. If you see a tinge of green on the edges of the plug, do not be alarmed - some petrol additives leave this marking.

Battery selection (general)

Due to recent advances in technology, there are new alternatives to choose from when selecting a battery for your Morgan. Standard automobile batteries have, for years, used antimony as a lead plate bonding agent. An unfortunate result of this, however, is that the reaction of antimony and sulphuric acid in the battery produces bubbles of gas. Therefore, batteries have a vent to atmosphere to allow this gas to escape but this results in having to add water to the battery from time to time.

However, the need for access to the cells for topping up creates the possibility of acid spills and damaged paint and metalwork.

There are two new types of batteries on the market:

1 *Low antimony, often advertised as low maintenance.*

2 *Calcium-based, often advertised as no maintenance.*

The low antimony type battery uses a smaller percentage of antimony to bind the lead in the plates together. This results in less gas production and a reduced need for topping up with water. The disadvantages of this battery are the continued possibilites of acid spills and corrosion in the engine compartment.

The calcium-based battery uses calcium instead of antimony as the bonding agent. Calcium produces no gas in its reaction with sulphuric acid. As a result, the battery can be sealed completely from the atmosphere, with only a one-way valve to allow boil-over in case of over-charging. This completely eliminates the need to add water, and greatly reduces the potential for leaks.

The calcium-based battery is the ideal battery for a Morgan. The lack of potential for leakage protects the engine compartment and paintwork and the no-maintenance feature is especially valuable for owners of two-seaters who would have to remove the luggage floor to check the battery. The calcium-based battery has another big plus for Morgan owners who do not drive their cars for extended periods of time, such as during the winter months. This type of battery will not normally discharge over time, as a regular antimony type will.

Be certain to use the felt washer terminal protectors, which are soaked in anti-corrosion fluid. These prevent any damage from terminal corrosion due to acid migrating up the battery posts, which can occur with any battery.

Finally, be certain to get a battery with at least 400-500 cold cranking amps. This will ensure that the starter on your Morgan will spin the engine over rapidly for quick starting, thereby reducing wear and tear on your starter motor.

• HINTS ON BATTERY MAINTENANCE
Keep your car's battery terminals clean. Dirty terminals prevent current being delivered to the electrical system and can give the impression of a dead battery. Lucas cup-shaped terminals can look perfectly good but be corroded inside sufficiently to reduce current flow and prevent operation of the system.

Purchase a battery charger. Twenty four hours on a 5-amp charger will restore any battery to full strength, if the battery is still good.

If you have an antimony battery, purchase an inexpensive hydrometer. This will enable you to check the specific gravity of the 'water' in the battery to determine if the battery is charged.

Headlight reflectors (general)

There have been many comments about Morgan headlight reflectors. They generally centre around whether to plate old reflectors with silver, chrome or nickel, all of which are far from perfect, or to change the whole system over to sealed beams for safe night driving. In fact, even with all original equipment you can meet sealed beam headlight standards! The most critical part of your headlight system is the reflectors.

Pre-1940 original equipment headlight reflectors were coated with silver. Freshly polished silver can reflect as much as 94 per cent of the bulb's light immediately after polishing. The reflectivity then declines as the silver tarnishes. To maintain silver at better than 80 per cent reflectivity, the generally accepted minimum for use in night driving, you would need to polish at least every other month and more often in some areas. Of course, there are only a limited number of times that you can repolish without replating.

Because of the problems with silver, many people have their car's reflectors plated with chrome. Chrome is certainly durable and never dulls. Chrome makes the car undrivable at night because the chrome reflects only about 65 per cent of the light available from the bulb, which is not enough for night vision. Other people have the reflectors plated with nickel because it has a softer, more yellow colour, giving the appearance of slightly tarnished silver. Nickel is even worse than chrome, reflecting only about 63 per cent of the available light.

In the late 1930s it became possible to coat moulded glass with highly reflecting surfaces of vacuum-evaporated aluminium. The aluminium surfaces, however, are very soft and decline in reflectivity if not properly protected from oxidation. This led to the development of the sealed beam headlamp, where the inside rear moulded glass surface is coated with an 85 per cent reflecting aluminium film in a vacuum coating system. The filament is added and the lens sealed on after filling the unit with inert gas. The process neatly protects the filament and protects the aluminium reflector from the ravages of oxidation.

It's now possible to vacuum deposit and seal aluminium directly on to the original equipment headlight reflector. The process begins with polishing the old tarnished reflector to remove the silver and underlying nickel layer. The brass is then buffed to an almost perfect sheen and the reflectors undergo a three-step coating process performed in a vacuum system. The outcome is a micro-thin layer of pure silica glass placed over a flash-evaporated pure aluminium coating. Just enough glass is used to obtain good durability while maintaining the highest reflectivity of the aluminium - 92 per cent. More glass would lower the reflectivity of the aluminium surface and add too much colour.

These overcoated aluminium coatings on your old reflectors yield reflectance equal to the best obtainable with silver and a reflector that looks like authentic, freshly polished silver in every way. This is a reflector that will never tarnish and will maintain its reflectivity as long as it is not mishandled or rubbed with any polishing or abrasive cleaning materials. These reflectors are durable enough to be cleaned carefully with mild soap and water or rubbing alcohol and cotton wool to remove fingerprints and dust.

Most companies offering this service undertake to recoat, at no charge, any reflector coated by them which fails to reflect at least 90 per cent for five years. With these reflectors and standard bulbs, you will have headlights essentially equal to sealed beams. Even better lights can be had by replacing the headlight sockets and using quartz-halogen bulbs. These bulbs are not recommended, however, because the lens may not be able to withstand the additional heat.

Throttle cable replacement (all models)

Morgans use throttle cables which vary in length from 12 to 30 inches. Since you cannot drive the car without using the throttle, these cables, naturally, are a high-wear item. It is common for the cable to wear, causing one (or more) of the strands to break and become entangled in the housing (sheath). The result is a 'notchy' movement which ultimately creates a cable that may jam or even break when the car is being driven.

The housing is also subject to great wear, the most common points of which are at each end of the housing where the cable enters and exits. This wear of the metal housing can also result in burrs being formed which can cause deterioration of the cable itself.

Cable replacements are readily available from an aircraft spares supplier or cycle shop. Aircraft cable is usually a stainless steel multi-strand type of higher quality and flexibility than the original equipment cable. Take your old cable with you and have the operator match it for diameter and length. The operator can also install the fitting at the end of the cable for you. But remember, replacement of the cable alone may not solve your problems; you may have to replace the housing as well. I's possible to obtain cables with a Teflon-lined housing. This ensures that the cable will slide within much more smoothly than the old one ever did. Since the Teflon surface provides the cable lubrication, the cable should be installed into the housing dry, with no oil or grease necessary or desirable.

One word of caution - double-check that the old housing was actually long enough to do its job correctly. I have seen instances of bends which are too sharp, as well as incorrect cable alignment. If you replace both the cable and housing, this is your chance to correct this fault.

Bowden cables (all models)

Bowden cables should be routed in gentle arcs, without any sharp bends. Most people oil their cables at either end and trust to luck that some actually finds its way to the centre of the cable. In fact, only about the first inch of each end receives any oil within the outer cable. This can be overcome by this simple solution. Obtain a corked bottle and drill the cork down the centre to the exact diameter of the outer cable. Then cut the cork in half down through the hole. Partially fill the bottle with oil; reassemble the two halves of the cork around the outer cable, the narrower end towards the free end of the cable; insert into the neck of the bottle, with the end of the inner cable extending inside. Then invert the whole assembly and support the bottle by some method. The oil will drain down the tube formed by the outer cable and lubricate the whole length of the cable. The process can be speeded up by periodically working the inner cable within the outer. In some cases it may be necessary to first use paraffin to penetrate the inner cable.

Location of chassis numbers (all models)

Chassis numbers are always stamped on the top of one of the chassis crossmembers on the offside. In the case of two-seaters, the Drop Head Coupés and the Plus 4 Plus Fixed Head Coupés, it will be found on the member immediately behind the gearbox. On the four-seaters it can be found on the member immediately behind the front seats. On the flat-radiator models of the 4/4 and the Plus 4 it can sometimes also be found on the reinforcing strip of the bonnet hinge about 6 inches from the front. Occasionally the number can also be found stamped on certain body panels, such as by the spare wheels and on the wood frame of the door hinge supports. These numbers are occasionally only lightly stamped onto the metal or the wood and are difficult to read.

It is a sound policy to make a note of as many numbers as possible in respect of the major components of your car - chassis, engine, gearbox, rear axle and so on - which will be most useful if ever you need to order or identify replacement parts. This is particularly important if you are not the first owner as the components may have been changed as a result of wear and tear or even an accident.

Whilst dealing with identification numbers, a brief explanation is necessary as to why the engine numbers on the Morgan do not follow on in sequence. The Morgan company has always purchased engines from outside manufacturers, meaning deliveries are made as stocks run low. When a delivery of 20, 30 or 40 engines is made, their numbers are, more than likely, consecutive, but when unloaded they are placed in front of the remaining stock from the previous delivery, not necessarily in order. When an engine is required, the first in line is invariably taken, and time is not wasted checking the sequence. Eventually all stock is used, but the engine numbers apparently have no logical sequence.

Model terminology (all models)

Over the years it has become apparent there is some confusion as to the correct way to denote the variations of Morgan models, particularly the four-wheelers. I include a short list of these designations as originally used by the Morgan Motor Company Limited, which are used in all references to the models either in the motoring press or by the motoring public in the UK and most other countries. This will, of course, assist the American reader should he have reason to write to the factory or any other UK Morgan organization.

4-4 or 4/4 - Refers to the model only and is said 'FOUR FOUR'.

Plus 4 - Refers to the model only and is said 'PLUS FOUR'.

Plus 4 Plus - Refers to the model only and is said 'PLUS FOUR PLUS'.

Plus 8 - Refers to the model only and is said 'PLUS EIGHT'.

To indicate which type of body the model has, the name of the model should be used as above, plus the type of body. For example:

PLUS FOUR, two-seater; PLUS FOUR, four-seater; PLUS FOUR, Drop Head Coupé.

In the case of a model which has had a range of engines, the practice is to indicate which model is being referred to by saying the series number or the engine type. For example:

'FOUR FOUR, SERIES TWO' or 'SERIES TWO FOUR FOUR'.

'PLUS FOUR FIAT-ENGINED' or 'FIAT-ENGINED PLUS FOUR'.

It is normally accepted that the above refers to a two-seater and the terms four-seater or Drop Head Coupé are only added if relevant.

Morgan

VI
RESTORATION ADVICE

Before undertaking any of the practical work described in this chapter it is essential to read the safety and working procedure notes given as an appendix to this book.

It is not my intention to deal with the dismantling and reassembly of every small part of all Morgan models, a task which would fill several books. Instead, I will confine myself to useful general methods, highlight common pitfalls and advise on the standard of finish for which an owner should aim.
Restoration is not a task to undertake without a lot of thought about what is involved. How many times does one come across an advertisement which reads: 'For sale ... insufficient time to restore', or, 'insufficient funds force sale'? Basically, one has to consider (a) the work involved, (b) the time involved, (c) the space required, (d) whether you are capable of carrying out the work or obtaining assistance from specific firms and, (e) can you afford the costs involved? If, having considered all these points, you still want to go ahead, I advise you to read and study every available book on the subject before you begin.

Remember, too, that a successful restoration job has invariably benefited from (a) a firm plan of action (b) patience (c) a tolerant partner (d) willing and reasonably strong friends to assist when required (e) the will to succeed against all setbacks (f) ample working and storage space (g) a good camera or an ability to sketch accurately (h) good tools of the right size (i) an electric drill (j) electric lighting and power points in the garage/workshop.

Safety
Whilst on the subject of accidents when restoring, never work under a car that is jacked off the ground without using axle stands. Never attempt to lift something by yourself that is too heavy or awkward for you. Always use protective clothing and, where necessary, masks and goggles. Always ensure that your power points really are power points and are correctly earthed. Check all power tools, inspection lights, etc, and ensure that they have good wiring and that they, too, are properly earthed. Never smoke or allow others to smoke in or near the garage, workshop or place where you are working. Always read, and read again, the manufacturer's instructions on any product you use and obey them. Always store inflammable material and products well away from the garage and the house.

Preparation for restoration
Before dismantling, some careful photography can save you hours of work at a later stage. It is advisable to take as many photographs - or make sketches of the car and its components - as possible, which should be taken from every angle to assist with correct reassembly. Photographs should also be taken at every stage of dismantling. Remember to make a list of the photos as they are taken and do so in a logical sequence. If possible, try to locate a member of the specialist club who has an identical model; most owners will be only too happy to allow you to take a series of photographs of their car and to give you various tips (some good, some bad ...). If possible, take these photographs in the same sequence as those of your own car; this will assist you when making comparisons.

Next, ensure that you have ample

boxes and containers to store the components and nuts and bolts from each particular section. You will also need labels to mark the containers; the self-adhesive type which can be purchased in most newsagents or deep-freeze centres are ideal. Remember to stick them on the side that will face forward when placed on shelves, not on the lid, as invariably there will be another container on top of the one you need.

Before you begin stripping components, be sure the car is situated in such a way that when it comes to removing the body you can do so without first having to remove the garage roof! Another common fault is to remove the wheels first which, of course, restricts every other job that requires changing the position of the car.

Dismantling

It is advisable first to dismantle the car into the minimum number of sections, each of which should be left complete until such time as you are ready to concentrate on that section by further dismantling, then cleaning, making good, painting and reassembling the various components.

Start by removing the spare wheel(s) and vulnerable parts like mirrors, headlamps, headlamp bar, side lamps, pass light, rear light and number plates - anything, in fact, that can easily be damaged. Remember to label the wiring so that you can identify it later and, if possible, disconnect wires at snap connectors. In general, first off is last on, so take care to store these items in a safe place where they cannot be damaged by other pieces falling on them.

Next, the windscreen, complete with wipers and motor (where applicable). This is best stored in the loft, under the bed or in the blanket chest. The most difficult section of the car, as far as handling and storing is concerned, is the bonnet, which should now be removed. The best method is to free both ends from their fixing points, then apply a liberal amount of penetrating oil to the centre hinge, no matter which type is used on your car. If it is the centre pin type, hammer it with a brass punch from one end and, as soon as enough appears the other end to get a grip with the mole wrench, slowly remove the pin by twisting it first one way, then the other. The centre looped spine type can be removed in the same manner, but care must be taken to avoid crimping the rolled edges. The bonnet is then in two manageable sections. A point arises here as to whether one should remove the bonnet fasteners ready to be sent off with other parts for re-chroming. From experience I have found it is better to send only a small number of pieces at a time to the plating works to minimize the chances of losing small but valuable parts. Therefore, leave them on until you are ready to work on the bonnet. This, too, should be stored in a safe place as the panels are very easily dented or bent; one suggestion is to secure them over the rafters of the garage.

Then comes removal of the radiator. First, drain it by removing the drain tap at the bottom, replacing the tap as soon as draining is completed. Unless you are extremely lucky, the hoses will have become vulcanized to inlet and outlet pipes. The best method is to put a hacksaw through the hoses (not the stubs!); at least then you are not tempted to put them back on when you replace the radiator, only to find they are cracked or that they burst as soon as you refill the cooling system with water. Store in a safe place (under the spare bed again?)

Wings and running boards are the next items for removal. First jack the car the side to be worked on and insert axle stands front and rear as a safety precaution. Next, remove the front and rear wheels but do not store them as they are going back on again. Follow the line of fixing screws along the wing edge and remove every other one, repeating this until all the screws are removed. If they are very rusty, they can be removed by careful use of a chisel. Replace the wheels, remove the axle stands and jack and lift the other side, repeating the complete operation.

All seats and carpets can now be removed, and this is an ideal time to take the battery out, too. Unless it is fairly new, make no attempt to save it, but if it is new, try to use it whilst the restoration is carried out; if it is to be of any use to you afterwards it must be charged and discharged regularly.

When dismantling any component always remember to replace the nut on the bolt from which it came, and wherever possible replace the bolt in its corect hole. This way, when it comes to putting everything back together you will know the correct size of bolt required should the original bolt need replacing.

Body removal

Before the body is removed be sure that it is completely free of all encumbrances. For example, the petrol fillers should be disconnected and, dependent upon which model you are restoring, the method will vary; a close examination of the filler and how it connects to the fuel tank will reveal the easiest way. It is not essential to drain the fuel tank at this stage as the drain plug is situated on the base of the tank, but whether or not you drain it you must plug the filler hole of the tank. Follow the fuel lines leading to the fuel pump and if these are fastened to the body by clips they should be disconnected.

The wiring should be cut away or disconnected and the ends marked appropriately. A new loom is by far the best course of action as there is no easy way of establishing the condition of the insulation throughout the whole system.

Next to be dealt with are the bulkhead and dashboard. If it is possible to detach them from the body it is better to do so; to be able to work on the wiring and other instrument connections without the encumbrance of the body is so much easier and quicker. On some models, where the dashboard is a part of the body, it may be necesary to remove the steering wheel and steering column before attempting to lift the body.

Check the whole body to see that nothing is attached to or passing through any part of it which would hinder or prevent the body removal from the chassis. If everything is clear, locate and loosen the nuts holding the body to the chassis. More than likely some of the nuts will be rusty and the coach bolts will turn with their nuts. There are several ways of dealing with this. First, give the bolts a good soaking with penetrating oil, then try tapping the head of the coach bolt further into the wood. However, if the nut is very rusty the whole thing will probably still turn. Another method is to cut a slot in the head of the coach bolt, wide and deep enough to take a screwdriver, but by far the best solution is to saw through the nut at the side or straight down the centre of the bolt, cutting the nut in half. In this way you avoid damaging the wood or metal behind the nut. If the nut is held in place by a split-pin as well, another alternative is to drill through the head of the coach bolt and use a tool of the 'Easyout' type to hold it. This type of tool is a tapered left-hand coarse tap made of hardened steel and can be supplied in different sizes; provided you drill the pilot hole to the correct size and hold the tool in a tap wrench it will grip nearly any rusted bolt. It is also invaluable for getting out broken bolts and studs. As a last resort, a sharp chisel and a 2lb hammer works wonders. If the last method is adopted, care must be taken to chisel on the nut side of the chassis, not at the end against the wooden frame. Always try to split the nut, rather than cut the bolt.

Now, after a final check, you are ready to remove the body. Firstly, obtain the assistance of four friends, one for each corner of the body (it is possible to lift the body with two people, but four is safer). Make sure that you have already prepared sufficient space in which to store the body and that access to the storage space has also been cleared: there is nothing worse than lifting the body, not knowing which way to move and then one of your helpers dropping and damaging his corner in the confusion. It will be a great help if you can arrange for the body to be rested across three or four trestles about 3ft to 3ft 6in in height. This saves moving the body again and is a convenient working height. While your assistants are manhandling the body, your own task should be to give them clear, positive instructions and, at the same time, adjust the position of the trestles so that the body will rest on them squarely, securely and correctly balanced. Having done this, you can then move on to removal of the flooring.

Marking components

I think at this point I should devote a paragraph or two to the marking of relative parts. There are two methods frequently used but neither is good practice and, where possible, should be avoided. The first is differing numbers of centre punch marks and the second hacksaw blade marks. Recently I assisted a friend in restoring a 1930 Lagonda and marking the place from where the halfshafts had been drawn was a total of 18 centre punch marks on both surfaces. It took some time to count them all, and just to help matters the opposite side had 17 punch marks. The simplest and cheapest method is to invest in a set of letter and number stamps, which may still be useful when you have completed the restoration; a set embracing A-Z and 0-9 provides you with a huge number of permutations.

First of all, ask yourself whether a particular part needs to be marked at all. If it does, then be sure to select a point for your punch where the component is least stressed. Also avoid placing it too close to the edge of a flange, as it will raise the metal on the face and you will spend hours trying to obtain a perfect mating surface; never allow a punch mark to spoil a working surface. By all means use the letter or number punches when necessary, but remember there are very few components that require such marking; masking tape of suitable size is ideal for the remainder.

Cleaning components

Most people restoring cars seem to want to remove the accumulated oil and grease from components as soon as possible, but this is not a good idea as the covering can be an effective rust

preventer. Only degrease a particular part when you are ready to start work on restoring it. The same may be said about draining the engine, gearbox and rear axle: leave them until you are ready to work on them.

Most areas of the country these days are covered by a travelling steam-degreasing company; they are not unduly expensive and can completely degrease a rolling chassis in under two hours. Before deciding to use such a service it is, of course, advisable to ensure that you have somewhere accessible where you can roll the chassis and where the resulting mess will not cause total war with your wife or neighbours!

The easiest manual method to adopt is to first scrape with an old knife or paint scraper and then, using an old paint brush, liberally apply paraffin (not petrol), washing off with a pressure hose or wiping off with an old rag. It cannot be over-emphasized how much care must be taken when working with petrol, paraffin, cellulose and any other preparations which have a high flash-point. I remember someone who for three years spent every minute of his spare time restoring his car. The time arrived to attempt to start it but, try as he might, it just would not fire. He then obtained some ether-based 'Easy Start' solution, but liberal applications of this in the carburettor still did not obtain the desired result. Meanwhile, unbeknown to him, one of the petrol feed unions was leaking. He then removed a plug and earthed it by holding it against the block, turning the engine over to check the spark. There was a spark! Result - his car, garage and workshop were completely destroyed and he spent nearly two years in hospital. You just cannot be too careful.

Paint removal

Chemical paint removers, of course, are convenient to use in the garage. However, chemical paint removers do not deal with rust, which will require special treatment, first with a wire brush (hand or powered) and then with rust inhibitors, although from experience I have found that no matter how thorough you are, somehow you never manage to remove all the rust. If you can afford to have the part sand-blasted you will save a lot of time and end up with a far more efficient job (if your classified telephone directory cannot help, remember that specialist service firms are often listed or advertised in car enthusiasts' magazines). However, this treatment is not recommended for Morgan body panels, due to the distortions which can be created by the process. By far the best way is to have the part chemically and professionally stripped in caustic soda vats, which is not nearly as expensive as you may think.

Replacement of worn or broken parts

To the amateur, this is the most worrying part of restoration, especially if he has no skills with a lathe or other machines. This is where the help of the specialist clubs can be so useful.

Most clubs have a spares register (the Morgan Sports Car Club being no exception), whose job is to locate new or secondhand spares and arrange for the special manufacture of otherwise unobtainable parts, providing there is sufficient demand from club members. When you discover a worn or broken part, do not panic; if you are restoring a Morgan you will find that the vast majority of parts can be replaced with the help of the spares registrar or by companies who specialize in Morgans.

Another source of supply is the numerous autojumbles which are so much a part of the classic car scene nowadays; not so many years ago an almost complete Coventry-Climax engine came my way for £35: it is amazing what can be found.

For the location of parts, services and materials covering all aspects of car restoration, I can also recommend the *Directory of Specialist Firms and Services* published by the British magazine *Classic Cars*.

Body panels that are damaged can be dealt with by a motor body shop that specializes in accident repairs, while damaged ash frames can be entrusted to any good carpenter and joiner, although in some cases the Morgan Motor Company may still have replacements in stock. Boat builders, of course, are a good source of well-seasoned ash for the frame and can also supply marine standard plywood for floors, scuttles, etc.

Restoration standards

No matter how much is writtenon a practical subject, there can be no substitute for experience. Remember that in restoring your car you will require skills in which, like as not, you have had little, if any, previous experience. By comparison the car manufacturer, or the professional restoration company is fortunate; they tend to employ people for various individual aspects of the job, which they do day-in, day-out, having first been taught their particular trade by highly-skilled colleagues.

Do not be deterred by lack of experience. Amazing results have been obtained by sheer determination, hard work and patience. In offering the advice which follows I hasten to add that I do not see myself as an expert, nor do

I claim that my advice is always based on the accepted correct way of tackling a particular job. But at least it is the result of watching and talking to talented experts; in applying their techniques, my own restoration work has led to many concours successes.

No matter what material you are working with - wood, metal, paint - the following basic rules must apply: (a) there are no short cuts; (b) never rush, no matter how slow you are or how long it takes to achieve a good standard; (c) never settle for second best; (d) have in your mind's eye the standard you are attempting to achieve; and (e) never lose patience with a particular job. If you feel you are doing so, leave it and return to it later. There is always plenty of other work to be done.

Before dealing with specific methods and materials, I think I should point out that over-restoring is just as bad as under-restoring. Do not have parts chromium-plated or polished if, when the car left the works new, they were painted. Examples of over-restoration are chromium-plated exhaust manifolds and down-pipes which were never chromium-plated when new; radiator header and floorboards made of mahogany or sapelia and then French-polished top and bottom. Chromium-plated steering gear is another popular example of 'restoring' a car to a finish it didn't have in the first place.

Paintwork

The question of whether to have the chassis, wheels, steering parts, rear axle, etc, stove-enamelled is a thorny one. If it is well done it produces a marvellous finish, but should it subsequently become chipped it can never be touched-in to look correct. Also, if stone chips go unnoticed, rust will spread under the enamelling. There is also the question of cost - stove-enamelling is not cheap.

On the other hand, really good finishes can be obtained using brushed-on coach paint or sprayed cellulose paint, providing you are patient and rub down well with varying grades of wet-or-dry abrasive paper between each coat. Whilst on the subject of rubbing-down paintwork, remember to use a rubbing-down pad or block for the wet-or-dry, also to continually rinse the paper to prevent clogging.

In an effort to remove a lot of the tedium of the job of painting a chassis and other so-called 'blackwork', a professional restorer will often use a twin-pack acrylic commercial vehicle chassis paint. First the chassis is cleaned and shot-blasted, then etch-primed and sprayed with a high-build primer. This is hand rubbed to a reasonably flat finish and then sprayed with the two-pack acrylic paint which gives a high-gloss finish comparable to any obtained by the usual slow and laborious methods. The more up-to-date method is to have the 'blackwork' epoxy powder-coated, which is the modern alternative to wet paint stove-enamelling. In this process, the part is electrically charged to become polarized and is then sprayed with minute plastic particles which have the opposite charge. Because opposite poles attract, the part becomes completely covered with an even coating of the powder. The part is then baked to provide a finish which, beside being very good-looking, is also very durable and chip-resistant. The process uses a wide range of plastics, epoxies, polyesters and polyurethanes, depending on the part to be coated. It can be carried out in a wide range of colours and finishes, including full gloss, matt, plain and metallic.

When painting the bodywork, start by applying several coats of rust inhibitor before using a filler coat. When you have applied the filler coat, cover it completely with a very thin coat of matt black, then rub it down gently. In this way you will discover all the high and low spots which must be dealt with before you go on to the primer and colour coats. Small indentations can be dealt with by using cellulose stopper which, when hardened, is rubbed down level with the surrounding metal or paint surface. Larger dents which cannot be reached by a panel-beater can be filled with plumber's lead; any plumber should be able to do the job for you if you cannot yourself. High spots, depending on size, can be flattened with the aid of a rubber mallet applied either to the spot or, if it is large, from the outer edges of the high spot, working towards the middle. Remember to hold a supporting pad or block to the underside of the metal and do not strike it too heavily.

Brass and copper

This should first be cleaned with fine wire wool and then polished with metal polish until all scratch marks disappear. Always renew any old brittle or badly-dented brass or copper tubing; nothing looks worse than a copper pipe wound in a close spiral, or bent to accommodate a change of direction, on which the bends have become flattened. Any good tool shop can supply bending springs which will avoid this.

Another solution is to anneal the pipe by heating to a cherry red and plunging it into cold water, which will soften the metal. Then block one end of the pipe with a small cork and fill the pipe with sand and seal. The bends can

This page and over, top: The four-seater Series I 4-4 started production at the end of 1937. The wheelbase was the same as the two-seater's; the rear seat was made possible by taking space from the luggage area and repositioning the fuel tank and batteries.

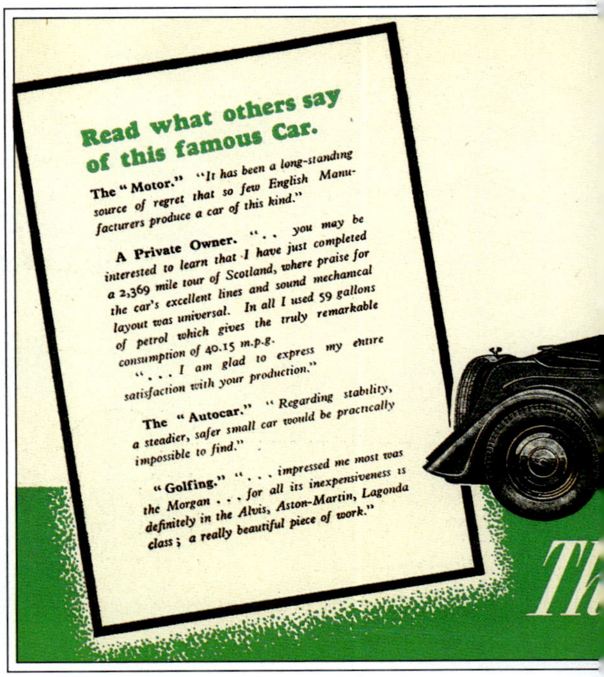

The Morgan factory is always ready to capitalize on press reports which praise their products. In this 1939 brochure they reproduce at length a report from the September 9th 1938 issue of The Autocar. *The brochure illustrates all three models, including the new Drop Head Coupé, arguably the most attractive of the Series I models. The dashboard photograph has been annotated by hand. The gearbox layout illustrates the four-speed Moss box, which has a reverse configuration to normal.*

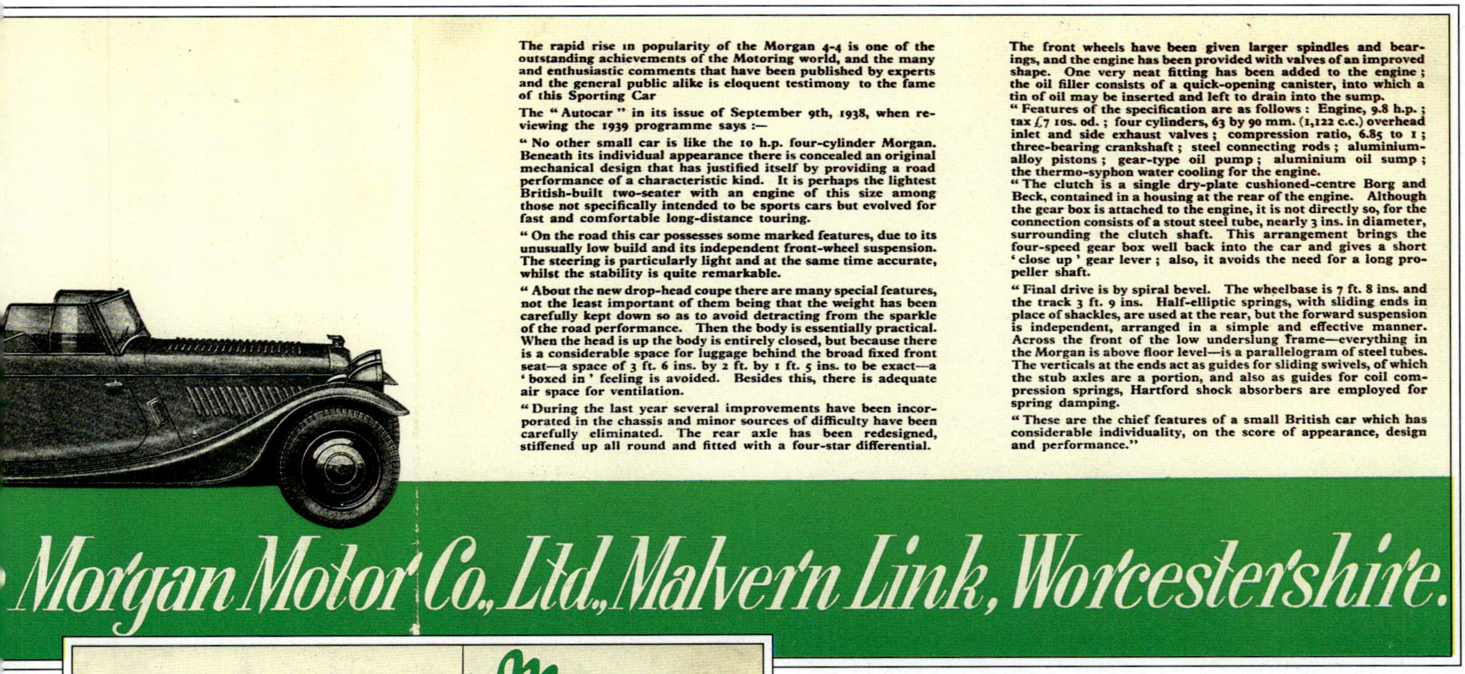

The rapid rise in popularity of the Morgan 4-4 is one of the outstanding achievements of the Motoring world, and the many and enthusiastic comments that have been published by experts and the general public alike is eloquent testimony to the fame of this Sporting Car

The "Autocar" in its issue of September 9th, 1938, when reviewing the 1939 programme says :—

"No other small car is like the 10 h.p. four-cylinder Morgan. Beneath its individual appearance there is concealed an original mechanical design that has justified itself by providing a road performance of a characteristic kind. It is perhaps the lightest British-built two-seater with an engine of this size among those not specifically intended to be sports cars but evolved for fast and comfortable long-distance touring.

"On the road this car possesses some marked features, due to its unusually low build and its independent front-wheel suspension. The steering is particularly light and at the same time accurate, whilst the stability is quite remarkable.

"About the new drop-head coupe there are many special features, not the least important of them being that the weight has been carefully kept down so as to avoid detracting from the sparkle of the road performance. Then the body is essentially practical. When the head is up the body is entirely closed, but because there is a considerable space for luggage behind the broad fixed front seat—a space of 3 ft. 6 ins. by 2 ft. by 1 ft. 5 ins. to be exact—a 'boxed in' feeling is avoided. Besides this, there is adequate air space for ventilation.

"During the last year several improvements have been incorporated in the chassis and minor sources of difficulty have been carefully eliminated. The rear axle has been redesigned, stiffened up all round and fitted with a four-star differential. The front wheels have been given larger spindles and bearings, and the engine has been provided with valves of an improved shape. One very neat fitting has been added to the engine; the oil filler consists of a quick-opening canister, into which a tin of oil may be inserted and left to drain into the sump.

"Features of the specification are as follows: Engine, 9.8 h.p.; tax £7 10s. 0d.; four cylinders, 63 by 90 mm. (1,122 c.c.) overhead inlet and side exhaust valves; compression ratio, 6.85 to 1; three-bearing crankshaft; steel connecting rods; aluminium-alloy pistons; gear-type oil pump; aluminium oil sump; the thermo-syphon water cooling for the engine.

"The clutch is a single dry-plate cushioned-centre Borg and Beck, contained in a housing at the rear of the engine. Although the gear box is attached to the engine, it is not directly so, for the connection consists of a stout steel tube, nearly 3 ins. in diameter, surrounding the clutch shaft. This arrangement brings the four-speed gear box well back into the car and gives a short 'close up' gear lever; also, it avoids the need for a long propeller shaft.

"Final drive is by spiral bevel. The wheelbase is 7 ft. 8 ins. and the track 3 ft. 9 ins. Half-elliptic springs, with sliding ends in place of shackles, are used at the rear, but the forward suspension is independent, arranged in a simple and effective manner. Across the front of the low underslung frame—everything in the Morgan is above floor level—is a parallelogram of steel tubes. The verticals at the ends act as guides for sliding swivels, of which the stub axles are a portion, and also as guides for coil compression springs, Hartford shock absorbers are employed for spring damping.

"These are the chief features of a small British car which has considerable individuality, on the score of appearance, design and performance."

Morgan Motor Co., Ltd., Malvern Link, Worcestershire.

Morgan

A very attractive two-seater on distinctive and sporting lines. The floor-boards are carried directly on the lower flange of the chassis frame, which results in a low and comfortable seating position. Each seat has its own pneumatic cushion with single back squab, behind which is a large space for luggage and a concealed hood, whilst behind this again is a 9-gallon petrol tank with quick action filler. A tail of registered design holds two spare wheels, is smooth and easy to clean and incorporates efficient stream-lined mudwings with deep valances.

TWO SEATER

190 GUINEAS

The essentials are exactly as the two-seater, with the body lengthened to provide two extra seats. The petrol tank on this model is below the rear seats, only one spare wheel is included in the equipment and a single 12-volt battery is under the bonnet. A "one man" hood is fitted, which, together with carefully arranged side curtains afford complete weather protection.

The bodywork is carried out in best style with DULUX Coach finish and is upholstered in Black leather; the standard colours are British Racing Green, Saxe Blue, Black or Carnation Red. Any deviation from the standard colours entails an extra charge of 5 guineas and a further 3 guineas for deviation from the standard leather upholstery.

FOUR SEATER

205 GUINEAS

In designing the Coupe (which is made throughout in our own factory) we have endeavoured to provide good weather protection and comfort combined with attractive appearance and good performance. A feature is that the windows above the 2 ft. 2 ins wide doors are arranged with twin sliding glasses. The object of these is three-fold. The first point is that each panel may be opened to give a gap at the front or the back of the window, so that the type of ventilation, either in-draught or out-draught may be chosen to suit the weather conditions. The second point is that the door is free of mechanism whereby a considerable increase of elbow room is obtained; the width is 44 ins. Thirdly, the whole frame carrying the glass panels can be removed entirely without much difficulty if in very warm summer weather a completely open car is wanted.

COUPÉ . . . . .

225 GUINEAS

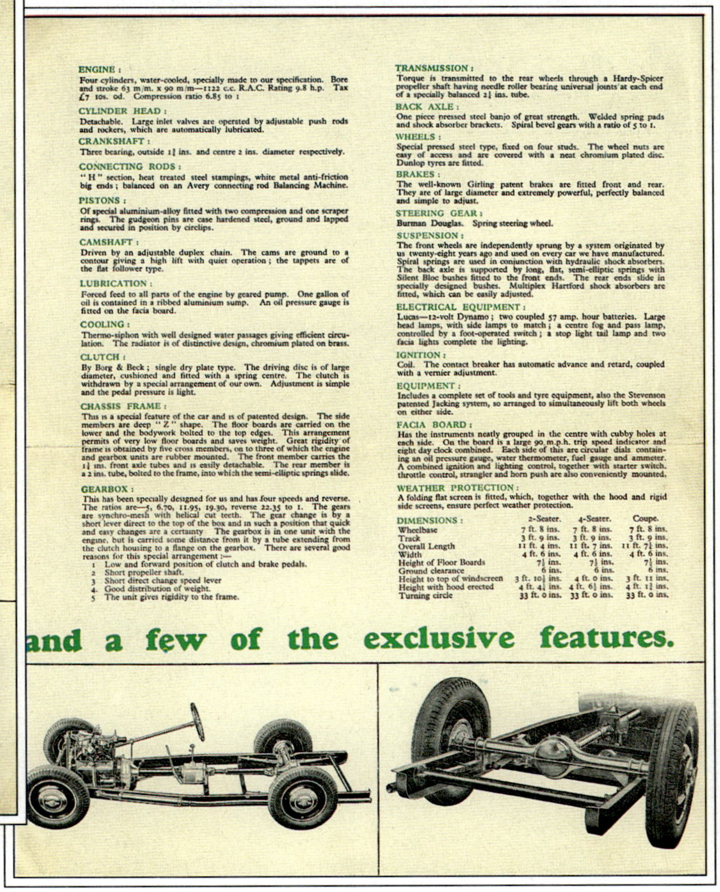

ENGINE :
Four cylinders, water-cooled, specially made to our specification. Bore and stroke 63 m/m. x 90 m/m—1122 c.c. R.A.C. Rating 9.8 h.p. Tax £7 10s. 0d. Compression ratio 6.85 to 1

CYLINDER HEAD :
Detachable. Large inlet valves are operated by adjustable push rods and rockers, which are automatically lubricated.

CRANKSHAFT :
Three bearing, outside 1¾ ins. and centre 2 ins. diameter respectively.

CONNECTING RODS :
"H" section, heat treated steel stampings, white metal anti-friction big ends; balanced on an Avery connecting rod Balancing Machine.

PISTONS :
Of special aluminium-alloy fitted with two compression and one scraper rings. The gudgeon pins are case hardened steel, ground and lapped and secured in position by circlips.

CAMSHAFT :
Driven by an adjustable duplex chain. The cams are ground to a contour giving a high lift with quiet operation; the tappets are of the flat follower type.

LUBRICATION :
Forced feed to all parts of the engine by geared pump. One gallon of oil is contained in a ribbed aluminium sump. An oil pressure gauge is fitted on the facia board.

COOLING :
Thermo-siphon with well designed water passages giving efficient circulation. The radiator is of distinctive design, chromium plated on brass.

CLUTCH :
By Borg & Beck; single dry plate type. The driving disc is of large diameter, cushioned and fitted with a spring centre. The clutch is withdrawn by a special arrangement of our own. Adjustment is simple and the pedal pressure is light.

CHASSIS FRAME :
This is a special feature of the car and is of patented design. The side members are deep "Z" shape. The floor boards are carried on the lower and the bodywork bolted to the top edges. This arrangement permits of very low floor boards and saves weight. Great rigidity of frame is obtained by five cross members, on to three of which the engine and gearbox units are rubber mounted. The front member carries the 1¼ ins. front axle tubes and is easily detachable. The rear member is a 2 ins. tube, bolted to the frame, into which the semi-elliptic springs slide.

GEARBOX :
This has been specially designed for us and has four speeds and reverse. The ratios are—5, 6.70, 11.95, 19.30, reverse 22.35 to 1. The gears are synchro-mesh with helical cut teeth. The gear change is by a short lever direct to the top of the box and in such a position that quick and easy changes are a certainty The gearbox is in one unit with the engine, but is carried some distance from it by a tube extending from the clutch housing to a flange on the gearbox. There are several good reasons for this special arrangement :—
1 Low and forward position of clutch and brake pedals.
2 Short propeller shaft.
3 Short direct change speed lever
4. Good distribution of weight.
5 The unit gives rigidity to the frame.

TRANSMISSION :
Torque is transmitted to the rear wheels through a Hardy-Spicer propeller shaft having needle roller bearing universal joints at each end of a specially balanced 2¼ ins. tube.

BACK AXLE :
One piece pressed steel banjo of great strength. Welded spring pads and shock absorber brackets. Spiral bevel gears with a ratio of 5 to 1.

WHEELS :
Special pressed steel type, fixed on four studs. The wheel nuts are easy of access and are covered with a neat chromium plated disc. Dunlop tyres are fitted.

BRAKES :
The well-known Girling patent brakes are fitted front and rear. They are of large diameter and extremely powerful, perfectly balanced and simple to adjust.

STEERING GEAR :
Burman Douglas. Spring steering wheel.

SUSPENSION :
The front wheels are independently sprung by a system originated by us twenty-eight years ago and used on every car we have manufactured. Spiral springs are used in conjunction with hydraulic shock absorbers. The back axle is supported by long, flat, semi-elliptic springs with Silent Bloc bushes fitted to the front ends. The rear ends slide in specially designed bushes. Multiplex Hartford shock absorbers are fitted, which can be easily adjusted.

ELECTRICAL EQUIPMENT :
Lucas—12-volt Dynamo; two coupled 57 amp. hour batteries. Large head lamps, with side lamps to match; a centre fog and pass lamp, controlled by a foot-operated switch; a stop light tail lamp and two facia lights complete the lighting.

IGNITION :
Coil. The contact breaker has automatic advance and retard, coupled with a vernier adjustment.

EQUIPMENT :
Includes a complete set of tools and tyre equipment, also the Stevenson patented Jacking system, so arranged to simultaneously lift both wheels on either side.

FACIA BOARD :
Has the instruments neatly grouped in the centre with cubby holes at each side. On the board is a large 90 m.p.h. trip speed indicator and eight day clock combined. Each side of this are circular dials containing an oil pressure gauge, water thermometer, fuel gauge and ammeter. A combined ignition and lighting control, together with starter switch, throttle control, strangler and horn push are also conveniently mounted.

WEATHER PROTECTION :
A folding flat screen is fitted, which, together with the hood and rigid side screens, ensure perfect weather protection.

DIMENSIONS :	2-Seater.	4-Seater.	Coupe.
Wheelbase	7 ft. 8 ins.	7 ft. 8 ins.	7 ft. 8 ins.
Track	3 ft. 9 ins.	3 ft. 9 ins.	3 ft. 9 ins.
Overall Length	11 ft. 4 ins.	11 ft. 7 ins.	11 ft. 7½ ins.
Width	4 ft. 6 ins.	4 ft. 6 ins.	4 ft. 6 ins.
Height of Floor Boards	7½ ins.	7½ ins.	7½ ins.
Ground clearance	6 ins.	6 ins.	6 ins.
Height to top of windscreen	3 ft. 10½ ins.	4 ft. 0 ins.	3 ft. 11 ins.
Height with hood erected	4 ft. 4¼ ins.	4 ft. 6½ ins.	4 ft. 1¾ ins.
Turning circle	33 ft. 0 ins.	33 ft. 0 ins.	33 ft. 0 ins.

and a few of the exclusive features.

Very few posters were produced for either the Three- or Four-Wheelers. This colourful example was used to illustrate the new Plus 4 and also, in 1951, in a brochure.

Below: The 1952 Plus 4 Saloon, produced to private order by Cooper Motor Bodies of Putney, London, arrives at Pebble Beach, California, to compete in the concours event of 1993. This was the first time a Morgan had been invited to enter an event: it achieved second in class.

Owners Gerry and Barbara Willburn, drive their one-off 1952 Plus 4 Saloon on to the prizegiving rostrum at the 1993 Pebble Beach, California, concours.

One of the rare Plus 4 four-seater Drop Head Coupés.

Simple and to the point: this 1955 brochure was the first to actually include the new rear-engined Series II 4/4 which was introduced in October 1955.

Right & opposite page, top: Throughout its years of manufacture the Morgan Motor Company has always tried to appeal to the deep-down sporting instinct of its customers. Although two years apart, these two brochures clearly illustrate this underlying aim. The 1960 example unknowingly predicts the international racing successes enjoyed by the Plus 4 in the hands of Chris Lawrence and his specially-tuned car, which became the forerunner of the Super Sports model that restored Morgan's competition reputation.

An under-bonnet view of the author's Le Mans Replica when it was supercharged. The low pressure, long shaft-driven Arnott supercharger was not actually supplied by the Morgan factory but was recommended by it. Only one or two cars were fitted with this extra, which made a remarkable difference in acceleration. However, if used on the Coventry Climax 1098cc full competition engine the centre main bearing had to be strengthened.

The author removed the supercharger when the Vintage Sports Car Club agreed to accept the car for its post-vintage thoroughbred register, providing it was returned to original specification. This was the first time a new marque had been admitted to the register and the first ever Morgan. It is regarded by the VSCC as 'a rare car of special interest'.

This 1963 Plus 4 wears wire wheels; an optional extra generally considered to greatly improve the car's overall appearance. A beautiful restoration job, the car pictured belongs to Edgar L Beere Jnr. of Ohio, USA.

As with all Morgans, this car is not simply a road-going machine. Owner Dixon Smith successfully drove his four-seater Drop Head Coupé in many competitive events. He is seen here competing in an Inter-Club Hillclimb at Shelsley Walsh in July 1964.

MORGAN + 4

SPECIFICATION

T.R.4. 4 cylinder engine of 2,138 c.c. developing 105 b.h.p. at 4,750 r.p.m. Compression ratio 9:1, twin Zenith Stromberg carburettors, 4 speed synchromesh gearbox; Girling front disc and rear drum brakes; independent front suspension; 5.60 x 15 high speed tyres on Dunlop wire wheels. Fuel tank capacity 10 galls. Maximum speed 110 m.p.h. Petrol consumption 26—30 m.p.g.

The Morgan Plus Four Plus is fully enclosed, providing complete weather protection. Wind-down side windows provide controlled ventilation and a heater is fitted as standard. The smooth flowing lines of the large windscreen are carefully designed to provide rain flow across the roof and away from the field of vision.

The deep fully-upholstered seats of the Plus Four Plus help to provide a comfortable ride at all speeds. The feeling of space inside the car is quite remarkable and the adjustable driving position is designed for arms-length comfort. The Plus Four Plus is a genuine Grand Touring Sports Car and ideal for tireless long-distance travel in comfort at high speed.

Standard finishes available: Westminster Green, Kingfisher Blue, Crimson or Broken White. Upholstery Black. Non-standard colours for paintwork and upholstery available along with optional extras, prices on application to the factory.

UNITED KINGDOM DISTRIBUTORS

BASIL ROY LTD.,
161 Great Portland Street,
LONDON, W.1

JOHNSON & BROWN,
268-270 High Street,
BROMLEY,
Kent.

BOLTON OF LEEDS LTD.,
14 Harrison Street,
Briggate,
Leeds, 1,
YORKS.

LIFES MOTORS LTD.,
32-36 West Street,
Southport,
LANCS.

WOODLANDS SERVICE STATION,
Cronton Road,
Tarbock,
Whiston,
LANCS.

CEDAR MOTOR HOUSE,
110-112 Bath Road,
Cheltenham,
GLOS.

BOWMAN & ACOCK LTD.,
Pickersleigh Garage,
Malvern Link,
WORCESTER.

HUXHAM'S (CARS) LTD.,
Penn Hill Garage,
Parkstone,
DORSET.

WESTLEIGH GARAGE LTD.,
1339 London Road,
Leigh-on-Sea,
ESSEX.

THE MORGAN MOTOR CO. LTD. Pickersleigh Road, Malvern Link, Worcester. Tel: Malvern 106.
Grams: Morgan Motors, Malvern Link.

Designed and Produced by Granville, Lang and Partners Ltd.

Only 26 examples were ever produced of this fibreglass-bodied Plus 4 Plus Fixed Head Coupé. It was a complete departure from the traditional bodywork, whilst the chassis, engine, running gear, etc., was the same as the Plus 4's.

Although the prototype, in the hands of Peter Morgan, proved successful in competition, it did not appeal to the motoring public. Manufacture was almost to order only and total production spread over a two year, four month period.

This fully enclosed 2-seater sports car is developed from the successful Morgan Plus Four. It shares with that model, tractability and ease of control, with the added refinements of deep luxurious seating and year-round weather protection.

The Plus Four Plus provides smooth, quiet travel at speeds up to 110 m.p.h. yet remains completely docile and manageable in heavy traffic. It is safely controlled by 11″ dia. front disc and rear drum brakes.

THE NEW CAR
FOR THE SPORTS ENTHUSIAST

Morgan
'PLUS FOUR PLUS'

Admire the smooth sweeping lines of the Plus Four Plus, then step into the spacious interior. There's more than ample room for two, in comfort, with luggage space, plus the useful boot. The seating position is low and purposeful, with excellent vision through the deep windscreen and winding windows. Full instrumentation is provided and controls fall easily to hand in the true sports car manner.

Price £1,055 plus p. tax £220 7s. 1d. **£1,275 7s. 1d.**

These fine examples of this very rare Plus 4 Plus model are owned and have been fully restored by American enthusiast John H Sheally II. The different angle photographs clearly illustrate the distinct lines of the model, the front being compared by the motoring press to a Jaguar XK150 and the rear with a Lotus Elite. Now very much a collector's car, this is a most sought-after model.

The TR4 and TR4A engines used in the Plus 4 Super Sports were sent to Lawrencetune at Westerham Motors, Acton, where they were stripped down, all internal moving parts balanced, the cylinder head polished and gas flowed and a modified inlet manifold and a pair of Weber carburettors fitted.

The classic Plus 4 lines! This original 1967 US export model had wire wheels and black-faced instruments (which were, by then, standard on the model). Note the tail and rear indicator lamps; these were later replaced by the stalk-type fittings used today.

The perfect picture of power and elegance combined is epitomised by this 1967 Plus 4 Super Sports.

In the mid-1960s the Morgan Motor Company switched its advertising style to the then-popular pin-up girl image, although it must be said that their ladies were well covered, as this 1967 brochure demonstrates.

75th Anniversary celebrations: just a few of the 1000-plus Morgans that attended the Morgan Motor Company's birthday party in 1984.

Twenty Morgans on parade are evidence of the interest in the marque in Italy.

The Morgan Sports Car Club meet at Catalonia in 1987.

Members of the Western New York Morgan Owners Group at the Victor Auto Festival in Victor, New York, site of the largest one-day car show in eastern USA (700 cars).

Perestroika comes to Estonia ... In 1989 this six metre high poster in Tallinn was announcing a new motoring magazine. This was before Estonia became an independent country again and illustrates the way thinking had changed in the USSR.

A South Eastern Finecast 1/24 scale model of a Morgan Plus 8.

BKL Autoreplica is the largest manufacturer of die cast white metal Morgan models in the world. The company manufactures, in all, fourteen different models in various scales, supplied in both fully finished and kit forms. The photographs above and left illustrate the latest additions to the range: 1/43 scale Morgan Three-Wheelers 'F2' and 'F4' with removable bonnets which reveal engine detail.

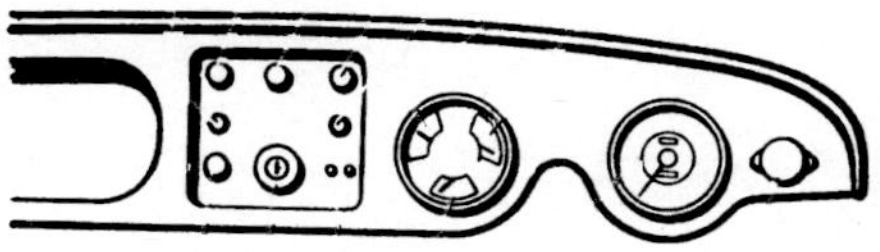

then be made without fear of flats being formed. When finished, empty out the sand and wash the pipe thoroughly under running water to ensure that not a single particle is left inside; blowing out with compressed air is insufficient as the sand tends to stick to the walls of the pipe.

Aluminium finishes

It is possible to obtain a finish on aluminium that can be mistaken for chromium-plating, but it involves hours of work. It is advantageous, wherever possible, to remove the part and work on it on a bench or table. The final finish that can be obtained must, by necessity, depend on the condition of the aluminium before you begin. Oxidized aluminium will have small pinhead-size holes over the surface. The depth of these will vary greatly, and the number that can be removed is largely dependent upon the thickness of the metal. It is better to leave some of the holes rather than reduce the thickness to such a level that the casting becomes fragile.

If the metal is in poor condition, begin by lightly filing the casting and then progress to the gentle use of engineer's scrapers. Then, by the use of emery cloth of diminishing grades of coarseness, or wet-or-dry paper impregnated with liquid metal polish, continue the polishing until you reach the desired standard. Care should be taken at all times to follow as closely as possible the grain of the metal. When you have achieved the standard required, polish finally with a small amount of metal polish such as 'Autosol' and shine with a soft duster.

Some castings should not be polished at all, particularly on sides exposed to the weather, such as the bottom of the sump. With these parts simply remove the forge marks by fettling with files and scrapers and finish off by wiping with a clean rag soaked in petrol (taking all necessary precautions to avoid a fire and skin contamination).

Pieces such as the base of SU petrol pumps, AC mechanical petrol pumps and carburettors look far better with a satin finish. This is best obtained by having them lightly bead-blasted, which has the added advantage of getting clean all those little crevices that are so hard to reach. Bead-blasting can also produce very good results on the exposed castings mentioned in the previous paragraph.

Remember that the largest proportion of any charge in connection with outside work is labour; therefore do as much preparatory work as you can before sending the part to a specialist firm.

Dashboard

The hardest part of dashboard restoration is obtaining the right type of well-seasoned timber. Boat-builders, timber yards and even undertakers are possible sources, but by far the cheapest way may be to visit auctions, house contents sales, or secondhand furniture dealers. It may take a little searching but you should find an old table, chest of drawers, bookcase, or even an old wooden fireplace surround. Any of these can usually be bought for a small sum, and by careful selection you will find a piece of ash, oak, teak, walnut or some other hardwood which is well-seasoned and can be cut to the size you require. However, be sure to inspect the item carefully before you buy it; woodworm, rot or cracks must be avoided for obvious reasons.

Having found your piece of wood, strip off the old paint or varnish - this can be done with any proprietary brand of chemical paint remover but be careful to scrape the old varnish off without damaging the wood. Then cut out the oblong you require and, using the remains of the old dashboard as a template, carefully mark out the exact shape you need. If the dashboard is missing completely, here is one of the instances when your photographic or sketching ability comes in handy. Having marked out the correct shape, and using an electric jigsaw, carefully cut round your marked-out shape, remembering to cut outside the line. The average saw cut is 0.12in-0.19in thick, and if you follow your marked line you will be that much out in the size you require.

The next stage is to cut the various holes, but always practise on a spare piece of wood before beginning work on the actual dashboard. If you think that you are unable to carry out this work, a carpenter or joinery shop should be able to tackle it for you. Also, they can usually be persuaded to machine-sand the surfaces for you, even if you have cut the shape yourself. This can save you hours of rubbing-down with sandpaper of diminishing grades of coarseness to achieve a good finish.

Next comes the staining, which can be done with wood dyes obtainable at most hardware stores. However, by far the cheapest method is to buy a few pence worth of permanganate of potash from your chemist, which is mixed into a thin paste and applied with a pad made of cotton wool about 2 inches square. Apply the paste using a gentle circular motion and, having covered the wood, allow to dry thoroughly before gently rubbing down with very fine sandpaper. (When staining it is advisable to wear a pair of old rubber kitchen

gloves to avoid staining your hands). Repeat if necessary until you obtain the desired colour.

The next stage is to paint the wood with a half-and-half mixture of white spirit and clear polyurethane varnish. Allow to dry thoroughly and then rub down well with fine sandpaper. The next coat should be undiluted polyurethane clear varnish. Two coats should be sufficient, but however many coats you use remember to rub down well with fine sandpaper between each one. The final coat should be left for several days before any other type of polishing is carried out. In practice, I have found that the finish obtained is so good that extra polishing is quite unnecessary.

The metal base of the dashboard is easiest made from 16-18 gauge aluminium sheeting, which must be polished, matt-finished or painted, according to the original specification of your car. A useful tip to remember when working with new aluminium rolled section, be it flat sheet, rounded or angled, is to wipe over with 'Jenolite' before polishing or painting. The reason is that part of the process of manufacture is to cover the rollers with lanolin. The 'Jenolite' removes this and leaves the surface clean. Castings of the same metal should not require this treatment as lanolin is not used in the manufacturing process. Some aluminium sheeting is coated with a plastic film before it leaves the factory; it is usually light blue in colour and can be removed with the aid of the manufacturer's recommended solvent.

Exhaust systems

Regardless of the type of finish you require, the only effective method to obtain the correct prepared surface is sand-blasting, which removes all the rust and old finishes, not only from the flat surfaces, but also from all the small indentations which will never be reached using a wire brush, even if it is powered.

The types of finishes vary from vitreous enamelling to brush-painted exhaust paints. In my opinion one of the best finishes is provided by using a VHT paint which is obtainable from most automobile paint shops. This is the paint that was developed for the re-entry module on all the manned moon rockets, which is capable of withstanding heat far in excess of anything that your engine will produce. However, one of its disadvantages is that if the system is sand-blasted before the paint is applied (it is supplied in aerosol form) it must be baked on. Although this can be done by running the engine, the result is usually unsatisfactory, and if you cannot find a company with a kiln the best method is to use a butane gas blow-lamp (not a paraffin lamp, as this leaves stains). Heat the system all over, slowly raising the temperature until it is red hot (dull cherry red), and then allow to cool. It is also advisable to apply the spray as best you can to the inside of the system as well.

Once it has been baked on, it protects the system very effectively from the ravages of the weather and road salts. It also has the capability of being restored back to its original colour by a hard brushing with a nail or scrubbing brush soaked in washing-up liquid, rinsed off with water, which should be done when the system is cold. A word of warning: most VHT paints are soluble in paraffin, even after they have been heat-cured. A more lasting and, arguably, better finish can be obtained by having the system 'aluminized' after sand-blasting and the VHT paint then baked on.

Whilst on the subject of exhaust systems, one must mention that stainless steel exhausts are now being supplied by various after-market manufacturers. There is a great deal to be said for them and they should not be dismissed without investigation. Originally, they were rather expensive, but now, in many cases, they are actually cheaper than the factory mild steel units.

The question of chromium-plating the manifold, *etc*, must be considered but, again, this is very expensive and has the tendency to 'blue'. In respect of originality, only a very few Morgans were ever supplied with such a 'luxury' as this; they were mainly the four-branch manifolds as fitted to the TT and Le Mans Replicas.

Electrical equipment and wiring

As mentioned earlier in the chapter, it is by far the safest policy to discard the old wiring. Depending upon the model being restored, it is possible that a replacement wiring harness will be available; here is where the club register could be of assistance to you. If not, and if you feel unable to carry out the work yourself, make contact with one of the specialist firms who advertise in motoring magazines.

Here are a few wiring tips that could be of use to you: (a) always solder the ends of wires into a solid piece to stop fraying; (b) where applicable, always solder connectors to the wires (terminals fitted to wire ends without without the special crimping tools always produce high-resistance joints and often work loose); (c) always use rubber grommets where cables pass through holes in metal; (d) ensure good connections and especially good earthing points (an intermittent or poor earth connection can produce the most peculiar faults which will take hours to locate); (e) always use junction boxes of a recognized manufacture (these can be obtained in all sizes); and lastly (f) use snap connectors (push-fit) at convenient places to assist cable runs. Always leave 2-3 inches spare on a run to avoid breakages due to vibration of taut cables. (See also the electrical section of Chapter 5).

Attention must be drawn to the fact that the lighting regulations have almost certainly altered considerably since your car was manufactured, especially if it is a pre-war model. Try a visit to your nearest reference library, which should hold copies of the lighting regulations. Alternatively, the local police station or an approved vehicle testing centre should be able to advise you.

Upholstery

The complete re-upholstering of a car is a very difficult job for the amateur and should, in the main, be left to the professional coachtrimmer and upholsterer. However, if the leather is not damaged and only looking sorry for itself, it can be brought back to its original lustre quite easily.

First, wash the leather with warm water, glycerine or non-caustic soap, or preferably with a proprietary cleaner, such as 'Connolly's Concentrated Cleaner', according to the manufacturer's instructions. If necessary, use a small tooth or nail brush, which is not too hard, to clean the ingrained dirt away from the seams and crevices. Try not to soak the leather and always ensure that the temperature of the water is no more than hand-hot. When dry, use a leather reviver and polish with a good leather polish.

The world-renowned firm of leather merchants, Connolly Brothers (Curriers) Limited, of London, are always most helpful and, in addition to supplying all the necessary cleaners and revivers, have a very good leaflet on the subject of renovation of leather and other car trimmings.

Fuel tank

Never use a blowlamp or any other form of naked flame on or near a fuel tank. Nor should you smoke or have naked lights anywhere near it, no matter how many years may have passed since the tank last had fuel in it.

All fuel tanks fitted to Morgans are made of tinned steel and, if old, generally have a considerable amount of rust inside. First remove the tank from the chassis and wash out repeatedly with water until all the loose rust is removed. The tank should then be dried immediately. This is best done by first draining and then standing in the sun. Remember to turn it end for end and side to side at regular intervals.

When the tank is thoroughly dry, remove the old paint with the aid of a chemical paint remover. Do not scrape the paint off dry as this removes the tinning as well as the paint. Dents, if any, can only be removed by unsoldering (with a soldering iron) the end giving best access to the dent, then hammering out the dent and resoldering the end plate back into place.

Carefully examine the tank for pin holes and cracks. The most likely place to find pin holes are on the bottom of the tank, especially if at any time it has stood with water in it. In this case it will be extremely thin and the best course is to solder a plate over the area. Cracks, of course, can be soldered providing they are not too wide.

When you have completed any necessary repairs, the best way of testing the tank for leaks is by sealing the inlet and outlet pipes with corks or rubber bungs, one of which should be drilled down the centre and an old inner tube valve inserted through it. Then, with the aid of a foot pump, pressurize the tank to 5psi (no more than that and never use a compressor!). Then immerse the tank in water, check for any air bubbles, mark their positions and remove and dry the tank. After depressurizing the tank, clean the area with a scraper and emery cloth before soldering.

Soldering on an object of this size requires a large iron which will not cool down too quickly. If you feel that you cannot manage this work, a good tinsmith will do it for you and even, if necessary, make a complete new tank to the original design using the original filler neck and other fittings.

Useful tips

Cleaning fuse clips: everyone who has tried knows just how difficult it is to clean corrosion off the little clips that hold the fuses: you just cannot get at the inside with a file, emery cloth or sandpaper.

Take a 0.25in wooden dowel (a pencil will do, but it must be a round one) and cut it to the length of a fuse. Glue fine emery cloth or sandpaper around the dowel. Drill a small hole in the side and insert a 2in nail as a handle. (Do not hammer the nail into the dowel without drilling as the wood will split).

Insert the dowel into the fuse holder clips and work the handle back and forwards two or three times. When removed, the inside of the clips will be found to be bright and clean.

Battery terminals: If you are still using the old-type battery which regularly corrodes its terminals, try glueing a copper coin to the top of the battery, close to but not touching the negative terminal. You will be surprised what a difference it makes.

Aluminium corrosion: to dissolve corrosion between aluminium components, try using vinegar instead of penetrating oil. All that the penetrating oil may do is to seal off the corrosion from other chemicals.

Water hoses and hose clamps: check hoses for unusual softness, cracks, drying out, etc. Any hose that looks less than 100 per cent should be replaced before it leaves you stranded.

When removing old hoses, it is wise to carefully slit them with a knife if they are stuck or bonded to the fittings. Putting too much pressure on the hose while attempting to force it off could cause thin radiator flanges to rupture and rusted steel piping to split. Inspect the pipe stubs for corrosion or rust damage at the ends where the hoses attach. All exposed aluminium or steel pipes should be painted to prevent rust or corrosion damage.

Hoses should be of the best reinforced designs. When cutting off excess length from a hose, exercise caution as they often contain a coiled spring. These springs are a good idea and should be retained if at all possible as they prevent the hose from collapsing due to water pump suction.

If your car will be a concours candidate, use hose clamps of the original design. Otherwise, any good quality stainless steel clamp will suffice. When installing most hoses, the fit should be tight. Again, avoid using undue pressure to force the hose on to the fitting. A silicone spray lubricant like WD40 will help ease the installation and prevent strain on the pipes. On corroded pipe stubs, or with looser fitting hoses, apply small amounts of silicone rubber to the pipe stub surface, then fit the hose and tighten the clamp as soon as possible. Wait a few hours before refilling the system.

Finally, check running clearances between hoses and fan blades, as hoses tend to move slightly when the engine is running. Check tightness of all the hose clamps after the first few hours of operation.

A good, tight water system will ensure proper operation and longer engine life.

Chain lubricant: motorcycle chain lubricant is available in a spray can under various brand names at your local motorcycle shop. The lubricant sprays at about the same viscosity as WD40 to penetrate cracks and crevices, and within minutes thickens to nearly the consistency of grease. Quite remarkable! It can be used with some success to rejuvenate semi-sealed bearings and is great for the rear springs. Unlike WD40, it will not disappear overnight.

Asbestos brake material: when you change your car's brake pads or shoes you probably notice lots of black residue (not the grease) on the backing plates or around the calipers. This black dust contains a lot of asbestos! The inhalation of this asbestos dust must be avoided completely. Do not use compressed air to blow it away or you will stir up an unpleasant cloud. Use water for cleaning it up. The wearing of a face mask is essential.

Brake repair notes: if the primary cup on your car's master cylinder is leaking, you will lose pedal pressure - that is, the pedal will travel all the way to the floor - but there may be little or no loss of brake fluid. If the secondary cup is leaking, you will lose brake fluid. Pinch the boot on the master cylinder and see if you get a drip of brake fluid from it. This is a good sign of a bad secondary cup.

Always check the brake lights when you are checking the master cylinder and if they are not working find out why and rectify the problem first. Another thing about brake systems: try to keep all the wheel and drum parts marked and together, so they go back in the same location on the car. Right front on the right front, and so forth.

Do not leave any twist in the brake hose if a wheel cylinder or hose has to be removed and replaced. Even the slightest twist will cause the hose to unscrew itself from the wheel cylinder as the car is being driven.

Always test fit brake shoes to the drums. The centre of the shoe should contact first while there is still clearance at the ends. The book says five or six thousandths of an inch at each end. A wood rasp works just as well as a brake shoe grinder. Do not breathe the dust - asbestos!

Tyre valves: a simple tool for removing valve cores is to epoxy a core-removing type valve cap on the end of an old valve stem.

Small oil leaks: stamped valve covers, oil pans, etc, always seem to leak. This is often caused by *over*-tightening of the bolts, causing 'sheet metal pull-down'. Additional tightening may temporarily stop the leak, but results in additional bent metal, causing the leak to return. The proper way to fix leaks is to straighten the sheet metal, install a new gasket (use a sealer) and torque fixings to specification.

Castellated nuts: next time you

remove a castellated nut, file a mark across the end of the axle, in line with the cotter pin hole. Lining up the nut will be a lot easier.

Windshield screws: those little screws holding your car's windshield in place must not touch the glass. They must be the exact length to hold the frame of your windshield but, if they touch the side of the windshield, expect the glass to crack.

Water leaks into car: one source of water leaks into the car is where the centre hinge is screwed to the firewall. If the sheet metal cracks, or the screws get loose enough to let water through, it comes down on your feet. Caulk this area with silastic putty. If the wood screws are stripped in the holes, dip a couple of wooden matchsticks in white glue and fit them in the holes before putting the screws back in. Be sure that the screws are flat head and seat flat in the end fitting. Round head, oval head, or oversize screws here are the major source of cracks in the bonnet, at the end of the hinge.

Steering column nuts: periodically check to see if the nut behind the steering wheel is tight! I have seen steering wheels with stripped hubs, resulting from excessive movement on the splines when the nut gets loose.

Emergency repair of threaded parts: since the need for special tools usually crops up at night, on holidays and other times when all suppliers are closed, here are a couple of tips to clean up threaded parts in an emergency.

If a stud or bolt has damaged threads, take a nut of the proper size and thread, saw it in half across the hole, clamp it around the stud or bolt (behind the damaged area if possible) with a pair of vice-grip pliers and unscrew the nut over the damaged section of thread. Cutting the nut in half does two things; it gives two sharp edges to clean up the threads, and it makes the nut a tight fit on the bolt to give a good cleaning action.

For a threaded hole that is damaged, take a bolt of the proper size and thread and file four grooves lengthwise in it with a triangular file. When this bolt is threaded into the hole, it will help clean out the threads.

None of the above is a substitute for taps and dies, but will work when nothing else is available.

Paint removal from plastic/vinyl: if you have ever attempted to remove paint from plastics or vinyl, you know that most paint thinners ruin plastic. Try model aeroplane fuel with about 10 per cent nitro base; it removes the paint with no damage to the plastic.

Gaskets: the standard mechanic's way of making a gasket when at least one of the parts has few if any protrusions, is to lay the gasket material on the part and let it (the part) be the cutting die. Gently tap around the edges of the holes and margins with the round end of a ball-peen hammer.

Oil/hydraulic fittings: use extreme care when replacing any oil or other hydraulic fitting. Threading properly makes a seal. Forcing new threads strips the seal and causes a leak. Always clean the threads and 'start' them by hand.

VII

MORGAN CLUBS

The Morgan Three-Wheeler Club Ltd

The oldest Morgan club in the world is without doubt the Morgan Three-wheeler Club Limited. The club as it is known today was first formed in 1944.

Before this there was a 'Morgan Club' which existed as far back as the early 1920s. This was a club exclusively for Morgan owners, organized very much on a local basis and predominantly of a social nature but the overall President of the club was HFS Morgan.

However, by the mid 1920s - immediately after EB Ware's crash at Brooklands in September 1924 - more and more pressure was brought to bear by the Junior Car Club and others to get Morgans banned from racing. As the pressure mounted so did the determination of Morgan drivers, particularly those interested in racing and record breaking, to find some way of ensuring they could continue. On what was almost certainly HFS's instigation, it was decided to re-constitute the club to allow membership to any Auto Cycle Union - defined cyclecar, (*ie* under 1100cc and 8cwt in weight - with three or four wheels).

With HFS as President, the new club, named 'The Cyclecar Club' was formed on August 4, 1927, and of course immediately became eligible under BARC rules to organise race meetings at Brooklands. Following objections from the Junior Car Club, which had previously been named the Cyclecar Club, the name was changed to 'The New Cycle Car Club', and by mid-1928 the racing committee organized its first Brooklands race meetings. As from January 1, 1930 the club became the Light Car Club and membership was extended to the owners of all cars up to 1500cc, whereas before it was limited

The first Morgan club was founded in the early 1920s. Although predominantly a social club, a few 'closed to club' competition events were organised.

THE MORGAN CLUB.

PROSPECTUS
OF
The Reliability Trial
ON
SUNDAY, AUGUST 23rd, 1925
(OPEN ONLY TO MEMBERS OF THE MORGAN CLUB)
FROM
"THE RED LION," HATFIELD,
FOR
THE MORGAN CUP
(Presented by H. F. S. MORGAN, Esq.)
TO BE WON OUTRIGHT.

GOLD, SILVER AND BRONZE MEDALS.

SPECIAL CUP (also presented by H. F. S. MORGAN, Esq.)
For the oldest machine gaining a gold or silver medal.

ENTRIES CLOSE AUGUST 17th, 1925.

All entries should be forwarded direct to :-
The Hon Trials Sec.—J. YULE, Esq., Kirkney, High Rd., Whetstone, N.20, to whom all communications should be addressed.

Arthur Maskell in a Super Sports accelerating hard away from the chicane at Brooklands at the New Cycle Car Club's first race meeting in 1928.

to 1100cc. This change came about mainly because the New Cycle Car Club membership comprised, almost exclusively, Morgan owners. The most notable innovation made by the club was to organise the 90-lap relay races held each year at Brooklands. In its latter years the club began to lose a lot of its impetus, reflecting the decline in the popularity of the cyclecar.

The Second World War curbed all forms of motor sport but the motoring

An early MTWC meeting, believed to be an AGM, at Brandon Hall near Coventry. The unusual Morgan in the foreground is a family model with a fixed head conversion.

magazines produced many articles concerning Morgans throughout those dark years; owners could read about their marque even if they could not get the petrol to use their cars. This lack of contact between owners was brought to a head when a letter was sent in 1944 to *Motor Cycling* magazine suggesting the formation of a club for three-wheeler Morgan owners. A regular contributor to the magazine on Morgan matters was WS Seaman who wrote under the pen name of 'Trepenpol'. It was therefore through him that the names of the owners were collected and the first club news sheet was distributed in September 1944. The first meeting was held at Carshalton, Surrey while Flying Bombs fell on London and one of the club's present Vice Presidents, Ernie Woods, attended. In 1945 petrol was made available, albeit in very limited and rationed quantities, for domestic and pleasure uses, and the club organized a few 'runs' for members. The Inaugural General Meeting was held in September that year and over 50 members attended. Slowly, over the next ten years membership increased to about 750 and has remained around that figure ever since.

Over the years the club's function has changed. Originally, most members' cars were in regular everyday use, with spares being readily available from commercial outlets. The club's role then was mostly social and competitive, with a few technical articles being published from time to time in the monthly magazine *The Bulletin*. However, as the value of the cars has increased, traffic conditions have changed and everyday use of the cars diminished, the club has had to fulfill a more technical capacity and, by 1972, found it necessary to organize a comprehensive spare parts manufacturing service for its members. This built to such a high annual turnover that, in 1980, it was decided to incorporate the club as a limited liability company.

In addition, over the years the club has gathered together a very extensive library, the contents of which are available to members, which helps to ensure that the owners of the surviving three-wheeler Morgans will be catered for well into the future.

The club has affiliated groups in both the USA and Australia, but most other clubs around the world are all willing to accept three-wheeler owners for membership.

Apply for membership in the first instance to the MTWC Membership Secretary, Morris Blease, Arden Cottage, Rocky Lane, Benllech, Anglesey, Gwynedd, Wales.

The Morgan Sports Car Club Ltd (Formerly the Morgan 4-4 Club & Morgan Sports Car Club.)

The publication of this book coincided with the 43rd anniversary of the Morgan Sports Car Club Limited, as it is now known. Any motor club which has been in existence for such a long time will have suffered setbacks and perhaps near-collapse. The Morgan 4-4 Club and Morgan Sports Car Club - both names by which the Morgan Sports Car Club Limited was once known - was no exception. But with the enthusiasm, and even on occasions financial assistance, of a few of the more devoted members, the club staggered its way through early crises to become today one of the largest one-make clubs in the world. It has a membership of almost 3800.

It might have been thought that

A few of the cars that turned up at the inaugural General Meeting of the Morgan 4/4 Club.

the introduction of the four wheeled Morgan would have been followed quickly by the creation of a club specifically for its owners, but this was not the case. At first, the Morgan Three-Wheeler Club decided that the owners of cars fitted with 'the extra wheel' would be allowed to join as associate members, but this idea was never a resounding success. It was not until May 1951 that positive steps were taken to establish a club for 4-4 owners. The turning point came when two Derby men, JS Atkins and DJ Whetton, placed an announcement in the May 18 issue of *The Autocar* inviting any Morgan owner interested in forming a 4-4 Club to contact Mr Whetton. One of those who did so was Lt-Col R (Dick) G Pritchard, TA, of Spondon, and later in the month a meeting of four or five Morgan owners was held at his home.

It was agreed that the number of replies to the invitation had been sufficient to warrent the holding of a further meeting, which was arranged to take place at the Flying Horse Hotel in Kegworth. Some 20 people attended and it was agreed that a club should be formed and an inaugural meeting be held at the same venue on the evening of June 30. Interest grew rapidly and

The first road rally organised by the Morgan 4/4 club was held in 1951. The event was won by Bill Allerton in his Le Mans Special. Note the twin upright spare wheels, painted radiator shell and missing 4/4 badge.

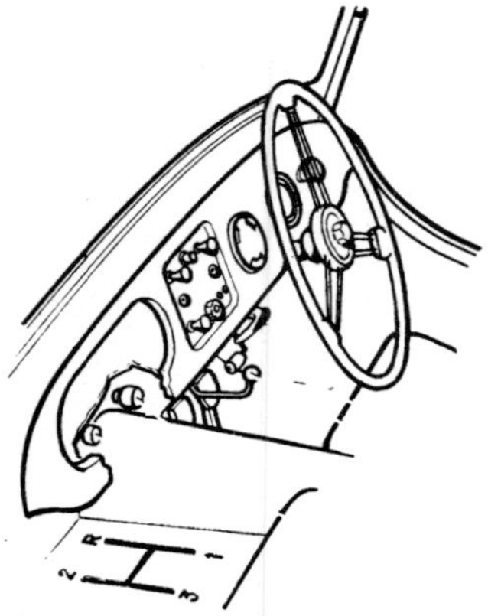

the club came into being with an initial membership of 63, including six ladies. Some of those present had travelled over 200 miles to attend.

The Morgan 4-4 Club was chosen as the official title of the club, and club colours of light blue, black and red were adopted. One of the first tasks was the appointment of officers and committee, and the club's first President was Jim Sparrowe, who was well known throughout the motor racing scene for his successes with his rapid 4-4, and who, in fact, had travelled with two other members to this inaugural meeting straight from a day's racing at Silverstone.

Dick Pritchard became the club Chairman , with LA Willsmere Vice-Chairman, J Sutton the Hon. Secretary and Treasurer and DJ Whetton the first Competitions Secretary. They were supported by a six man committee comprising RD Hadley, W Allerton, B Carroll, Flt. Lt. RC Bridges, IA (Tim) Parkes and EP Huxham, although the last-named had to decline office owing to pressures of business.

Although the club's name had been linked with the Morgan 4-4, the Plus 4 model had been in production for several months, so it was decided that owners of the more powerful model should also be eligible for membership, while non-Morgan owners would be admitted as associate members at the same £2.00 annual subscription that applied to full members. It was also agreed the club should apply for affiliation - subsequently granted - to the Royal Automobile Club, which enabled it to run its first officially recognized event the following October.

Meanwhile, the club's first competitive event - a 70 mile road rally from Burton-on-Trent to Banbury - took place on August Bank Holiday, August 5. This seems to have been a peculiar date to choose which, no doubt, accounted for the somewhat meagre entry. It was very much a case of more Chiefs than Indians, for there were about 15 officials present to attend to a mere five entries! The event ended with a series of driving tests in which most of the officials and marshals tried their hands after the competitors had completed their official runs.

The distinction of winning the club's first event went to Graham Stallard, of Worcester, who drove a TT Replica reputed to have once been the property of HFS, although this was never really established. Second place went to Tim Parkes, of Pershore, and third to Bill Allerton, from Chester, who was driving a Le Mans Replica and also took the award for the best performance on the road sections.

Somewhat to the surprise - and certainly the delight - of the committee, the late Holland Birkett asked the club to enter a team of cars for the first Six-Hours Relay Race which his 750 Motor Club was to organize at Silverstone on August 25, and a team assembled comprising of Sparrowe, Parkes, Allerton and Atkins.

Any number of cars up to six could be nominated as a team and the objective was to carry a coloured sash around the circuit as many times as possible in the six hours, the sash being passed from driver to driver and worn by each in turn as their car did its stint on the circuit. A handicap system was applied by means of credit laps, the Morgan team being credited with 34 laps over the Jaguar XK120 team, which started on scratch. The race was not without incident for the Morgan team, but they held on grimly to average 61.5mph and finish second overall to a team of vintage Bentleys. This was the beginning of a long association with the 'Six-Hours', an event in which the club has by no means disgraced itself.

The club's own first RAC-recognized event was a driving test meeting on October 7, 1951, at Queensford Aerodrome, near Dorchester-on-Thames. It was a well attended competition which was won by Graham Stallard. Then came the club's first night rally, on December 1/2, which attracted 24 entries to starting points in Manchester, Gloucester and Hereford. The route took competitors around Lincoln, Gainsborough and the Derbyshire Dales and back to Leicester, a total of 300 miles.

A few treasure hunts and similar social runs kept the club active for the next few months, culminating in an August Bank Holiday anniversary run over the route of the 1951 inaugural event, but in this instance much better supported. Up to now all the club's events had been restricted to members, but during the winter of 1952 a 200-mile night rally was observed by the RAC, as a result of which they granted the club a permit to organize invitation events of a similar kind. Further progress was made when the 1953 Spring Rally, also over 200 miles, and starting and finishing at Kidlington Aerodrome, Oxfordshire, was again officially observed by the RAC, as a result of which another permit to run invitation events was granted. The RAC observer on this occasion was Holland Birkett, who expressed particular interest in the club's operation of dumb (unmanned) checkpoints along the route and asked if he might copy the idea on events he organized for his own club which, later that year, he did.

One of the features of the club's

Club Driving Tests are always well supported. These three photos show one which was held on the car park of the Metal Box Company in the 1960s.

administration in those early days was that because officials and committee members were so widely dispersed, their regular monthly meetings were arranged by each member in turn. This meant that venues ranged over an area from Chester in the west to Sheffield in the north and Malvern in the south and many hundreds of miles were motored to attend. Some of the people who did so tell that it was great fun, but I believe the high mileage and amount of time involved led indirectly to the setting up of different centres of the club which now operate throughout the country.

The advent of the first cowled-radiator Plus 4s in 1954 caused a certain amount of heart-searching by the committee and quite a few rude remarks by the more committed 'flat-rad' people.

But, of course, the owners of the cowled-radiator models were in due course welcomed to the club, membership of which has grown to 3800, with more than 1000 active overseas members as well. The end of the flat-rad production, however, marked the end of an era, for the club as much as for the company; it was, in a sense, the end of the formative years.

As the club grew, so did its reputation in all forms of motor sport, and in the hey-days of rallies, sprints, driving tests, *etc*, hardly a week passed without the Secretary receiving invitations to enter teams in some event or another.

In order to link the membership, a club magazine was founded which took the form of a monthly newsletter called *The Miscellany,* with a larger quarterly

edition under the title *Motoring and Morgans*. Unfortunately, in those early days, production proved to be quite a struggle; the problem was the shortage of printable material. Gradually the membership responded and can now look back on over 30 years of regular publications, while the quality of the production has increased as well;*The Miscellany* now looks a very professionally produced magazine, complete with illustrations.

The membership of the club is divided into full, associate and overseas members, whilst, following an Annual General Meeting of the club on April 30, 1955, family membership became available under which the benefits of membership were extended to a member's immediate family on payment of the appropriate subscription .

As the membership increased so did the complaints about the long distances which some members had to travel in order to attend the noggins 'n' natters held in the Midlands. To overcome this problem Richard Upshall, with the committee's blessing, organized a meeting in the London area late in 1959 with the idea of forming a London Centre. However, the project was shortlived, due to lack of support, and it was not until October 1965 that a committee was formed at an inaugural London Centre meeting held in the garage premises of W Griffiths, in Willesden. This time it was a great success and the London Centre, now one of 30 covering nearly all the UK, has never looked back. Each centre has its own secretary and can organize social and competitive events independently of the main club, while all centre activities are open to any MSCC member, regardless of where they may live - visitors from one centre are always most welcome at others.

The first attempt to change the name of the clubs was made in August 1962, when a member from Canada proposed that the name be changed to the Morgan Owner's Club, which seemed to echo the thoughts of a number of home-based members, mainly Plus 4 owners. However, when the committee decide to hold a postal vote of members, a large majority opted to keep the original name and the subject was not raised again until 1968. Members were invited to air their views through the pages of *The Miscellany*, but opinion seemed to be evenly divided, so the question was deferred again for 18 months. Then, in November 1969, a postal referendum took place in which 317 of the 451 voting papers sent out were returned completed. To the question whether or not members wished the club's name to be changed, 195 replied 'Yes' and 122 'No'. To the subsequent question, which of three names would members prefer in the event of a change, the scoring was: Morgan Car Club - 121 votes; Morgan 4-Wheeler Club - 39 votes; Morgan Sports Car Club - 136 votes. The remaining 21 papers were either spoilt or unanswered. The club was renamed - from January 1, 1971 - The Morgan Sports Car Club.

The next landmark for the club was the formation of the 4-4 Series I and Plus 4 Registers in mid-1972 to provide, 'a means for owners to contact each other, a source of advice, data and information, help in obtaining spares, *etc*'. John Orton was the first Series I Registrar and Tim Cree the first Plus 4 Registrar. This was the birth of the club's comprehensive spares and advisory service for members which embraces a Series I Registrar and Technical Adviser, a Plus 4 Registrar and Technical Adviser, a used spares Registrar and a Librarian. In recent years this format has been changed and the club now has a Technical Services section offering a more comprehensive service to members with technical advisers covering the whole four-wheeler production years, and a new and second-hand spares service. The Registers and the Librarian are now entirely separate.

In 1978 David Stretton-Smith put forward the idea of MOGAID, which is an international register of club members who are willing to assist any other member who breaks down in their area. Each member is issued with a list of names, addresses and telephone numbers which he can call upon for assistance if required. It is a splendid idea which other one-make clubs might do well to copy.

As from May 1991 the Club became a Limited Company, its full title now being The Morgan Sports Car Club Limited. Great care is taken with the way the board of directors of the company are elected, and also the length of time they can serve before they must put themselves forward for re-election. The maximum period is three years. All other officers (except for President and Vice President who are elected 'for life') will be elected each year. The only nominated positions are those which pay an honorarium or salary, which are appointed by the board as and when appropriate.

Finally, each officer whose position is elected annually now has a full job description, clearly laying out the responsibilities of each office and the requirements from the company of anyone elected into such office.

I have no doubt that the founder fathers of the original Morgan 4-4 Club did not envisage its humble beginnings would develop into the present day structure. However, having known at least three of the founders I am sure they would be very proud if they were alive today.

All correspondence should be addressed to Mrs Christine Healey, Registrar, The Morgan Sports Car Club Ltd, 41, Cordwell Close, Castle Donington, Derbyshire, DE74 2JL.

Morgan clubs around the world

To own a Morgan is to own an international calling card. If a Morgan owner visits another country, with or without his car, and spots another Morgan, he will invariably seek out its owner and often the resulting discussion is the start of a lasting friendship.

Morgan ownership is rather like an addiction; once hooked the addict has little chance of being cured. How many times have you heard these familiar words, 'I used to own a Morgan and I wish I'd never sold it'? Through this particular addiction, Morgan owners have helped each other out in many parts of the world and, as the circle's widened, have formed themselves into an exclusive club. Some remain very small, still others have grown into large and very active organizations.

In this chapter the clubs are listed, together with a brief history and rundown of their activities. Also included is a latest known contact name and address of a club official (usually the Secretary). In some cases I have explained the methods adopted by clubs in awarding trophies and staging events in case they may be of interest to other Morgan clubs.

I think the very essence of a Morgan club is epitomized by the wording printed inside the register of members of the Morgan Owners Group - Great Lakes. It reads:

The Morgan Owners Group was established in 1965 to promote the Morgan automobile and to provide technical advice and social companionship for those fortunate enough to have found their very own world's last true sports car.

MORGAN OWNERS MORGAN CLUB AUSTRALIA

Morgan Owner's Club of Australia

The club was founded in August 1958 in Sydney, on the instigation of the Ward brothers, Ken and Barry, with 10 members attending the first meeting. Early meetings were informal gatherings at members' homes, but by 1960 increasing membership caused the meeting places to be changed to various halls or club premises in the Sydney area. That year the club joined the Confederation of Australia Motor Sport (CAMS) which is the representative of the FIA in Australia.

By 1963 the membership had grown to 44, by 1968 it was 82 and at the end of 1988 there were 180 fully paid-up members. The club fought and won a five-year battle with CAMS as a result of which all Australian clubs with at least 40 members, and which had been in existence for at least four

The Morgan Owners Club Dawn Dash at Mount Wilson in 1987.

years, were granted full membership with voting rights; hitherto, clubs with less than 100 members had only been granted associate membership with no voting rights.

In 1970 the club joined the Council of Veteran, Vintage and Throughbred Motor Clubs (VCCTMC), an organization which brings together about 30 clubs in the Sydney area and stages a static annual display of between 500 and 700 vehicle and motorcycles at Warwick Farm race course, including a Concours.

Communication between distant members is largely dependent upon the club magazine, which is a 20-page publication named *Morgan Ear* and is published monthly. This is a well-produced informative magazine. Full membership is open to all Morgan owners and associate membership to others who are *simpatico*.

The club organizes about 30 events a year. Competition events include circuit sprints and hill-climbs, there being a vigorous championship in conjunction with the Combined Sports Car Association. Social types are very well catered for, too, with events such as barbecues, dinners, tours to the country and interstate early morning 'dawn dashes', often ending with a picnic somewhere like a distant beach or in the mountains. A useful bridge between social and competition events is the Driver Training Day usually held at Oran Park racing circuit.

There is an extensive array of club regalia available, including T-shirts, windcheaters, caps, car badges, lapel badges and transfers. Club morale is high and members help each other with technical problems and maintenance. The range of cars owned by members is quite extensive and includes several three-wheelers restored and running, flat-rad 4-4s and Plus 4s - four-seaters and roadsters, modern 4/4s and Plus 8s.

For the name and address of the club contact see the Affiliated Clubs list at the end of the chapter.

The Morgan Sports Car Club of Austria

Upon the initiative of Max Bulla, the importer and dealer of Morgans in Austria, a club was formed on June 26, 1977 with 12 enthusiasts. Statutes were written by Hans Schmolzer, the first president of the officially registered club.

The main purpose of the club is to hold competitive as well as social meetings, and to advise and help other members on technical problems with their Morgans. Membership can be obtained by owning a Morgan car, taking part in club activities at least once a year and the payment of the dues, which have remained the same since the founding date (ASh 500 annually).

At present the club has a membership of some 80 cars/persons, which include three-wheelers, 4/4 four-seaters and Plus 8s, as well as Plus 4s. Most of the members' cars are from the late 1960s, the 1970s and 1980s and are registered in Austria; however, they also have members from Switzerland and Bavaria.

The club is governed by a board which is responsible to the annual general assembly. Present President is Fred Myer, the Vice-President is Max Bulla, Treasurer and Regalia Officer is Peter Taucha, Sporting Events Commissioner Hans Jachim, Club Secretary is Janet Jesch assisted by Sissy Marine and the Editor of the *Morganeer Gazette* is Christine Maschl.

The activities of the MSCC-A include two bi-annual rallies according to the statutes, one at Whitsun and one in early autumn (September or October); the Thanksgiving Goose Eating Festival, with the new wine; the Christmas party with club championships celebration and the presentation of cups; participation at other Morgan rallies in foreign countries or participation at veteran car rallies in Austria due to the acceptance of Morgans as a veteran car in its own class at these events; publishing the *Morganeer Gazette* as well as irregular news bulletins and rally announcements; the Vienna Sections 'Noggin' Round Table, the second Thursday every month.

Production and sale of regalia peculiar to and designed for the MSCC-A such as pins, badges, (cloth or metal) and key chains, can be ordered from the Regalia Officer or the President.

Members welcome visitors from England, France, Switzerland, Germany, Italy and Belgium at club rallies and also participate on an individual

The Belgian Morgan Sports Car Club's stand at the Classic Car Show at Antwerp in 1989.

basis at Morgan meets in foreign countries. Due to having membership of the OMVV they can also participate in FIVA sanctioned rallies in Austria. The Morgan Sports Car Club of Austria has been associated for five years with the Morgan Sports Car Club (GB).

Anybody can join the MSCC-A as long as they own a Morgan, pay the annual dues, participate at one or more functions and adhere to the club's constitution. For name and address of the club contact see the Affiliated Clubs list at the end of the chapter.

Morgan Owner's Club, Belgium

In the summer of 1973 Dr Albert De Mey and Patrice Libiez met at the latter's home along with two other Morgan owners and two Jaguar owners, with the idea of forming a Morgan club in Belgium. From this modest beginning the club had grown to 65 members by 1978 and caters for French-speaking Morgan owners.

Members meet on the second Wednesday of each month at the Café Delta in Brussels, when owners of other sports cars are also welcome, as are any Morgan owners who happen to be in Brussels on that date. The club tries to organize five events each year, comprising of a social run with a visit to a place of historic interest. The main rally of the year must be unique in the Morgan club world, being held in conjunction with an international hot-air balloon meeting. The balloons lift off from the market place in the town of Saint Niklaas, the first one carrying a trophy. The Morgan owners then follow this balloon in their cars and the first one to reach it when it lands claims the trophy. The idea is not as dangerous as it might sound due to the slow progress of the balloons, and the event has the co-operation of the local police.

Membership is open to all Morgan owners or enthusiasts. For name and address of the club contact see the Affiliated Clubs list at the end of the chapter.

Morgan Sports Car Club, Belgium

This club was started in 1987 when the French-speaking group went its separate way. At this time the club had about 60 members, most of which were French-speaking. However, in just two years the Flemish-speaking club had grown to 60-plus members.

The club organizes at least two events a year with the regular concours and rally tests. The annual dinner is held in January. The main event is the club stand at the Belgium Classic Car Show at Antwerp. Here the club is well known to other clubs for its hospitality - and especially its free champagne! The club also hopes to be able to organize an international event every two years.

The Belgian club's annual meet in 1988 was the Vlammog 88 at Lier.

Members of the Canadian club tune-up outside the motel where they stayed when attending the annual American meet.

There are three board members; President, Secretary and Public Relations Officer, who are all elected by the members. However, most club problems, *etc* are discussed at the monthly 'Noggins'.

The club also produces the well-produced, bi-monthly magazine *Belmognews*.

For name and address of the club contact see the Affiliated Clubs list at the end of the chapter.

Morgan Car Club of Canada (elect) (Toronto Morgan Owner's Club)

The golden era for Morgans in competition in Canada was the 1950s when they competed with great success at the Harewood Airport circuit and on the unpaved hill-climbs at Rattlesnake and Hockley, and dominated the ice races at St. Mary's Lake. On the rallying side they were less successful, perhaps because these events were sometimes 1300 miles long and held in temperatures as low as minus 40°C. However, in November 1961 three Plus 4s won the team prize in the Press-on-Regardless rally.

The first attempt to form a Morgan club occurred in 1962, when 18 Morgan owners met at the Jolly Miller Tavern in Toronto. However, Alan Sands, then 'between Morgans', was concerned that the proliferation of one-make clubs in Canada was undermining the sports car movement and he persuaded the Morgan owners to stay with their existing general purpose sports car clubs. Consequently, the idea of a Morgan Club of Canada faded for five years.

Then, in 1967, Doug Price was driving his Plus 4 when he spotted two Morgans parked at the kerb. A quick U-turn resulted in him meeting Ian Campbell and Ken Miles, as a result of which the first post-1962 meeting was held at Alan Sand's farm, Piper's Hill, where, with the unexpected publicity boost from radio CFRB, 12 Morgans assembled. The next few months were spent locating other Morgan owners, and by 1968 some 25 cars had been traced, including five three-wheelers. Contact was made with the Great Lakes Group in Detriot, and joint meetings arranged and an effort made to form a group in Vancouver.

As with so many clubs, interest waned after the formative years, and whereas some 36 cars were on the register in 1974, two years later, when Reg and Audrey Beer staged a barbeque, there was only one present apart from their own. Most of the others were rotting in garages under the pretext of undergoing restoration. The club then decided to take action and another meeting was held at Piper's Hill Farm for a 'Spring Tune-up'. A mixture of Plus 4s of older members and 4/4s and Plus 8s of younger and more recent owners turned up, enthusiasm was rekindled and 10 cars were present for

the next barbeque. By 1978 club membership had risen to 89, events had become well supported and at last the club was a viable movement.

The newsletter. published irregularly and called *The Blurb*, covers an area from the Artic Circle to the Great Lakes and carries an illustration of a pig driving a Morgan, This has its origins in the fact that Toronto is sometimes unkindly referred to as Hogtown, and that the MSCC often chooses a name for its centres appropriate to its location, hence Taffmog for South Wales, Hopmog for the hop-growing area of Kent and Hogmog for Canada. Despite the long distances involved, many Canadian members manage to attend the national meeting of the American clubs each year, and some even collect prizes.

At the time of writing the club is in the process of being officially renamed The Morgan Sports Car Club of Canada. Co-incidental with the name change will be the production of a new badge.

For name and address of the club contact see the Affiliated Clubs list at the end of the chapter.

MORGAN CLUB OF DENMARK
D
MORGAN
K

Morgan Club of Denmark

The club was founded in October 1973 at a meeting held at the home of an owner, at which six Morgans were present. Since than the membership has grown to over 50, which is thought to represent three-quarters of the total of Morgans in the country. In addition to normal social gatherings and competitive events - including race meetings in Denmark - members of the club (which is now affiliated to the Morgan Sports Car Club and is known as Danmog) also compete in events held by the Swedish Club.

The club is open to all Morgan owners and enthusiasts. For name and address of the club contact see the Affiliated Clubs list at the end of the chapter.

Morgan owner - Finland

Until recently the only Morgan in Finland was the Plus 4 drophead coupé owned by Nick Marchan, so it is not surprising there is no Morgan club in that country. However, Mr Marchan is a member of the MSCC, as well as belonging to the Sports Car Club of Helsinki, which covers all makes and has approximately 100 sports cars within its membership.

The Plus 4 has the chassis number 6851 and engine number CT 80080 and was purchased direct from the factory in 1969. The car is absolutely standard, but was supplied with four extra 72-spoke wire wheels for use with winter tyres. Recently, a 1972 Plus 8 was imported into Finland.

As can be seen from this picture, the annual meet of the Morgan Club de France is very well supported. This is Mog Avignon-Ventoux in 1987.

Nick Marschan will always be happy to talk to other Morgan owners should they happen to be in Finland. His address is Bertel Jungs väg 8, 00570 Helsinki 57, Finland.

Morgan Club of France

The club, which is now affiliated to the MSCC, was founded in 1973 by Patrick Boisvieux and another Morgan owner who lived in Paris. From this humble beginning - and an initial lack of interest by other owners - it has grown steadily and now has more than 180 members throughout the country. Because of the large distances involved in attending meetings, the club has been divided into seven regions, each of which runs its own events, including concours and rallies, In addition the club organizes one international event each year, which started in 1981. These events are very popular and attract well over 120 Morgans every year.

The club publishes a quarterly magazine named *News*, which reports on meetings, technical advice and general Morgan news.

The first President of the club, Jaqueline Frot, has been succeeded as President by her son Jean-Christophe.

Four particularly interesting Morgans are owned by club members: Maurice Louche's very rare left-hand drive genuine Plus 4 Super Sports Drophead Coupé, which is fitted with a high grille and twin spare wheels; Marin Frot owns the 1921 speed record-breaking Darmont single-seater; Jean-Mare and Francoise Taboutet own an international concours-winning 1957 Plus 4 two-seater and Mme Vian still owns the car brought by her late husband, Boris Vian, who is considered to be one of the most important post-war French authors.

For name and address of the club contact see the Affiliated Clubs list at the end of the chapter.

Morgan Club of Germany

The Morgan Club of Germany, which is now affiliated to the MSCC, was founded in 1970 by six Morgan enthusiasts at a meeting in Frankfurt, where a committee was elected, the first President being Nick Borsch. Slowly, the club grew in numbers and in 1972 Pius Kuhlmann took over as President, who in turn was succeeded by Jurgen Bell in 1976. In 1975 the club adopted a formal constitution and was registered as an association by the city of Essen. Potential members join as guests for a year, then can be granted full membership at a General Meeting after having been proposed by two sponsors, and if elected can display the official club badge on their Morgan. Total membership had risen to 47 by 1976 and to over 300 in 1985.

The club is divided into four regions - North, Central, South and Berlin - each organized by a section leader, who arranges monthly meetings to enable members to meet each other and cultivate friendships. Sharing experiences and technical tips and the restoration and rebuilding of Morgans are recurring popular topics, and the success of the club in helping to forge friendships can be seen in the number

Celebrating the 15th anniversary of the Morgan Sports Car Club Holland with a scenic tour; the cars are parked in front of the Royal Palace in Amsterdam.

of private meetings which take place in members' homes.

The club has had its own magazine, *The Morgan-Post*, since 1975. Its first editor was Heiner Giersberg, from Berlin, who handed over after one year to Reiner Wandert, originator of the magazine's current illustrated style and presentation. Pressure of work caused him to pass the editorship to Helmet Kuhlemann, who has made the magazine an interesting, informative and colourfully presented source of information to members.

Between 60 and 70 per cent of member's cars are Plus 8s, about 15 per cent are 4/4 four-seaters and the remainder are 4/4 two-seaters. Two flat-radiator Drop Head Coupés are owned by club members, but are in need of rebuilding, and the only registered three-wheeler is a 1934 Super Sports with a water-cooled 998cc Matchless engine, owned by Eberhard Schirdewahn from Delmenhorst.

The club's membership also includes Morgan drivers from Holland, Belgium, Switzerland and Austria, so it can claim to be truly international. For name and address of the club contact see the Affiliated Clubs list at the end of the chapter.

Morgan Sports Car Club - Holland

This club was formed in April 1972 through the enthusiasm of one man, Hans Dee, who, at the time of writing, is Treasurer. The first meeting was attended by 15 Morgan owners, but the club has since gathered momentum and become one of the most active in Europe, with membership of over 130 and affiliation to the MSCC. Of the original 15 members, 14 still have the same car.

The club organizes two big events each year, one in May and the other in October, comprising of driving tests, sprints, concours, social get-togethers and, last but not least, good food. As in nearly all large countries, the club has local centres which organise monthly noggins 'n' natters, local tours and rallies.

Although most Morgans in Holland are stored for the winter to avoid the salt on the roads, the club always holds a New Year's cocktail meeting arranged by the Dutch Morgan agent, who also happens to be the proprietor of the Dutch National Automobile Museum, so the surroundings are ideal.

A quarterly magazine called *Fanta Morgana* and its editor, Dick de Bruyn, excels in obtaining material from all over the world for inclusion. Membership is open to all Morgan owners, as well as 'sympathisers' although the basic rule of the club is No Morgan, No Vote. For name and address of the club contact see the Affiliated Clubs list at the end of the chapter.

Morgan owners - Hong Kong

There is no official Morgan club in Hong Kong , partly because of the strict governmental regulations relating to the formation of clubs, and partly because the majority of Morgan owners travel overseas a great deal and are often transferred to other countries. However, during 1988 and 1989 there was an incredible resurgence of interest in the Morgan in Hong Kong, so much so that the formation of an official club was then being seriously considered. It wass envisaged to be the Morgan Sports Car Club with a constitution and proper legal registration. At the time of writing the 'membership' boasts some 60 enthusiasts with 23 cars in the colony. Among the events being planned is a 'Morgan Drive to China'.

The club functions at the moment through the Classic Car Club of Hong Kong. It may be slightly cheating to include the Classic Car Club of Hong Kong in a list of Morgan clubs because it is not a single marque club. However, it has such important Morgan associations it cannot be passed over!

The CCCHK was born as a result of an abortive attempt in 1970 to found a Morgan club in the Crown Colony. Today the club is almost 100 members strong and it is interesting to note that although there are only four Morgan-owning members, three of these are on the club committee. The four cars are interesting examples in fine condition. The oldest one is known as Betsy and is a 1953 flat-radiator Plus 4 four-seater. This has the most interesting history as it took part in the first Macau Grand Prix in 1954. Its owner, Club Secretary, Dick Worrall, bought the car in late 1967, sold it in April 1970, regretted doing so and bought it back again in July 1973. The President, Tom Surrency, has a 1958 Plus 4 four-seater which he restored to such perfection in the period 1975-8 that it was the overall winner of the 1980 Hong Kong Automobile Association's concours for cars manufactured after 1945. Tom North, the club's Publicity Officer has a 1969 4/4 four-seater and T Gibbs has a 1981 Plus 8.

The bad road conditions of Hong Kong tend to restrict usage of the cars, and have necessitated the introduction of modifications aimed at curing steering-wheel shake, one such alteration being the fitment of a small shock absorber into the steering. Another hazard for Morgan owners is parking, and it is best to leave your car in a multistorey car park hidden beneath a specially made nylon cover, otherwise a parked Morgan will immediately attract a crowd which may not be content just to look, but will clamber all over the car to have photographs taken!

Despite the lack of a club, the list of Morgan owners and the literature held by them is maintained, and any Morgan owner moving to, or passing through the colony is assured of a warm welcome from Derek Hinings, Banyan Villas, House B7, 9, Stanley Village Road, Stanley, Hong Kong.

Morgan Club - Italy

The official club was formed on June 10,1981, and, from its small beginnings, has grown to become a very well organized and well supported club with over 110 members. Membership is strictly confined to active Morgan owners.

The number of events and meetings vary from year to year but include taking part in historical motor events by special arrangements with the promoters. Many other events are open to all Morgans, especially if the cars are 'late built'. The best competition result achieved by the club was in 1986 when it was placed first, second and third at the Coppa delle Alpi and first overall at the two day meeting at Monza.

Besides issuing newsletters throughout the year, the club also publishes an illustrated year book, which is exceptionally well produced.

For name and address of the club contact see the Affiliated Clubs list at the end of the chapter.

Morgan Sports Car Club of Japan

It is very doubtful whether any Morgans were imported into Japan before World War Two, and after the war the import of foreign-made cars was prohibited until 1958. When this ban was lifted very few Morgans were imported, but the one or two seen in Tokyo in the early 1960s were owned by US serviceman or British college lecturers.

In 1967 Mr Toshio Takano was appointed Morgan distributor in Japan and took delivery of four new cars. Few people knew about the Morgan car at that time and it took several months to sell them. Now, as nearly everywhere else, orders for new Morgans far outnumber availability.

In late 1972 Masahiro Naito took delivery of a 4/4 two-seater after a two-year wait, and was so impressed with his car that he contacted other Morgan owners and formed the Morgan Sports Car Club of Japan. For the first meeting in December, there were 18 members and, by 1978 (by which time the club had staged 34 events from touring rallies and hill-climbs to gymkhanas, sprints and races), membership had grown to 80. At the time of writing it has risen to 145. National meetings are held each year with the Tokyo Auto Club Sports and the Japan British Automobile Society.

It is believed that about 12 three-wheelers survive in the country, the rarest known being a 1937 barrel-back model with lowered body and an air-cooled Matchless engine, which is owned by Yutak Ando, while the rarest four-wheeler is a 1938 Coventry Climax engined 4-4 owned by Masahiro Naito.

The club now has nine branches throughout Japan, with each branch having its own chairman and organizing its own local events. The branches are located as follows: Central Branch; Tokyo; Chiba; Kanagawa; Kofu; Nagoya; Shizuoka; Kansai; Kyoto; Osaka; Wakayama; Hiroshima; Okayama; Saitama; Tohoku; North Area.

Contact amongst the branches is maintained by the organisation of the annual national meeting and through the club's twice-yearly magazine, *Great Runabout*.

Article 2 of the club constitution says : *The object of the club is to enjoy auto sports through Morgan cars, promote friendship among members, ob-*

serve traffic rules while trying to improve driving skill and retain the spirit to endure every difficulty experienced with an automobile 'Morgan'. Anyone agreeing to support these objectives can be admitted to the club, and Morgan enthusiasts visiting Japan are assured of a warm welcome.

Morgan Owners Club Luxembourg

Current membership totals 24 and is open to anyone interested in the marque. The club holds one international event a year, plus some four or five local events, which include the MOCL Tennis Tournament.

The club was started in May 1978 when six Morgan owners held a social meeting, since when membership has increased steadily.

The club produces a quarterly magazine entitled *The Chronicle*. Technical advice is also given. For name and address of the club contact see the Affiliated Clubs list at the end of the chapter.

The club stand at the Luxembourg Autojumble in 1988.

Morgan Sports Car Club of New Zealand

This club began in 1971 as the result of a chance meeting of Morgan owners on the road, held its first meeting at Easter 1972 and then was run informally for three years until becoming officially established in 1975. On its formation the total membership of 35 represented 80 per cent of Morgans in the country, while at the time of writing the number has risen to around 100. There was a club magazine, *Borrowed Time* (the title was copied from an article on Morgans by Peter Garnier in *Autocar* in September 1965) which was published monthly, but this since ceased publication and has been replaced by an occasional newsletter.

Marsden Robinson (Club Secretary) and Martin Adams (President) keep the flag flying in the Auckland area - especially in the 'intermarque challenge' (a series of races/sprints/concours culminating in the 'Le Mans Relay' race run by the Porsche Club on the New Zealand Grand Prix circuit.) 'Team Morgan' - always the favourite with the

The Christchurch area of the New Zealand Morgan Sports Car Club staged a member's run to Hammer Springs in 1987.

Club member, Lloyd Gleeson, competing in the Thoroughbred & Classic Car Club's sprint meeting in February 1981 on the Pukekoho (New Zealand) Grand Prix circuit.

spectators - usually acquits itself quite well at the Le Mans event. It took the chequered flag in 1986, but concours are generally not taken too seriously in New Zealand Morgan circles, so at the time of writing the Challenge Cup has so far eluded them.

Amongst their members' cars are P2038 - the very first Plus 4 ever made in 1950 originally as a DHC, but rebodied by the factory in 1968 with the then current 'cowled roadster' style; P2412 - a one-owner Plus 4 roadster with total mileage of less than 40,000, most of them on racetracks and hillclimbs (and a dislay case of trophies to prove it!); and the 1972 Earls Court Motor Show Plus 8 (R7536) still in the hands of its original owner. And from the 'Old Morgans never die' department - a 1951 example that was 'totalled' by a train some 30 years ago and left to rot in a paddock. It was subsequently uplifted by a keen enthusiast desperate for a Morgan and has recently risen from the dead, with the help of just about everyones' spare parts bins. It proves the basic philosophy of the New Zealand Morgan enthusiast that 'No Morgan is ever beyond repair!'

Club member, John Rock-Evans, put both the club and the marque on the map in 1978 by appearing in the *Mastermind* competition on New Zealand television and answering questions on the history of the Morgan Motor Company. He reached the semi-finals before falling on the question: 'Has the Morgan Motor Company ever produced a model, other than the Plus 4 Plus, which could be classed as an enclosed vehicle?' (They did, but can you name it?)

Another keen owner collected his car, a 1977 4/4 four-seater, in England and drove it to South Africa, all without trouble, before shipping it the remainder of the journey.

Membership of the club is open to all Morgan owners or 'Morgan nuts'.

Morgan owner - Peru

Just before the outbreak of World War Two, three Englishmen each imported a 4-4, but because of the war they had to return to England, leaving their cars behind them. One of these cars has disappeared but the other two were found by A Gardillo, who is now in the process of restoring them. The only other known Morgan in the country, a 4/4 two-seater, was imported new in 1963.

Mr Gardillo, who is a member of the MSCC, would be delighted to meet any Morgan owner who visits Peru and can be reached at Luis Felipe Villaran 395, San Isidro, Lima 27, Peru.

Morgan owners - Sicily

Six Morgan have been traced in Sicily, two 4/4 four-seaters, three 4/4 two-seaters and a Plus 8, the owners of which are all known to each other. Although they do not meet on a regular basis or have an official club to link them, they offer each other the usual Morgan mutual assistance whenever necessary.

In 1974 strict exhaust regulations came into effect, and at the same time a 35 per cent VAT was applied to cars over 2000cc, so the owner of the Plus 8 counts himself very lucky to have

The very first meeting of the South African Club held on July 4th 1971.

brought his car just before these rules came into effect.

Anyone visiting Sicily and wanting to talk Morgans should contact R Bonomo, Via Alcide de Gasperi 203, Palermo, Sicily.

Morgan owners - Singapore

Although there is no official Morgan club in Singapore, the five known Morgan owners keep in regular contact with each each other. The earliest of their cars is a pre-war 4-4, which is owned by an English couple who really cherish it. A New Zealander owns a 1954 two-seater, there are two four-seaters dating from 1963 and 1964, and an Indian Singaporian keeps a two-seater in excellent condition.

Several local factors affect the ownership of Morgans and certain other cars in Singapore, one being a rather strange government taxation system which increases tax on all cars over ten years old (regardless of condition) and almost doubles it by the time the car is 15 years old. The weather is also a major influence, for it rains on average once a day and most other times there is very hot sunshine. This has led to hood modifications by Morgan owners, one arrangement being to provide a sedanca hood position, with the top folded back from the screen halfway and fastened to the hood irons. Another way of avoiding being drenched by one of the sudden rainstorms is to drive the car with the hood up, but with a zipped flap for the rear window which can be operated from inside the car.

The five Morgans have proved very popular in Singapore, and their owners are approached almost weekly by people offering to buy their cars. Any visitor wishing to meet a local Morgan owner should first contact WJ Lawrence, PO Box 57, Tanglin Post Office, Singapore 10.

Morgan Owner's Club of South Africa (SAMOG)

Morgan owners in South Africa seem to be independent-minded individuals, so much so that they do not feel the need for a formal Morgan club. Nevertheless, they have formed themselves into an association so that the 30 or so cars spread throughout this large country remain identified and are well looked after.

The group was started by Angela Heinz, an adventurous woman whose hobbies include flying her own aeroplane and selling Morgans and motorcycles, who is a familiar sight to the Police in Johannesburg driving her very potent Plus 8 into town with her plaited pony tails streaming in the wind. In the days of free import of Morgans into South Africa she acted as the official agent, but now import regulations make it virtually impossible to get a new Morgan into the country.

Her first gathering of Morgan owners was held on July 11, 1971, at her oil executive husband's prestigious house in a select Johannesburg suburb, where seven Morgans gathered together. Since then broadcaster Ronnie Wilson and his wife, Colleen, have organized a number of social events, including

Members of the Morgan Owners Club committee dressed to contend with the very cold Swedish weather but, in true Morganeering tradition, the hood remains down.

drives to gracious old mine managers' homes on the Golden Reef, barbecues, and so on, and recently Ronnie was chosen to be Chairman of SAMOG. As a result of a letter asking him for details for inclusion in this book he arranged a special gathering of Johannesburg-based cars at the Jukskei Motor Museum for a photographic session, which drew approximately a third of all Morgans known in South Africa.

In motor sport the Morgan name is upheld in South Africa by former MG owner Steve Spencer, who now races a particularly fast Plus 8, and by John Frewin, from Middleburg, who is a keen rally competitor. South Africans are great travellers, and several owners have made the pilgrimage to Malvern. Likewise, a warm Morgan greeting awaits anyone visiting South Africa. For name and address of the club contact see the Affiliated Clubs list at the end of the chapter.

Morgan Sports Car Club of Spain

The founder and first President of this club in June 1984 was Mrs Maribel Puig. Originally, there were six members of the club (two of whom actually attended the 75th Anniversary celebrations at Malvern that year). Membership has now grown to 25 members from Spain and one from Portugal.

Once a year the club holds a general meeting which is used for planning forthcoming activities for the year. Normally there are six excursions for one, two or three days and normally no less than eight Morgans take part.

Morgan Owner's Club of Sweden

The formation of this group came about as a result of a chance meeting of two Morgan owners in the car park of a sports car exhibition in Gothenburg in April 1966. When Arne Holmstrom, now Chairman of the group, parked his 1952 Plus 4 Drop Head Coupé, he was approached by another Morgan owner and in the ensuing conversation they discovered that, between them, they knew of 10 Morgan owners in Sweden, and so the idea of forming club was born. The first meeting was held at the Gothenburg home of one of the owners on January 11, 1967.

A provisional name of Morgan Register of Sweden was proposed for the club at a subsequent meeting a few weeks later, but in April that year it was changed to allow the use of the letters MOG.

The first competitive event tackled by the members was a club race at the Dalslandsring, run by the Gothenburg Sports Car Club, and in July 1967 nine members took part in a two week holiday to England with their cars, the nine Swedish numbered cars attracting a great deal of attention throughout their tour, which included a factory visit at Malvern, where they were warmly greeted by Peter Morgan. The following month, when the first officers and committee of the group were elected at a meeting at Granna, membership had

grown to 20. With the help of an active competition programme, including hillclimbs and rallies, and a good social calendar, membership has grown steadily ever since. The most recent count is 61 members, who own 75 Morgans between them.

The club produces its own magazine and has two sub-sections covering the north and south of the country. For

The Swiss Club's summer meet.

name and address of the club contact see the Affiliated Clubs list at the end of the chapter.

Morgan Club of Switzerland

Following a gathering of Morgan owners in July 1976 the Morgan Club of Switzerland came into being the following June, since when the membership has grown from 45 to 135. Each year the club organizes a race, two two-day get-togethers which include a gymkhana, treasure hunt, *etc*, as well as monthly regional gatherings. Nearly all the cars of club members are extremely well kept and original, most with low mileages as very few are used as everyday transport, especially during winter when salt can remain on the road for anything up to six months. About 80 per cent of the cars within the club were manufactured between 1966 and 1976 and, of these, 60 per cent are Plus 8s, 20 per cent 4/4 two-seaters and 10 per cent 4/4 four-seaters. There are also five three-wheelers (two V-twins and three F models).

Not long ago a Swiss member of the club became the proud owner of a rarity, namely the racing Morgan MMC 4. He also owns the original Morgan 4 built in 1936, chassis no. 037, one of the few vehicles where the doors open backwards. It seems this is the only original 4-4 of this type on the Continent. There are also known to be a Plus 4 Plus, a Plus 4 Super Sports and five other three-wheelers in the country, the owners of which are being encouraged to join the club.

Although the club has very detailed rules (which fill a 12-page booklet), nobody seems to take them too seriously. The club facilities include technical advice and the supply of cheap parts, while the club magazine, *MCS Organ,* is edited by the Secretary and issued twice a year. A Morgan register is being compiled which, it is hoped, will contain full details and specifications of every Morgan in Switzerland.

Unfortunately, it has not been possible to import new Morgans officially since 1974. Because of this, each owner

A fine assortment of Morgans gathered together for the Southern California Club concours.

tries to keep his car in good, if not very good, condition.

The club and its members are always happy to receive visits from Morgan owners from other countries. For name and address of the club contact see the Affiliated Clubs list at the end of the chapter.

Morgan car clubs in the USA
Morgan Plus 4 Club of Southern California

Founded in 1954, this club has the distinction of being the oldest Morgan club in America, and second in the world, after the Morgan Sports Car Club Limited. Without doubt the club is also one of the most active. Regular meetings are held in the rear of Me 'n' Ed's Pizza Parlor in Long Beach, where it is not unusual to find more than 40 enthusiasts present. Two meetings are held every month; one a general

A few of the three-wheelers owned by club members lined up at the Morro Bay Show in 1987.

business meeting - really a euphemism for a social gathering - and the other a board meeting. A good balance is maintained between social events, rallies and slaloms, with the occasional wine-tasting tour, baseball game or picnic thrown in. All events are well supported, despite the large distances often involved.

The President and officers adopt parliamentary procedure at their board meetings, business being conducted in a reasonably formal fashion, the size of the group permitting good delegation of responsibility. Particularly impressive is the roll of Spares Chairman, which is filled by a member who manages a very sizeable inventory of Morgan spares. At the December 1978 meeting I was privileged to be extended Honorary Membership of the club in recognition of my 'extensive contributions to the cause'.

The membership is 250-plus with over 30 events a year. A monthly magazine called *The Format* is issued to all members and contains a great deal of information on club meetings and events, as well as technical articles of a very high standard. Also, in spite of the name, membership is open to all Morgan owners regardless of which model they own.

In the past, an annual social event was held, often at a place called the Sky Ranch, but recently this has given way to something more ambitious, patterned after the 'Luray' event on the east coast of the United States. After having driven her 4/4 across country to Luray in 1981 (and back, with a broken arm and throttle cable), Evelyn (Lynn Bird) Willburn put together a similar west coast event co-hosted by Northern California's Bob Schmidt. The event mushroomed from around 80 Moggies the first year to well over 100 at the 75th Anniversary celebration at Laguna Seca Spring vintage races, Monterey, California in 1984. The event was first called Morgans at Morro Bay, but has since been held at Pismo Beach and Cambria; the goal is to find a mid-point between southern and northern California Morgan clubs. Lately, the event is taking over the old MOGWEST name formally associated with the north/south meet at the annual August Laguna Seca historic races.

A significant event occured at the historic races at Laguna Seca in 1978 when three Morgans, one a rare SLR, competed in different events. In the fifth race, Brian Howlett, a past President of the club driving a 1962 Plus 4, utterly humiliated two Porches to win the most exciting race of the series, ensuring the Morgan name would be imprinted in Southern California motor racing history for some time to come.

A more recent triumph for a Morgan in California happened in August 1993 when Gerry Willburn, a past President of the club, was invited to enter his recently restored one-off 1952 Plus 4 Saloon Morgan in the prestigious Pebble Beach Concours, the first time a Morgan had ever been invited to compete. Although he did not win his class he was placed second, which is a magnificent achievement when one considers that the Pebble Beach concours is the world pinnacle in this type of motoring competition.

The club has a very extensive range of trophies including the Ken Hill 'Close Encounters' award which is awarded by the committee to the member who, in its opinion, has had the closet brush with the law in the previous year. In 1989 it was awarded for the Best Traffic Ticket Evasion.

For name and address of the club contact see the Affiliated Clubs list at the end of the chapter.

Morgan Sports Car Club (formerly Morgan Plus 4 Club) of Northern California

The history of the Morgan Sports Car Club of Northern California is synonymous with the availability of the marque in the San Francisco Bay area. The California club was formed in 1955, shortly after Rene Pellandini established, in Los Angeles, a Morgan distributorship for the Western United States - Worldwide Motors.

Berkeley and Marin County claim the honour of having been residences for the first Morgans in the Bay area, both purchased from Worldwide Motors. Later, a third Morgan was purchased at Malvern Link, which also settled briefly in the East Bay. As enthusiasm for Morgan motoring began to take hold in Northen California, Morgans were imported from such far away places as Morocco, Venezuela and the Malay States.

By late 1956 there was sufficient interest among Bay area Morganeers to form a Northern California Club. Minutes of the club record that the inaugural meeting was held on March 24, 1957. Four of the members owned 1953 vintage Morgans and the fifth car is said to have been a TR3-engined model. At about the same time new Morgan dealerships were established: Fergus in New York in 1959 and Triumph Inc., in San Francisco. However, the impact of increasing Morgan imports on the club's membership was felt immediately; nineteen new members had joined the club by the end of 1957. Membership was soon touching the 200 mark and the diversity of models led the club

A general view of the annual meeting of Morgans from all over America and Canada. This is MOG 13 (1983).

to vote to change its name, which took effect from December 1988.

The club has a very extensive social and competition calendar which caters for all members, no matter what their interests. Originally, the club magazine was named *The Morgan Plus 4 Format*, then in 1973 it was changed to *Morganasm*. This had a very short life and the next title was the *Morgazette*. In more recent years it has become known as *The Flexible Flyer*.

For name and address of the club contact see the Affiliated Clubs list at the end of the chapter.

Morgan Car Club, Washington DC
Founded in March 1959 by a small group of enthusiasts in the area, the club is the third oldest in the USA.

Membership has grown to over 300 and is now run on a national basis. The club holds a monthly social 'noggin' linked with a brief business meeting. Also there is a monthly driving tour to an event or location of interest. In addition the club organizes and co-ordinates the annual Morgan Owners' Gathering (MOG).

In 1971 the first MOG was held to coincide with the inaugural Ponoco 500 at the Ponoco International Raceway, a venue which the club used for the next four years before moving to Luray, Virginia in 1976. Morgan owners travel hundreds of miles to attend these gatherings, and Mr and Mrs Peter Morgan were guests of honour in 1972 and 1978, and myself and my wife, Janet, in 1980. Since then, due to a fire at the Luray Hotel, the venue has been changed each year.

When the club was formed, club racing was just beginning in the USA and most of the venues were within easy striking distance of Washington DC. A club team of Morgans, called Team Revel, competed at all of these events, but the showcase was the 1.2-mile road circuit at Marlboro, which was to prove an ideal training ground for future stars like Mark Donohue, Peter Revson and Bob Tullus. A highspot of the club's early days was when 50 Morgans attended Patuxent Naval Air Show and associated spring meeting.

When the economic conditions of the mid 1960s caused closure of most of the convenient road circuits, the club turned its attention to rally and autocross events, where its autocross and sprint teams nearly always finished amongst the top three whenever entered.

The policy of the club is to encourage other Morgan clubs to co-sponsor

A few of the 3/4 Club members pose for a photograph specially taken for this book.

annual MOGs. It is responsible for MANNA (Morgan Aid Network of North America), and is affiliated to the Morgan Sports Car Club. The club issues a magazine aptly-named *Rough Rider*, which is edited by Edmund (Ed) Zielinski and is of good quality with technical bulletins, a swap-and-sell column and news of club activities, illustrated with photographs by one of the club's past Presidents, John H Sheally II.

Morgan owners the world over are always welcome. For name and address of the club contact see the Affiliated Clubs list at the end of the chapter.

3/4 Morgan Group of America

A group of 15 Morgan owners from New York area who attended the first MOG event at Ponoco, Pennsylvania, decided to form their own club, membership of which has grown to approximately 100. Most live in the north-eastern United States, but the club does have members from as far south as Virginia and as far west as Chicago, while the easternmost member is Peter Morgan, who was presented with honorary membership of the group by the late President Harry J Carter, when he visited the USA in 1977.

As the title infers, the club caters for both three- and four-wheeled Morgans; as a registered corporation it has comprehensive articles which extend to 12 typed pages. The clear objectives of the group are:

The purposes for which this corporation has been organized are to promote fellowship and extend acquaintanceships among Morgan car enthusiasts by means of social gatherings and discussions to promote interest in the ownership and enjoyment of the operation of such cars by means of such activities as, but not necessarily limited to, rallies, concours d'elegance, autocross, gymkhanas, races, time trials, technical workshops, meets, noggins and tours; to engage generally in any causes or objects similar to the above-mentioned in order to promote the welfare of the membership and for 'maintaining the breed' (which phrase shall be deemed to mean the promotion or the preservation, and the regular use, of the three- and four-wheeled Morgan cars which were originally designed by the late HFS Morgan and by his son Peter HG Morgan and which were, and are being, produced by the firm which they built, the Morgan Car Company Limited of Pickersleigh Road, Malvern Link, Worcestershire, England.

The club membership owns a fine assortment of models but the rarest car is the 1938 4-4 TT Replica owned by John Artley. The club magazine, *Morganeer*, is published monthly.

Membership is open to all Morgan enthusiasts. For name and address of the club contact see the Affiliated Clubs list at the end of the chapter.

Ohio Morgan Owners

This is an informal regional group of Morgan owners dedicated to keeping their cars on the road by sharing technical information, trading spare parts and getting together two or three times a year for various Morgan-related events. It was started in 1973 when three Morgan owners invited all 15 known owners in the area to an initial picnic meeting in Columbus, Ohio, to which 10 of them responded. Mainly by word-of-mouth contact, membership had risen to 72 inidividuals at the last count, owning 82 Morgans between them. All but 20 of the members have become fee-paying since 1978 in order to finance expanded mailings and various club activities.

Ohio developed a good population of Morgans in the early 1970s when Don Simpkins' MM & M Imports of Derby brought many cars into the USA, when they were not as fashionable as they are today. He now runs Morgan

Fab Industries (Morgan restoration shop) in Buford, Georgia,and has probably handled half the cars now owned in Ohio.

The group has always been well represented at national Morgan events sponsored by the Washington DC club, laurels having been won by Walt Squires' 1971 Plus 8, Jerry Boston's 1961 Plus 4 Drop Head Coupé, Jim Bessts' 1958 4/4 Roadster, George Brokaw's 1961 Plus 4 Drop Head Coupé, Charles Miller's 1970 4/4 Roadster and Bob Burrows'1963 Plus 4 Roadster. The group also participates in local classic and vintage car shows and sports car rallies to promote the marque, and a typical recent season included combined events with Jaguar, MG, Triumph and Austin-Healey clubs involving rallies, sports car shows and swap meets.

Facilities for members include the group's own car badge, lists of all Morgan spare parts available in Ohio, together with lists of recommended parts suppliers and specialist shops for mechanical body and upholstery work.

On the members' roster there are 35 Plus 4 Roadsters, 13 Plus 4 four-seaters, 10 4/4 Roadsters, four 4/4 Competition Roadsters, six Plus 4 Drop Head Coupés, seven Plus 8s, three three-wheelers and two 4/4 Series 1 Drop Head Coupés.

Interesting cars which have been or are being restored include Jim Bessts' 1932 Family Model three-wheeler, Bob Burrows' 1949 4/4 Series 1 Drop Head Coupé, Dutch Junge's 1928 Super Sport three-wheeler, Dick Nelson's 1955 dual spare (long-nose) Plus 4 Drop Head Coupé, Winn Rosch's 1934 three-wheeler and 1964 Plus 4 Plus and Pail Schodorf's 1955 Plus 4 four-seater Drop Head Coupé. An interesting newcomer is Bruce Edwards' Issis Imports 1978 Plus 8 propane-powered roadster.

It is not necessary to be a Morgan owner to enjoy membership of this group. For name and address of the club contact see the Affiliated Clubs list at the end of the chapter.

Morgans of Philadelphia (MOPS)

Formed on 13 March 1977 when 25 enthusiasts responded to an advertisement in the city newspaper, within two years the club membership had reached almost 40. Monthly gatherings and visits to local auto shows are held, and in the autumn the club hosts a gymkhana which is well supported by other clubs. In a recent event the crowd were treated to the sight of President Gus Spahr's concours-winning 1932 Aero three-wheeler doing very well while competing against a field of 4-4s, Plus 4s and a Plus 8.

There are some interesting cars amongst the membership, and Alex Smith, who used to work for a local Morgan dealer back in the days when the cars could be imported legally, has one of the few Plus 4 Plus models in the USA. Fellow member Bob Lehr, who is also Vice-President of the Washington DC Club, has a right-hand drive Plus 4 Super Sports which was probably the last Super Sports to be imported into the USA and is one-of-a-kind in that it is not entirely aluminium-bodied. Factory correspondence indicates that in 1967, when the car went into production, the supply of aluminium was low and, rather than wait, the owner elected to have his care built with steel. Bud Bernard's 1967 Plus 4 is a perfect example of a Morgan transition car in that it has the rear treatment of the 1959-66 Plus 4 with other features, including the lockable glovebox, of a 1967, and was one of the first to enter the United States during that last year of legality.

A 1966 Plus 4 Super Sports has been undergoing restoration since 1971 by Dick Larrick, who is so fanatical about detail that some of his restored parts may soon need restoring! Jim Schrader's 1964 Plus 4 has a TR transmission with overdrive to go with its TR engine, this near conversion having been done before he bought the car. In addition to his multi award-winning 1962 Plus 4, member Dick Smith has the last Drop Head Coupé to be imported into the USA, an early 1968 car in concours condition. Alex Knight's 1933 barrel-back three-wheeler Super Sports should be back on the road by now, and Peter Gunshor's 1975 4-4 four-seater is a recent addition to the club strength.

Membership of MOPS is described as 'open to anyone who is tolerant enough to put up with the inexplicable affection they have for Morgans, and the driving pleasure they bring them', and a monthly newsletter, *MOPS Ma-*

nia, is supplied to all members.

For name and address of the club contact see the Affiliated Clubs list at the end of the chapter.

Morgan Owner's Group - Great Lakes

This group began informally in 1964 or 1965, its principal founders being Don Frantz and the late Jack Markley, long-time Detroit Morgan enthusiast, who would get together with about 10 other Morgan owners when the mood struck them to socialise a little and discuss their mutual Morgan interests. By chance meetings on the road and word-of-mouth, new members were attracted, but the group remained informal under the unofficial presidencies of Don Frantz and Pat Kobielski until it was decided around 1970 to bring some organisation into the club, with the payment of modest dues, the preparation of a calendar of events and the regular mailing of notices to members.

This led in 1975 to the official incorporation of the Morgan Owners Group as a Michigan non-profit corporation, and the club had grown to a membership of more than 65 by 1979, most of whom lived in the Detroit, Michigan and surrounding Great Lakes areas of the mid-West, but some were from all over the country and even as far away as Australia. More than 75 Morgans are owned by the membership, including three-wheelers, pre-war 4-4s, a Plus 4 Plus and what is probably the only 1976 4/4 presently in the United States.

A full calendar of events includes monthly meetings at a local restaurant and at least one other monthly function. In the winter events tend to be mainly social as few members are willing to expose their cars to the harsh Michigan weather. The outdoor season begins in the early spring with the annual English day road rally held in conjunction with the Austin-Healey, Jaguar, MG and Triumph clubs, followed by the spring car clinic, a gathering with a technical emphasis to inspect members's cars and prepare them for summer driving, which includes an English day at the Waterford Hills SCCA track and various picnic and cross-country outings.

Highlight of the year is the annual mid-West meet, a weekend affair held at a northern Michigan resort, involving competition events such as a road rally and Concours d'Elegance, as well as a banquet. The first of these was held in 1976, when it was felt that the long journeys to national meets discouraged many local members from participating and that the club had grown large enough to support such an event alone. They have proved a huge success, usually with over 30 cars attending. However, in recent years the location of this popular gathering has been extended to include some outstanding resorts in Ohio, and in 1989 the meet was held at Waterloo, Ontario, Canada.

In addition to sponsoring activities, the club also puts out publications

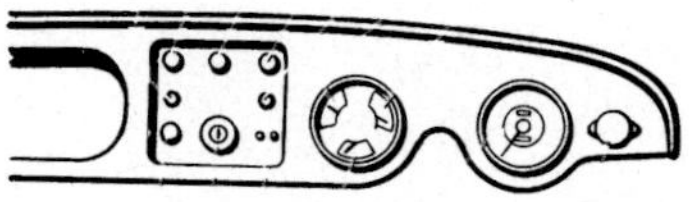

of interest to Morgan owners, including event notices and general news in the monthly *Flexible Flyer*. The club is also putting together a technical manual of hints and suggestions on how to maintain the Morgan car, based on the experiences of members and other Morgan owners. This is a very active and enthusiastic group.

For name and address of the club contact see the Affiliated Clubs list at the end of the chapter.

Morgan Owner's Group - Northwest

Morgan Owners Group Northwest was founded in early 1975 in Portland, Oregon, with about a dozen members. As it grew in number it also grew geographically. The club now holds business meetings quarterly in centrally-located Olympia, Washington. They have a monthly gathering for socialising and sharing technical knowledge in Portland, Oregon for the southern membership and in Seattle, Washington for the northern contingent.

The club has participating members in Washington, Oregon, Idaho and British Columbia. This represents a large territory and most members think nothing of driving one or two hundred miles to arrive at the start of a tour of several hundred miles! Considering that this area of the USA and Canada is blessed with incredible scenery, it is not surprising that the group has evolved into a touring club that enjoys one- or two-day explorations of picturesque coastline, arid desert or snow-capped mountain ranges. Interspersed among the six to eight tours each summer are three or four social gatherings at members' homes.

Additionally, the club supports one long-distance tour each year and, a few years ago, spent 10 days covering the

Sixteen Morgans on the lawn at Deep Bay, the house of George Sterne who, for 15 years, was the Morgan agent on the west coast of Canada. This was just one of the long-distance trips organised by the Morgan Owners Club Group North West.

2500 mile round-trip to Cambria, California for the west coast Morgan Meet, then a further week touring British Columbia.

At the time of writing, the club has 95 paid-up members. Between them they own 115 Morgans and the club's database keeps records on an additional 35 to 40 cars in the north-west USA and British Columbia. Membership is open to anyone interested in the marque and not limited to Morgan owners, although currently they have only two non-owners on the roster.

The club produces a monthly magazine entitled *Northwest Mogazine* which is supplied to all members. For name and address of the club contact see the Affiliated Clubs list at the end of the chapter.

Minnesota Morgan Club

This group had its beginnings in 1978 when Morgan owners, Tom Smisek and David Fellman, ran an ad in a Minneapolis newspaper exhorting fellow Mog-persons to appear with their Morgans at a picnic to be held at Minnehaha Falls. Even though the picnic never took place, the advertisement served to identify enthusiasts. Membership now totals 38 and is open to all Morgan owners and their families.

This is an informal group of Morgan owners dedicated to keeping their cars on the road by sharing technical information, trading spare parts and getting together several times a year for Morgan-related events.

Club activities include vintage racing at Brainerd International Raceway, British 'Kick-off' tour and picnic, autumn drive to Classic Motorbooks in Osceola, Wisconsin and Freeze-Your-Gizzard Morgan Gathering - a late fall gathering where snow accumulation in the cockpit of attending Morgans is measured and duly recognised!

Member's cars of interest include Dr Bill Nelson's 1953 Plus 4 transition model which recently underwent a lengthy restoration project. Paul Massnick owns the only three-wheeler

Members inspect Kathy and George Fink's 1962 Plus 4 Super Sports at the Club's 1985 'Freeze Your Gizzard' event. That year there was no snow due to unseasonably warm weather.

in the area; it's undergoing a complete rebuild in the kitchen of his historic home, located on Nicollet Island. Fred Nairn owns the only flat-rad in the area, a 1952 Plus 4.

Kathy and George Fink's 1964 Plus 4 Plus Coupé was the 20th car built out of a total production run of 26 cars; only 19 of this model are known to have survived. Ginny and Gordon Spartz draw lots of attention in their 1962 Plus 4 Drop Head Coupé.

While no Plus 8 Morgan is counted among the group, speedy Morgans abound in Minnesota, there's Ken Kauffman's 1967 Weber-carburetted 4/4, Pat Starr's 1958 Plus 4 racing Morgan and George Fink's 1962 Plus 4 Super Sport, which campaigned at Sebring in 1963.

Morgan Motor Car Club - Dallas, Texas

The organisational meeting of the Morgan Motor Car Club was in December 1974, in Dallas, Texas. The primary force behind the organisation was Jim Woodyard, who started making a list of Morgans of which he became aware soon after acquiring his own Plus 4 in 1970; the list included licence numbers of Morgans he saw on the street and checked the ownership of through the State office. The first attempt at a get-together was in June 1974 and resulted in the appearance of nine Morgans of various descriptions, including one three-wheeler. Thereafter, a meet of some type was held every month prior to the organisational meeting of the club itself. The quartet of Jim Woodyard, Chuck Harris, Bob Moses and Bill Boyless made up a nucleus around which the club was initially formed. Six of the nine owners who joined at the initial meeting remain members today. At the first meeting, Woodyard was elected President, the sole officer.

An interest in the marque is the only criterion for club membership. There are no written rules, nor have any been necessary. The aim of the group is to share the troubles and triumphs of Morgan ownership. The club newsletter, *Mog Log*, is a monthly publication.

An informal 'Flying Mog Squad' has been formed to go to the aid of any Morgan owner whose car needs attention or assistance. The Mog Squad provides not only camaraderie but also the combined expertise of many members.

Whilst the Morgan Motor Club was the initial club founded in Texas, and all its badges feature the Texas Lone Star, the club was not meant to be a local, provincial organisation. Rather, the organisers hoped to include and serve all Morgan enthusiasts in the southwest portion of the United States that were unrepresented by a formal organisation. Hence, the non-regional official naming of the club as the Morgan Motor Car Club.

Among the more active members there is a lack of appreciation of obstacles to Morgan driving. Four Morgans travelled together on a 3500-mile round trip trek to Washington DC for the 1982 meet, followed the next year by a 4500-mile round trip with five cars to California's initial Morro Bay event. The Plus 4 Plus of Bill and Judi Boyles was shipped to the Zandvoort Holland event in 1987 and logged some 3000 miles while on the continent of Europe.

At one time, three of the known 19 Plus 4 Plus models still existing were owned by MMCC members, including the last one ever built. Also unique is that the Morgan Motor Car Club was the first Western Hemisphere club accepted as an official centre of the home club in England, the Morgan Sports Car Club.

For name and address of the club contact see the Affiliated Clubs list at the end of the chapter.

Western New York Morgan Owner's Group

This group was started in the winter of 1978 by Alan Isselhard and Dean Meyer. The club has only two 'officers', who have guided it since the beginning: Alan Isselhard is the group organiser, who keeps the club on an informal basis and attempts to keep the treasury from becoming too fat, and Ed Kowalski - who purchased his Plus 4 flat radiator roadster new in 1953 and still owns it - is the treasurer. There are no elections for club officers and, in spite of offers by Alan and Ed to stand down in favour of someone else, no change has taken place as yet. Membership currently stands at 40 and activities include attending Watkins Glen races and social events.

Most of the membership is from the upstate New York area, mainly between Buffalo and Syracuse, with Rochester in the middle. Several of the members own sports/collector cars in addition to their Morgans. As is true with the majority of American Morgan people, most of their member-owned cars are pre-1968 - thanks to the safety-minded bureaucrats in Washington.

A club magazine entitled *Morgandats* is published five times a year. For name and address of the club contact see the Affiliated Clubs list at the end of the chapter.

Morgan Owner's Group South

The club was founded in 1976 at a weekend event held at Myrtle Beach, South Carolina, and has remained active ever since. The current membership totals approximately 60, and membership is open to anyone, without restriction.

MOG South has a broad geographic coverage, with members from the states of Georgia, South Carolina, North Caro-

Members of the Morgan Owners Group South at the Club's 1989 Spring Meet at Highlands, North Carolina.

lina, Tennessee, Alabama, Florida and Virginia. As a result, get-togethers are less frequent but are held in more varied locations.

The club has four events each year. The first get-together is at the Walter Mitty Challenge at Road Atlanta (in Georgia) the last weekend in April. This is an all-makes vintage racing event. The second event of the year is Spring Meet, which is usually held in May. The site varies from Highlands, North Carolina (which is just across the border from Georgia), to picturesque Charleston, South Carolina. The third get-together is another full weekend affair, traditionally held in the North Carolina Mountains in October when the leaves are in full colour. Fourth and last event of the year is a Christmas Party held in the home of one of the members, usually in the Atlanta, Georgia area. All of the events are informal and generally always include a car show but not a formal concours. The club magazine is called *Southern Fours and Eights*. The current editor is Lance Lipscomb, who lives in Atlanta, Georgia.

For name and address of the club contact see the Affiliated Clubs list at the end of the chapter.

Morgan Three-Wheeler Club, USA Group Organiser

There is no actual club in America catering solely for three-wheelers, but all are welcome as members of the previously listed clubs. However, the Morgan Three-Wheeler Club of England does have a group in the USA. For name and address of the contact see the Affiliated Clubs list at the end of the chapter.

The World Morgan Regalia Collector's Society

This club is truly international and came about as a direct result of the amount of regalia sold each year at the American MOG meets. A chance conversation between Melvyn Rutter, who was visiting MOG12 in 1982, and Win Sharples, the recently appointed Regalia Chairman of the Washington Club (MCCDC), planted the seed of an idea. After further discussions it was decided that Win Sharples should investigate the possibility of forming such a club. Typical of Win, he was not just content to see if there was any interest in the idea, he spent a year researching Morgan club regalia worldwide, with particular emphasis on badges. Then in June 1983 he sent out invitations to prominent Morgan people all over the world to become charter members of the proposed society.

The inaugural meeting was held at the Abbey Hotel, Malvern, during the 75th Anniversary celebrations, and was attended by 27 of the then 30 members. Today, membership is getting very close to the 100 maximum agreed by the charter members. Membership is not automatically granted to anyone who cares to apply. Before even being considered the applicant must be able to prove that he is a serious collector by producing evidence that he owns at least ten badges and is also already a member of an established Morgan club.

The Society caters for all collectable Morgan-related items, including

Part of the badge collection of Win Sharples, founder and chairman of the Morgan Regalia Collectors Society.

The pictures on these two pages are of a Series I 4-4 which was found and restored by its Latvian owner. At the time, he had never seen a Morgan and his initial restoration resulted in the car receiving the wrong back-end treatment.

Over the years the owner made contact with various other Morgan owners, particularly in England, and studied books and photographs. As a result, he was able to rectify the car's rear end. As can be seen, he made a remarkable job of the restoration.

toys, models, photographs, brochures, books and handbooks - in fact, anything pertaining to the marque. It produces a quarterly journal and its own regalia which are sold strictly to members only.

For name and address of the club contact see the Affiliated Clubs list at the end of the chapter.

Morgan owner - Latvia

The only known Morgan in the former USSR is the pre-war 4-4 in Riga, which was featured in my first book on the flat-radiator Morgans. Naturally, the owner does not use the car during the winter months, and stores it in a novel way by winching it up and suspending it in the roof space of his garage, above another car.

Modern cars are few in Latvia and collector vehicles are even rarer. So besides being extremely lucky to own this Morgan, he has started what is believed to be the only motor museum in the region and has some very interesting vehicles on display.

He would be most interested to hear from other Morgan owners and his name and address is G Michailow, Lenin Str 121-58, Riga 225012 ind, Latvia.

Morgan clubs affiliated to the Morgan Sports Car Club

AUSTRALIA
Colin Davidson
PO Box 44
Sutherland 2232
Australia

AUSTRIA
Peter Taucha
Morgan Sports Car Club of Austria

The engine bay complete with Coventry Climax engine, which is readily recognisable even if the layout of the ancillaries is different.

Weidlinger Hauptstrasse 154
A-3400 Kloster Neuburg
Austria

BELGIUM
Fred de Coster
Morgan Sports Car Club Belgium
Belgium

Leon Giet
Morgan Owners Group Belgium
Rue a la Forge
5170 Arbre
Belgium

CANADA
Mrs Audrey Beer
Morgan Owners Group Toronto
R.R. 3
Bolton
Ontario L7E 5R9
Canada

DENMARK
Alan Hall
Morgan Club Denmark
Mollegaardsvej 16
Gl Hjortkaer
6818 Aarre, Denmark

FRANCE
Mme. Jaqueline Frot
Morgan Club De France
20 Rue Daguerre
75014 Paris
France

GERMANY
Klaus Harmut Scholz
Morgan Club Deutchland
Dagoberstrasse 90
5000 Koln
Germany

Rudolf Linke
Morgan Sports Car Club
Deutchland Bonkelaar 15
Burgwedeler Str. 25
D-3002 Bissendorf
Germany

HOLLAND
Machiel Kalf
Morgan Sports Car Club Holland
1633 Gn Avenhorn
Holland

ITALY
Ambbrogio Macchi
Morgan Club Italia
Via V. Foppa 6
20144 Milano
Italy

Paolo Lazzari
Sun, Sex, Sea & Sound
Via Cuneo 57
16038 S. Margherita Ligure
Genova
Italy

JAPAN
Morgan Sports Car Club of Japan
Satoru Araki
23-8 Narashinodia
2-Chome, Funabash9-Shi
Chiba 274
Japan

LUXEMBOURG
Merens Fred
67 Rue de la Money
498 Reckon/Mess
Luxembourg

NEW ZEALAND
The Morgan Sports Car Club of New Zealand
PO Box 14-104
Auckland
New Zealand

NORWAY
Fred Ljone
Norsk Sportsvogn Klubb
M. Register Fjordmog
Arnulf Overlands vei 216
0764-Oslo 7
Norway

SOUTH AFRICA
Kevin Bolon
South African Morgan Owners Group
20 Wantage Road
Parkwood
Johannesburg 2193
South Africa

SPAIN
Xavier Romeu
Morgan Sports Car Club Espana
Mas Canyelles
08211 Castellar del Valles
Barcelona
Spain

SWEDEN
Ingvar Fredriksson
Morgan Owners Group of Sweden
P.O. Box 102
S-762 23 Rimbo
Sweden

SWITZERLAND
Daniel Zoller
Morgan Club Schweiz
Via al Chioso
CH-6929 Gravesano
Switzerland

UNITED STATES OF AMERICA
Alan Marsh
Morgan Car Club of
Washington DC
2854 Connecticut Ave. 24
Washington DC 20008
USA

Larry & Patty Moss
Morgan Owners Group Great Lakes
3377 Timber Crest Ct
West Bloomfield
Mi 48324
USA

Bill Boyles
Morgan Motor Car Club
PO Box 50392
DallaS
Texas 75250
USA

Win Sharples
Morgan Regalia Collectors Society
Rrl Box 537a
Round Hill
Va 22141
USA.

Bob Chamberlain
Morgans On The Gulf
Morgan Owners Group
3304 Sunset Blvd.
Houston
Texas 77005
USA

Bob Koupal
Morgan Sports Car Club
of Northern California
919 Mowry Avenue
Fremont
Ca 94536-4141
USA

Chas Wasser
Ohio Morgan Owners Group
7663 Windy Knoll Drive
Cincinnati, Ohio 45241
USA

Craig Seibert
Morgan Owners of Philadelphia
PO Box 242k
Martins Creek
Pa 18063
USA

Lance Lipscomb
Morgan Owners Group South
2932 Melton Court
Lilburn
Georgia 30247
USA

Michael Hattem
Morgan +4 Club
10823 Oregon Avenue
Culver City
CA 90232
USA
Spence Young,
Morgan Three-Wheeler Club, USA GO
Morgan
Three-Wheeler Club
6614 E Corrinne Drive
Scottsdale
AZ 85254
USA

Paul Levitt
3/4 Morgan Group Ltd.
112 Mercer Street
N.Y. 10012
USA

Perry Nuhn
The Gator Mogs Morgans of Florida
9067 S.E. Star Island Way
Hobe Sound
Florida 33455
USA

Wynell Bruce
2721 Acworth Due West Road
Kennesawi
GA 30144
USA
Tony Frederick
Morgan Owners Group
3024 Jarrard Street
Houston
Texas 77005
USA

Nancy Dice
Morgan Owners Group Northwest
13802 S.E. 3rd Place
Bellevue
WA 98005
USA

Bill Warren
Western NY Morgan Owners Group
22 San Rafael Dr.
Rochester
New York 14618
USA

John & Evelyn Willburn
Morgan Plus 4 Club of Southern California
Mog Cottage
11423 Gradwell
Lakewood
California 90715
USA

Denis Glavis
Morgan Sports Car Club
1322 Redwood Drive
Santa Cruz
California 95060
USA

Ed Zielinski
Morgan Car Club Washington DC
616 Gist Avenue
Silver Spring
Maryland 20910
USA

Dave Bondon
3/4 Morgan Group of America
88-89 24 N. Point Road
Colts Neck
NJ 07722
USA

Robert T Burrows
Ohio Morgan Owners
7080 Westview Drive
Worthington
Ohio 43085
USA

George Fink
Minnesota Morgan Club
2608 West 45th Street
Minneapolis,
Minnesota 55410
USA
Morgan Owners Group -
Great Lakes
Webb Drive
Farmington Hills
Michigan 48018
USA.

Other international Morgan contacts

BRAZIL
Richard J Flynn
Rua Natingui 494
Sao Paulo SP
Brazil

DENMARK
Henning Thyrre
Sommerlyst 26
8500 Grenaa
Denmark

PORTUGAL
Manuel Ramos
Av Das Rosas 152
Francelos
4400 VN GA1A
Portugal

TASMANIA
Rod Macey
20 Taronga Road
Taroona
Tasmania 7053

APPENDIX I

ORIGINAL SPECIFICATIONS, PRODUCTION CHANGES, GENERAL DATA

Series I 4-4

• 4-4 SERIES I SPECIFICATIONS

Engine:	Coventry-Climax (pre-war Standard Special - see post-war data for details).
Type:	Four cylinders, in-line, water-cooled (thermo syphon); cast iron block and cylinder head.
Bore and stroke:	63mm x 90mm.
Displacement:	1122cc.
Valves:	Overhead inlet and side exhaust.
Brake horsepower:	39 at 4500rpm.
Chassis:	'Z' section steel pressings, underslung, with inverted 'U' section crossmembers.
Clutch:	Borg & Beck single dry plate, connected to gearbox by short shaft enclosed in tube.
Gearbox:	Meadows four-speed with synchromesh on third and top. Ratios 5.0, 7.0, 12.1 and 17.1 to 1. Moss four-speed with synchromesh on second, third and top. Ratios 4.72, 6.7, 11.42 and 21.3 to 1. Engine and gearbox rigidly attached to chassis at six points. Initially no form of rubber or other type of flexible mounting, then fitted with Silentbloc mountings.
Final drive:	Open Hardy Spicer propeller shaft with needle bearing universals to conventional back axle with spiral bevel gears in pressed steel banjo case.
Front suspension:	Independent with sliding stub axle assemblies on vertical pillars, upper enclosed and lower exposed coil springs with Newton telescopic shock absorbers.
Rear suspension:	Half-elliptic laminated springs, underslung, 40in between centres. Andre Hartford frictional shock absorbers.
Brakes:	Girling, 8in drum, rod and cable operation.
Wheels:	Pressed disc type (spoked or flat) with hub caps, four stud fitting.
Tyres:	5.00 x 16in Dunlop '90' extra low pressure tyres.
Fuel:	9 gallon tank, rear mounted.
Steering:	Originally a reduction gear mounted on steering column; changed to Burman worm and nut.
Jacking system:	Stephenson, simultaneously lifting two wheels on one side.
Dimensions:	Wheelbase 7ft 8in; length 11ft 8in; width 4ft 6in; height 3ft 8in to top of screen. (4ft 2in with hood up); track 3ft 9in.
Weight:	1512lb.

• 4-4 SERIES I SPECIFICATION CHANGES

Below are listed the more important changes, which all take effect from the chassis number or date quoted.

44 First 4-4 exported (Sandford, Paris). (July 1936).

50 Ceandess clip-type radiator fillers and necks replaced treaded pattern on and after this chassis number, except for nos. 051 and 065. (July 1936).

59 New-type Coventry Climax engines fitted. Valve clearance adjustment made more convenient by inverting the balljoint between the pushrod and the rocker so that it would retain oil and the ball was attached to the rocker. Engine and gearbox were mounted on rubber bushes. Also the cooling system was fitted with thicker tubular-type radiator block, and a large diameter connection taken from the bottom of the radiator to the offside of the cylinder block at its lowest point. Also at about this chassis number the electrics were improved: the dynamo was mounted separately from the distributor and driven by belt or chain; the distributor now being driven independently by skew gears. *Note: August 1936. A few cars were produced with magneto ignition up to the outbreak of the war.*

75 New-type front wings fitted. (September 1936).

78 New-type front frames with straight tie rods. The front suspension was altered so that the fixed spindle, previously vertical, was inclined backwards and the sliding member became a steel forging with a phosphor-bronze bush at the top and bottom and incorporating the steering arm with a face to which the brake plate was bolted at four points. The spindle diameter was increased from 0.75-1 inch. By mounting the brake plate to a large rigid face the plate itself was effectively stiffened, enabling the brakes to operate more smoothly and consistently. (September 1936).

79 Change from rear to front hinged doors introduced. (circa September 1936).

80 Steering altered. In place of the reduction gear housed half-way down the steering column a Burman Douglas steering box was mounted on the end of a continuous column extending from the Ashby sprung steering wheel. The drag link ran across the front of the chassis from the drop arm to a balljoint on the track rod, close to its connection with the nearside steering arm. (circa October 1936).

92 Engine no. MA79 up to this number B998 HERAL 21 pistons known as Mark IV fitted. After this Mark V stamped C15 fitted. (October 1936).

155 Amal-type bonnet clips fitted, except for chassis 158. (January 1937).

166 Higher ratio Burman Douglas steering columns fitted. Also K.L.G. type L.K.1 sparking plugs fitted as standard. (February 1937).

206 New-type bellhousing fitted. (March 1937).

238 Works experimental Ford 22hp 'V8' engine. (April 1937).

239 Sidescreens changed to higher type with frame extension. Also from Meadows gearbox no. 10337 shorter driving end employing longer transmission primary shaft. Also longer speedometer drive gearshaft. (April 1937).

257 Windscreens fitted with two wiper blades. Plus chassis nos. 226 and 253. (April 1937).

291 Knock on hub caps available as option. (June 1937).

303 The four-seater introduction announced. Weight 1642lbs. This car works experimental. (August 1937).

312 Sliding axle covers fitted. Also rear axles strengthened. (June 1937).

317 New-type track rod ends with bosses on end to take balljoint pins. Also on chassis nos. 297, 302, 306 & 307.(June 1937).

325 Chronometric speedometers fitted except for chassis nos. 382 & 385.

356 Burgess silencers replaced by Carbjector type. (July 1937).

395 New-type radiator sides fitting outside valance boards at bottom. (August 1937).

410 New-type bonnet with end clips fitted, except for chassis no. 458. (November 1937).

421 Small-type bonnet knobs. (December 1937).

427 First production four-seater. (December 1937).

493 1.25in brake shoes; slightly tapered spindles; RMS 5 & RMS 8 bearings fitted. (January 1938).

522 New-type rear axle B.A.8.A. fitted. (February 1938).

525 New-type rear hubs. (March 1938).

545 1.5in brake shoes fitted as standard. (March 1938).

596 Full taper spindles with RLS 9 inner bearings. Also on 578, 585, 589 and 590. (May 1938).

604 Spark plugs K.L.G. Corrundite Type F50X fitted as standard. (circa June 1938).

648 Drop Head Coupé introduction announced. First production car. Weight 1736lbs. Fitted with vertical slat radiator grill. These also supplied as an option on all models from here on. (October 1938).

677 Primary driveshaft 0.12in shorter due to the alteration to spigot bearing. Also quick opening oil filler introduced. October 1938).

723 Bellhousing fitted with Silentbloc mountings. (December 1938).

724 All Drop Head Coupés fitted with Moss gearboxes and Silentbloc mountings. (December 1938).

747 Last Meadows gearbox fitted. (December 1938).

748 Moss gearboxes only fitted. (January 1939).

785 Radiator tops recessed for flush fitting bonnets. (December 1938).

March 1939
Le Mans Replica model announced to be fitted with 1098cc Coventry Climax competition engine. Not put into production; three only produced.

829 No toe-in of front wheels. (May 1939).

880 Chassis stays shortened 1/2in to compensate for the setting back of bottom front axle. (June 1939).

894 Standard Special engine introduced as standard. (June 1939.)

896 New pattern frame front with flanged lugs; larger centre pins; longer main springs with rebound springs and spring covers fitted to each spring. Also on chassis nos. 872 and 892, plus works cars 585 and 724. A few cars were fitted with old-type springing after 896. (June 1939).

900 New-type wing stays, not in pairs. (July 1939).

904 New-type radiator caps, plain top less mascot. (July 1939).

• 4-4 PRE-WAR CHASSIS NUMBER SEQUENCE

Numbers 1-999	Last number used: 930.
Chassis 1, 2 & 3	Works experimental cars from about May 1935.
Chassis 3 & 11 same car	Sold via Bowman & Acock on March 3, 1936, having been rebuilt on chassis no. 11 after crash.
Chassis 4	Not used.
Chassis 5	First production car. Sold

	March 27, 1936.
Chassis 6 & 7	Not used.
Chassis 44	First car exported. Sold to Stewart Sandford, Paris. First car to be coloured other than green - Carnation Red with black wings. Sold July 6, 1936.
Chassis 892	Used twice.

• ENGINE NUMBER SEQUENCE

Coventry Climax:	1122cc & 1098cc.
From chassis number:	1-88 prefixed by letter 'M'. Highest number: 106.
From chassis number:	89 prefixed by letters 'MA'. Highest number: 898.
Standard Special:	1267cc. From Q1E Q69E.

Note: engines were not fitted in number sequence.

• PRODUCTION DATES

March 1936 - September 1939.

• PRODUCTION BY ENGINE

Coventry-Climax 1122cc	713 (one chassis no. used twice).
Coventry-Climax 1098cc	58
Coventry-Climax 12hp	1
Standard Special 1267cc	37
Ford various	3
Total cars	812
Chassis only 1122cc	82
Chassis nos. not used	105
Total car & chassis	**1000**

• PRODUCTION BY BODY AND ENGINE

Two-seaters 1122cc	577 (one chassis no. used twice).
Two-seaters 1098cc	47 (all models).
Two-seaters 1267cc	15
Two-seaters 12hp	1
Two-seaters Ford engine	3
Total two-seaters	**643**
Four-seaters 1122cc	102
Four-seaters 1098cc	5
Four-seaters 1267cc	4
Total four-seaters	**111**
Coupés 1122cc	34
Coupés 1098cc	6
Coupés 1267cc	18
Total Coupés	**58**
Chassis only 1122cc	82
Chassis only 1267cc	1
Chassis nos. not used	105
Total cars & chassis	**895**
No. cars & chassis exported	**121**

• 4-4 OTHER USEFUL INFORMATION

Borg & Beck clutch springs

The clutch springs fitted to the 4-4 were colour coded as follows:
BLUE coloured springs were fitted as standard practice.
ORANGE coloured springs had a poundage of 140-146lbs.
BROWN coloured springs had a poundage of 153-160lbs.
GREEN coloured springs had a poundage of 175-180lbs.
The brown and orange springs were used for reliability trials where the owner did not object to a fierce clutch. The pressure for the green springs was exceptional and was used for racing, pure and simple.

Rear axle oil seals

Type B.A.8. was fitted from the first 4-4 until early in 1938 (chassis no. 522), when it was changed to B.A.8.A. The rear axle and seal numbers are:

B.A.8. type rear axle
2 off no. 237120 oil seals for rear hubs.
2 off no. 21914 oil seals for axle shafts.
1 off no. 20018 oil seals for bevel pinion.

B.A.8.A. type rear axle
2 off no. 237120 oil seals for srear hubs.
2 off no. 21914 oil seals for axle shafts.
1 off no. 237112 or no. 237114 oil seals for bevel pinion.

Gearbox oil seals

Meadows-type gearbox 4 4
1 off no. 20010 oil seal for front end.
1 off no. 225112 oil seal for rear end.

Moss type gearbox 4-4
1 off no. 1831 oil seal for front end.
1 off no. 237112 oil seal for rear end.

Rear axle type B.A.8 bearings 4-4

1 off no. SKEFKO 6305 bevel pinion spider end.
1 off no. N.M.25 bevel pinion, pinion end.
2 off no. 1.11885/1.112016 or TIMKEN 26112T-26283 rear hub end.
2 off SKEFKO 6207 crownwheel.

Rear axle type B.A.8.A. bearings 4-4

1 off no. Hoffmann M10CD or DUFLEX MS10 bevel pinion outer ballrace.
1 off no. Hoffmann RMS11 or SKF. CRM9 bevel pinion inner roller race.
2 off no. Ransomes & Marples LJT/35 or SKF 7207 or SKF 6207A differential crown wheel ballraces.
2 off no. Timken 26112T-26183 hub roller races (taper).

Moss gearbox bearings

No. SKF. RLS8 front bearing.
No. SKF. RLS11 primary shaft.
No. SKF. 6306 mainshaft.
Two phospher-bronze bushes for the layshaft.

Moss gearbox speedometer drive

Driving gear on mainshaft has four start and ten driven gears.

Coventry-Climax dynamo pulleys
The pulley fitted to the Morris 10HP Series II is almost identical to the above and can be fitted without modification.

4/4 Series (Postwar)

• 4/4 SERIES POSTWAR SPECIFICATIONS

Engine:	Standard Special.
Type:	Four cylinders in-line; water-cooled; cast iron block and cylinder head.
Bore and stroke:	63.5mm x 100mm.
Displacement:	1267cc.
Valves:	Overhead, offset from centre of cylinder.
Brake horsepower:	40 at 4300rpm.
Chassis:	'Z' section steel pressings, underslung, with inverted 'U' section crossmembers.
Clutch:	Borg & Beck single dry-plate, connected to gearbox by short shaft enclosed in tube.
Gearbox:	Moss four-speed with synchromesh on second, third and top. Ratios: 4.72, 6.7, 11.42 and 21.3 to 1. Engine and gearbox attached to chassis at six points, Silentbloc mountings.
Final drive:	Open Hardy Spicer propeller-shaft with needle-bearing universals to conventional back axle with spiral-bevel gears in pressed-steel banjo case.
Front suspension:	Independent with sliding stub-axle assemblies on vertical pillars, upper enclosed and lower exposed coil springs with Newton hydraulic dampers.
Rear suspension:	Half-elliptic laminated springs, underslung, 40 inches between centres. Andre Hartford frictional dampers.
Brakes:	Girling, 8 inch drum, rod and cable operation.
Wheels:	Pressed disc-type (spoked or flat) with hub caps, four-stud fitting.
Tyres:	Dunlop, 4.50 x 17 inch, two and four-seaters. 5.00 x 16 inch, Drop Head Coupés.
Fuel:	9 gallon tank, rear-mounted.
Steering:	Burman worm & nut.
Jacking system:	Stephenson, simultaneously lifting two wheels on one side.

Dimensions:

	Two-seater	Four-seater	Drop Head Coupé
Wheelbase	7ft 8in	7ft 8in	7ft 8in
Length	11ft 4in	11ft 7in	11ft 7.5in
Width	4ft 7in	4ft 7in	4ft 7in
Height (hood up)	4ft 4.5in	4ft 6.5in	4ft 3in
Track	3ft 9in	3ft 9in	3ft 9in
Weight:	1624lb	1624lb	1680lb.

• 4/4 SERIES SPECIFICATION CHANGES

1000	4.50 x 17 inch spoked disc wheels fitted to both two-& four-seaters; 5.0 x 16 inch disc wheels on Drop Head Coupé. Also three-spoke steering wheels fitted. (March 1946).
1046	Voltage control and handbrake release knob now plated. Side screens fitted with external buttons. (December 1946).
1201	Bluemell steering wheels fitted. (March 1947).
1232	New-type front stays and squarer front wings fitted. (November 1946).
1278	100mph speedometers fitted as standard. (June 1947).
1480	First left-hand drive, four-wheeled Morgan. (September 1946).
1518	New-type wheel studs and nuts fitted (14mm). (September 1947).
1589	New-type instruments fitted. (January 1947).
1666	9 inch front brake drums fitted. (March 1948).
1715	(circa) New Standard Engine Instruction Book dispatched with all cars. (April 1948).
1732	New-type cut-out fitted. (May 1948).
1754	14mm wheel nuts fitted as standard. (Hubs the same as used on a German milk float). High melting point grease in front hubs. (September 1948).
1754	(circa). Steering damper blades fitted as standard. (September 1948).
1855	9 inch front and rear brake drums fitted. (April 1949).
1952	Small bushes fitted to stub axles. (November 1949).
2026	'S'-type stub axles fitted. (September 1950).
2038	Chassis used for experimental Plus 4. (July 1950).

• 4/4 POSTWAR PRODUCTION FIGURES (1098CC COVENTRY-CLIMAX-ENGINED CARS)

Two-seaters (Roadsters, including Le Mans replicas)	4
1267cc Standard Special-engined cars	
Two-seaters (Roadsters)	243
Four-seaters	140
Drop Head Coupés	101
Sold as chassis	54
Chassis numbers not used	539
Total production	**542**
Total number of cars and chassis exported	256
Total home market	**825**

Note: certain dates and chassis numbers would appear to be out of sequence and a large number of chassis were not used. Immediately after the war ended many different companies placed orders for new cars. This, in spite of the fact that the factory was unable to produce cars even for export, owing to the lack of sheet steel or aluminium. All orders received were entered in strict rotation. Cars were sold and the chassis number stamped on the car relative to the actual chassis number allocated to the company at the time of ordering. Full production restarted from January 1, 1948, at chassis number 1700.

• CHASSIS NUMBER SEQUENCE

Numbers	1000-2081
Last number used	2081

• ENGINE NUMBER SEQUENCE

Coventry-Climax = MA645; MA665; MA674; MA675
Standard Special = Q27E (lowest number)-700 (highest number used)
Note: engines were not fitted in number sequence

• PRODUCTION DATES

March 1946-February 1951

• PRODUCTION BY ENGINE

Coventry-Climax 1098cc	4 (3 Le Mans Replicas)
Standard Special 1267cc	488
Total cars	**492**
Chassis only1267cc	54
Chassis nos. not used	539
Total car & chassis	**546**

• PRODUCTION BY BODY AND ENGINE

Two-seaters 1098cc	4 (3 Le Mans Replicas)
Two-seaters 1267cc	240
Total two-seaters	**244**
Four-seaters 1267cc	139
Coupés 1267cc	101
Total postwar car production	**484**
Chassis only	58
Chassis nos. not used	539
Total cars & chassis	**302**
Number of cars & chassis exported	**256**

• TOTAL PRODUCTION BY BODY: PREWAR AND POSTWAR

Body	Prewar	Postwar	Total
Two-seaters	633	241	874
Four-seaters	110	139	249
Coupé	58	101	159
TT Replica	4	-	4
Le Mans Special	7	-	7
Le Mans Replica	-	3	3
Chassis only	83	58	141
Chassis not used	105	539	644
Total exported	121	256	377
Total cars & chassis	**895**	**542**	**1437**

• POSTWAR MOSS GEARBOXES

Gearboxes marked with 'H.2.' have an altered (higher) second gear ratio.

4/4 Series II

• 4/4 SERIES II SPECIFICATIONS

Engine:	Ford 100E.
Type:	Four cylinders, in-line.
Bore and stroke:	63.5mm x 92.5mm.
Displacement:	1172cc.
Valves:	Side.
Brake horsepower:	36bhp at 4400rpm.
Chassis:	'Z' section steel pressings, underslung, with inverted 'U' section crossmembers.
Clutch:	Borg & Beck single dry-plate.
Gearbox:	Ford three-speed with syncromesh on second and top. Ratios 17.29, 8.91 and 4.44 to 1.
Final drive:	Open Hardy Spicer propeller shaft to Salisbury hypoid-bevel axle unit.
Front suspension:	Independent with sliding pillars; coil springs with Armstrong telescopic shock absorbers.
Rear suspension:	Live axle with semi-elliptic springs and Armstrong piston-type shock absorbers.
Brakes:	Girling, 9in diameter drums, hydraulic, two leading shoes at front, 1.75in wide shoes.
Wheels:	Pressed disc type with hub caps, four-stud fitting.
Tyres:	5.00 x 16in, 18psi front and rear.
Fuel:	8.5 gallon tank, rear-mounted.
Steering:	Cam gear, 2.5 turns lock to lock.
Jacking system:	Stephenson, simultaneously lifting two wheels on one side.
Dimensions:	Wheelbase 8ft; Length 12ft; Width 4ft 8in; Height 4ft 2in; Track 3ft 11in.
Weight:	1568lb.

• 4/4 SERIES II SPECIFICATION CHANGES

3242	Engine no. 3601. Experimental Series II 4/4 built on this Plus 4 chassis. (March 1955).
A347	New-type bumper front, old type rear. (October 1957).
A349	New-type bumpers fitted as standard. (October 1957).
A353	All stub axles now E57-142 392859. (November 1957).
A403	Large stop lamps and Plus 4 radiator grille fitted. (May 1958).
A517	Windscreens as fitted to Plus 4s. Four-inch wider body now fitted (in fact, a Series III body). Flashers now fitted as standard. (November 1959).
A557	New-type steering column with shorter drop arm introduced.

• CHASSIS NUMBER SEQUENCE

Numbers	A200-A586
Last number used	A586

• ENGINE NUMBER SEQUENCE

Engines were not fitted in number sequence.

• PRODUCTION DATES

October 1955-November 1960

• PRODUCTION BY BODY

Two-seaters	341

Four-seaters	1 (experimental; chassis no. A296)
Coupés	1 (chassis no. 553)
Competition model	43
Chassis only	1
Total cars & chassis	**387**
Number of cars exported	**225**
Total exported to USA	**157**

• SERIES II COMPETITION MODEL

Chassis no.	*Engine no.*	*Colour*	*Date*	*Destination*
A299	353737	Green	25.5.57	UK
A351	391021	Black	14.2.58	UK
A356	392210	Red	6.2.58	UK
A371	118076	Green	27.3.58	N. Zealand
A374	114500	Green	25.4.58	UK
A376	118149	Green	25.2.58	USA
A378	118122	Grey	5.3.58	USA
A384	118006	Ivory	17.3.58	France
A397	123055	Green	1.5.58	UK
A399	123065	Red	6.5.58	UK
A414	128459	Red	31.5.58	UK
A427	128460	Green	16.7.58	USA
A429	128462	Red	5.7.58	UK
A449	128946	Red	31.10.58	UK
A450	128941	Red	5.11.58	USA
A456	128945	Green	17.3.59	UK
A475	142234	Blue	17.2.59	Canada
A483	142201	Red	24.4.59	France
A489	142229	Ivory	11.5.59	Canada
A490	140225	Red	13.5.59	USA
A496	142194	Blue	8.6.59	USA
A499	154537	Green	26.6.59	UK
A506	154535	Red	30.7.59	Canada
A515	154762	Blue	23.9.59	USA
A527	158461	Grey	6.11.59	UK
A529	158462	Green	13.11.59	UK
A540	1612325	Red	22.12.59	UK
A544	S153693	Red	25.1.60	UK
A551	S153674	Red	24.2.60	USA
A557	S172882	Grey	21.4.60	Cyprus
A561	S172874	Blue	20.5.60	France
A563	S172883	Blue	13.5.60	UK
A570	S172880	Broken White	16.7.60	UK
A572	S172575	Green	15.6.60	USA
A574	S172574	Green	16.7.60	UK
A578	S176010	Turq. Blue	10.10.60	UK
A579	S176011	Turq. Blue	26.9.60	UK
A584	S176012	White	14.10.69	Cyprus

• SERIES II FITTED WITH AQUAPLANE HEAD (NOT COMPETITION MODELS)

A272	262838	Ivory	1.3.57	USA
A280	353707	Green	17.5.57	USA
A284	353717	Blue	17.5.57	USA
A285	3533711	Red	31.5.57	USA
A321	353706	Grey	17.7.57	USA
A327	353709	Red	2.9.57	UK
A347	390992	Red	6.11.57	USA

• SERIES II FITTED WITH WIRE WHEELS

A370	A378	A425	A427	A428	A435	A437	A446
A455	A467	A468	A469	A470	A560	A514	A550

• SERIES II FITTED WITH FULL DISC WHEELS

A395

(All other cars fitted with spoked disc wheels similar to those used on the Series I 4/4).

4/4 Series III

• 4/4 SERIES III SPECIFICATIONS

Engine:	Ford 105E.
Type:	Four cylinders, in-line.
Bore and stroke:	88.86mm x 48.41mm.
Displacement:	997cc.
Valves:	Overhead.
Brake horsepower:	39bhp at 5000rpm.
Chassis:	'Z' section steel pressings, underslung, with inverted 'U' section crossmembers.
Clutch:	Single dry-plate.
Gearbox:	Ford four-speed with syncromesh on second, third and top. Ratios 18.1, 10.54, 6.21 and 4.4 to 1. Reverse 23.7 to 1.
Final drive:	Open Hardy Spicer propeller shaft to Salisbury hypoid-bevel axle unit.
Front suspension:	Independent with sliding pillars; coil springs with Armstrong telescopic shock absorbers.
Rear suspension:	Live axle with semi-elliptic springs and Armstrong piston-type shock absorbers.
Brakes:	Girling, 9in diameter drums, hydraulic, disc front brakes optional extra.
Wheels:	Pressed disc type with hub caps, four-stud fitting.
Tyres:	5.20 x 15in.
Fuel:	8 gallon tank, rear-mounted.
Steering:	Cam gear, 2.5 turns to lock.
Jacking system:	Stephenson, simultaneously lifting two wheels on one side.
Dimensions:	Wheelbase 8ft; Length 12ft; Width 4ft 8in; Height 4ft 2in; Track 3ft 11in.
Weight:	1568lb.

• 4/4 SERIES III SPECIFICATION CHANGES

A614	All 4/4s fitted with 7HA rear axle. Plus A611 & A612. Except A632 which was fitted with 6HA type axle. (March 1961).
A634	1.5in brake shoes introduced. (July 1961).

• CHASSIS NUMBER SEQUENCE

Numbers	A590-A648
Last number used	A648

• ENGINE NUMBER SEQUENCE

Engines were not fitted in number sequence.

• PRODUCTION DATES

October 1960-November 1961

• PRODUCTION BY BODY

Two-seaters	59 (+ 1 Competition Model A664).
Experimental	1
Total cars	**61**
Number of cars exported	**43**
Total exported to USA	**32**

4/4 Series IV

• 4/4 SERIES IV SPECIFICATIONS

Engine:	Ford 109E.
Type:	Four cylinders, in-line.
Bore and stroke:	80.96mm x 65.07mm.
Displacement:	1340cc.
Valves:	Overhead.
Brake horsepower:	62bhp at 5000rpm.
Chassis:	'Z' section steel pressings, underslung, with inverted 'U' section crossmembers.
Clutch:	Borg & Beck single dry-plate.
Gearbox:	Ford four-speed with syncromesh on second, third and top. Ratios 18.8, 11.00, 6.00 and 4.56 to 1.
Final drive:	Open Hardy Spicer propeller shaft to Salisbury hypoid-bevel axle unit.
Front suspension:	Independent with sliding pillars; coil springs with Armstrong telescopic shock absorbers.
Rear suspension:	Live axle with semi-elliptic springs and Armstrong piston-type shock absorbers.
Brakes:	Girling, 9in diameter rear drums, hydraulic, 11in disc front brakes.
Wheels:	Pressed disc type with hub caps, four-stud fitting.
Tyres:	5.60 x 15in.
Fuel:	8.5 gallon tank, rear-mounted.
Steering:	Cam and peg, 2.25 turns lock to lock.
Dimensions:	Wheelbase 8ft; Length 12ft; Width 4ft 8in; Height 4ft 2in; Track 3ft 11in.
Weight:	1456lb.

• 4/4 SERIES IV SPECIFICATION CHANGES

B651	Disc brakes fitted as standard. (November 1961).
B657	Engine no. 229748 works experimental car. (December 1961).
B850	Engine no. 283445 used for prototype Series V 4/4. (December 1962).

• CHASSIS NUMBER SEQUENCE

Numbers	B650-B855
Last number used	B855.

Note: chassis nos. B710 - B799 were not used. The works records show B709 followed by B800. It is normal practice to leave a gap in the chassis number sequence when changing to a new engine and model. It was thought at the time that the next delivery of engines would be the new Ford 116E engine for the Series V. Whilst awaiting delivery, the chassis were stamped with the new numbers starting at B800, but the delivery consisted of 56 more Ford 105E engines.

• PRODUCTION DATES

October 1961-March 1963

• ENGINE NUMBER SEQUENCE

Engines were not fitted in number sequence.

• PRODUCTION BY BODY

Two-seaters	15 (including A664 Competition Series III)
Chassis only	1
Chassis nos. not used	90
Total cars	**206**
Number of cars exported	**76**
Total exported to USA	**44**

4/4 Series V

• 4/4 SERIES V SPECIFICATIONS

Engine:	Ford 116E.
Type:	Four cylinders, in-line.
Bore and stroke:	80.97mm x 72.74mm.
Displacement:	1498cc.
Valves:	Overhead.
Brake horsepower:	65bhp at 4800rpm. GT 83.5bhp at 5200rpm
Chassis:	'Z' section steel pressings, underslung, with inverted 'U' section crossmembers.
Clutch:	Single dry-plate.
Gearbox:	Ford four-speed all-syncromesh. Ratios 16.2, 10.92, 6.44 and 4.56 to 1. Reverse 18.1 to 1.
Final drive:	Open Hardy Spicer propeller shaft to Salisbury hypoid-bevel axle unit.
Front suspension:	Independent with sliding pillars; coil springs with Armstrong telescopic shock absorbers.
Rear suspension:	Live axle with semi-elliptic springs and Armstrong piston-type shock absorbers.
Brakes:	Girling, 9in diameter rear drums, hydraulic, 11in disc front brakes.
Wheels:	Spoked.
Tyres:	5.20 x 15in or 155 x 15.
Fuel:	8.5 gallon tank, rear-mounted.
Steering:	Cam and peg.
Dimensions:	Wheelbase 8ft; Length 12ft; Width 4ft 8in; Track 3ft 11in front, 4ft 1in rear.
Weight:	1848lb.

• CHASSIS NUMBER SEQUENCE

Numbers	B856-B1495
Last number used	B1495.

• PRODUCTION DATES

February 1963-March 1968

• ENGINE NUMBER SEQUENCE

Engines were not fitted in number sequence.

• PRODUCTION BY BODY

Two-seaters	367
Two-seater GT models	270
Two-seater Special body to order (B1384)	1
Experimental 1600	1
Experimental 1600 GT	1
Total cars	**640**
Number of cars exported	**442**
Total exported to USA	**235**

Plus 4 Vanguard-engined

• PLUS 4 FLAT RADIATOR SPECIFICATIONS

Engine:	Standard Vanguard.
Type:	Four cylinders, in-line.
Bore and stroke:	85mm x 92mm.
Displacement:	2088cc.
Valves:	Overhead.
Brake horsepower:	68bhp at 4300rpm.
Chassis:	'Z' section steel pressings, underslung, with inverted 'U' section crossmembers.
Clutch:	Borg & Beck single dry plate, connected to gearbox by short shaft enclosed in tube.
Gearbox:	Moss four-speed with synchromesh on second, third and top. Ratios 4.11, 5.4, 8.0 and 13.5 to 1. Reverse 13.87 to 1. Optional ratios 3.73, 4.24, 7.38 and 12.85 to 1. Reverse 12.58 to 1.
Final drive:	Open Hardy Spicer 1300 series propeller shaft with needle bearing universals to hypoid bevel axle unit (Salisbury 3HA23).
Front suspension:	Independent with sliding stub axle assemblies on near vertical pillars, coil springs with Girling hydraulic telescopic shock absorbers.
Rear suspension:	Half elliptic laminated springs, Girling piston-type shock absorbers.
Brakes:	Girling hydraulic 9in diameter drum and two leading shoes.
Wheels:	Pressed disc type with hub caps, four stud fitting.
Tyres:	5.25 x 16in.
Fuel:	11 gallon tank, rear mounted, changed to 10 gallon on four-seater models.
Steering:	Burman Douglas worm and nut, two turns from lock to lock.
Jacking system:	Stephenson, simultaneously lifting two wheels on one side.
Dimensions:	Wheelbase 8ft; Length 11ft 8in (Four-seater and Coupé 11ft 10in); Width 4ft 4in (Four-seater 4ft 6in, Coupé 4ft 2in); Height to top of screen 3ft 10 in; Track 3ft 11in.
Weight:	From 1764lb.

• VANGUARD ENGINE SPECIFICATION CHANGES

The detail specification of the 2088cc Standard Vanguard engine was changed progressively during the period it was fitted to the Plus 4, the most significant alterations occurring from the engine numbers quoted below:

Engine	**Comment**
V56ME	Fitted in chassis no. P2158. Petrol pump changed from type 'U' to type 'UE'.
V82ME	Fitted in chassis no. P2185. Oil gauge feed moved from gallery to filter body.
V113ME	Fitted in chassis no. P2216. Exhaust valves fitted with thinner springs. (This modification will fit earlier engines).
V169ME	Fitted in chassis no. P2258. Thermostatically controlled hot spot inlet manifold introduced.
V198VE	Fitted in chassis no. P2304. Dynamo pulley lugs machined off to clear fan.
V210ME	Fitted in chassis no. P2351. New cylinder head with redesigned combustion chambers fitted, also new pistons.
V288ME	(and above this number). Fitted in chassis no. P2328. Disc in top of rocker cover.
V244ME	Fitted in chassis no. P2381. Breather restrictor location changed from manifold to valve cover.
V400ME	Fitted in chassis no. P2490. Diameter of exhaust valve stems increased from 0.31in to 0.34in.

• PLUS 4 FLAT RADIATOR MODELS SPECIFICATION CHANGES BY CHASSIS NUMBER

2038 (4/4)	First car to Plus 4 specifications, built on 4/4 chassis. Works experimental car. Later rebuilt with cowled radiator.
P2100	Prototype Plus 4. Works experimental. Registration no. HUY 982. Fitted with high ratio rear axle (circa September 1950).
P2109	First four-wheeled Morgan produced with left-hand drive (February 1951).
P2140	Export-type headlamps (055941B) fitted on all cars for home and export markets, except for USA (May 1951).
P2150	Double dipping headlamps introduced (May 1951).
P2156	Drilled wheels, 15 holes now fitted (June 1951).
P2165	Top water hose connection changed from single to double (June 1951).
P2182	All export cars other than USA, continental lamps fitted (July 1951).
P2227	First experimental four-seater Drop Head Coupé

	(circa October 1951).
P2234	New Girling dampers fitted (October 1951).
P2260	Radiator core changed from film to tubular type (December 1951).
P2268	Gearbox no. Moss 144. From this number all gearboxes were fitted with new baffle behind ball race at speedometer end (December 1951).
P2270	New pattern pass lamp (no. 053127A) now fitted (December 1951).
P2275	Rear wheels slotted and drilled for snow grips (not coupe model) (December 1951).
P2293	Radiator grille modified; badge plate fixed outside bars (January 1952).
P2306	Clutch operating pawl (eye bolt) strengthened from 0.37 inch to 0.5 inch BSF thread. Push connectors added for headlamp wiring (January 1952).
P2312	New pattern exhaust manifold introduced (February 1952).
P2360	Headlamp block lenses introduced, also smaller sidelamps fitted (no. 52733A) (April 1952).
P2400	Steering box drop arm shaft adjusted, set screw added (June 1952).
P2410	Steering dampers modified to enable damper blades to be renewed without dismantling the springs (July 1952).
P2500	Front suspension rebound springs strengthened. (November 1952).
P2587	Steering dampers fitted with distance tubes. (April 1953).
P2612	Experimental four-seater Drop Head Coupé (circa April 1953).
P2662	Changed by works to interim radiator shape with straight shaped headlamp housing (March 1954).
P2669	New pattern seats on all four-seaters (June 1953).
P2710	'T' type front hubs introduced (September 1953).
P2725	Thompson track rods first fitted (October 1953). Then various cars until P2756.
P2750	First TR2 engined car. Drop Head Coupé. (Only flat radiator TR engined car) (October 1953).
P2756	New-type track rods now fitted to all cars. Shock absorbers on all four-seaters 0.5inch higher (November 1953).

• CHASSIS NUMBER SEQUENCE
Numbers P2100-P2756

• ENGINE NUMBER SEQUENCE
VI1ME (engines were not fitted in number sequence).

• PRODUCTION DATES
February 1951-June 1954

• PRODUCTION BY BODY

Two-seater	345 (including one TR2-engined car).
Four-seater	141
Drop Head Coupé	
Two-seater	117
Drop Head Coupé	
Four-seater	2
Sold in chassis form	50. Experimental: 3 (one from Series I, 4/4).
Total cars & chassis	**658**
Number of cars exported	**437**
Total exported to USA	**183**

Note: some cars were returned to the works by dealers, unsold. These were rebodied as cowled radiator models and resold. Known chassis numbers are: P2662, P2737, P2739, P2740, P2741, P2742, P2745, P2747, P2750, P275 5, P2762.

• PLUS 4 VANGUARD-ENGINED INTERIM/COWLED RADIATOR

First chassis number:	P3000
Last chassis number:	P3922 Last car to be fitted. General production ended at chassis no. 3450.
Lowest engine no:	V18ME
Highest engine no:	V786ME (Engines are not fitted in number sequence).

• PRODUCTION DATES
December 1953-May 1958.

• PRODUCTION BY BODY

Two-seater	60
Four-seater	30
Drop Head Coupé	
Two-seater	11
Drop Head Coupé	
Four-seater	27
Chassis nos. not used	2
Total cars & chassis	**128**

Chassis number 3238 used twice.

• TOTAL PRODUCTION FIGURES FOR VANGUARD-ENGINED PLUS 4S

Two-seater	405
Four-seater	171
Drop Head Coupé	
Two-seater	128
Drop Head Coupé	
Four-seater	29
Sold in chassis form	50
Chassis nos. not used	2
Total cars & chassis	**783**

Chassis number 3238 used twice.

Plus 4 4-seater Drop Head Coupés - factory record (September 1951 to February 1956)

Date left factory	Chass. no.	Eng. type	Eng. no.	G'box no.	Axle ratio	Coachwork col.	Upholstery col.	Dest.	Notes
Sept 1951	P2227	V	V122ME	113	4.1	Blue	Black		1
March 1952	P2612	V1	V420ME	435	4.1	-	-		1
July 1954	P3144	V2	V730ME	822	4.1	Blue	Black	UK	
Oct 1954	3179	V2	V686ME	799	4.1	Ivory	Black	UK	2
Nov 1954	3207	TR2	TS5810	898	4.1	Ming Blue	Beige	USA	3
Nov 1954	3211	V2	V762ME	871	4.1	Blue	Black	UK	
Dec 1954	3214	V2	V749ME	840	3.73	Ivory	Black	UK	
Dec 1954	3222	V2	V765ME	809	4.1	Black	Red	UK	
Jan 1955	3230	V2	V732ME	825	4.1	Green	Black	UK	
Jan 1955	3234	V2	V767ME	842	4.1	Red	Black	UK	
Jan 1955	3235	V2	V713ME	930	3.73	Green	Black	UK	
Jan 1955	3237	V2	V740ME	962	4.1	Green	Black	UK	
Jan 1955	3238	V2	V719ME	958	3.73	Blue	Black	UK	
Feb 1955	3243	TR2	TS5138	837	N/A	Ivory	Red	USA	
Feb 1955	3250	V2	V761ME	950	3.73	Blue	Black	UK	
March 1955	3254	V2	V760ME	929	4.1	Blue	Black	UK	
March1955	3256	TR2	TS5028	921	3.73	Red	Black leather	USA	
March 1955	3259	V2	V698ME	956	4.1	Green	N/A	UK	
March 1955	3261	TR2	TS4195	935	3.73	Ivory	Black leather	USA	
March 1955	3264	V2	V664ME	957	4.1	Green	Black	UK	
March1955	3266	V2	V769ME	992	3.73	Ming Blue	Grey leather	UK	
March 1955	3267	V2	V783ME	968	3.73	Green	Beige	UK	
April1955	3274	TR2	TS4925	940	3.73	Black	Red	USA	
April 1955	3283	V2	V741ME	976	4.1	Blue	Black	UK	
May 1955	3289	TR2	TS5598	967	3.73	Red	Black Leather	USA	
May 1955	3290	V2	V721ME	961	N/A	Blue	Black	UK	
May 1955	3298	TR2	TS4602	N/A	3.73	Black	Red leather	USA	
June 1955	3300	V2	V734ME	849	4.1	Green	Black	UK	
June 1955	3302	TR2	TS4927	941	3.73	Blue	Black leather	USA	
June 1955	3307	V2	V781ME	1018	3.73	Ming Blue	Beige	UK	
July 1955	3319	TR2	TS5684	944	3.73	Black	Red	Belgium	
July 1955	3322	V2	V706ME	1037	4.1	Green	Black	UK	
July 1955	3323	TR2	TS4824	982	3.73	Off-white	Red leather	USA	
Aug 1955	3332	TR2	TS6018	996	3.73	Maroon	Black	Australia	
Aug 1955	3335	TR2	TS5251	1006	3.73	Red	Beige leather	USA	
Aug 1955	3339	TR2	TS5466	1003	3.73	Black	Red leather	USA	
Sept 1955	3342	V2	V693ME	1023	4.1	Leaf Green	Black	UK	
Sept 1955	3346	TR2	TS5698	1014	3.73	Ming Blue	Red leather	USA	
Sept 1955	3348	V2	V759ME	1045	4.1	Primer	Beige leather	UK	
Sept 1955	3354	TR2	TS5866	1050	3.73	Black	Beige leather	USA	
Sept 1955	3356	V2	V775ME	1053	4.1	Blue	Black	UK	
Oct 1955	3368	TR2	TS7775	1046	3.73	Ivory	Red leather	USA	
Nov 1955	3372	TR2	TS6060	1054	3.73	Red	Black leather	USA	
Nov 1955	3375	V2	V18ME	1055	4.1	Blue	Black	UK	
Nov 1955	3393	V2	V707ME	1154	4.1	Green	Black	UK	
Dec 1955	3394	TR2	TS8199	1092	4.1	Blue	Light Grey	Spain	
Dec 1955	3398	TR2	TS7901	1063	4.1	Ivory	Red leather	Spain	
Dec 1955	3400	V2	V778ME	1065	3.73	Ivory	Black	UK	
Feb 1956	3419	TR2	TS829ME	1107	3.73	Blue	Black leather	USA	
Feb 1956	3438	TR2	TS8287	922	4.1	Red	Black leather	USA	

Notes:
1 'Works' cars - both were subsequently updated from 'Flat Rad' prototypes to cowled fronts. 2 1954 show car. 3 Originally built with a Vanguard engine but changed to a TR2 unit before delivery.
Engines:
V: Early type standard Vanguard Phase 1 engine. V1: Later type standard Vanguard Phase 1 engine. V2: Standard Vanguard Phase 2 engine. TR2: Triumph TR2 engine

Plus 4 TR2-engined

• PLUS 4 TR2-ENGINED SPECIFICATIONS

Engine:	Triumph TR2.
Type:	Four cylinders, in-line.
Bore and stroke:	83mm x 92mm.
Displacement:	1991cc.
Valves:	Overhead.
Brake horsepower:	90bhp at 4800rpm.
Carburettor:	Twin SU.
Chassis:	'Z' section steel pressings, underslung, with inverted 'U' section crossmembers.
Clutch:	Borg & Beck 9in single dry plate, connected to gearbox by short shaft enclosed in tube. Gearbox: Moss four-speed with synchromesh on second, third and top. Ratios 12.51, 7.43, 4.9 and 3.73 to 1. Reverse 16.09 to 1.
Final drive:	Open Hardy Spicer propeller shaft with needle bearing universals to hypoid bevel axle unit.
Front suspension:	Independent with sliding stub axle assemblies on near vertical pillars, coil springs with Girling telescopic shock absorbers.
Rear suspension:	Semi-elliptic laminated springs, Girling hydraulic shock absorbers and Silentbloc bushes.
Brakes:	Girling hydraulic 9in drums with two leading shoes front and rear.
Wheels:	Pressed steel with hub caps.
Tyres:	5.25 x 16in.
Fuel:	11 gallon tank, rear mounted.
Steering:	Cam and peg, 2 turns from lock to lock.
Jacking system:	Stephenson, simultaneously lifting two wheels on one side.
Dimensions:	Wheelbase 8ft; Length 11ft 8in; Width 4ft 8in; Height 4ft 3 in; Track 3ft 11in.
Weight:	1848lb.

• COWLED RADIATOR PRODUCTION BY ENGINE: TRIUMPH TR2

First chassis number:	P2750. This is the first car fitted with a TR2 engine, but general production commenced from chassis no. 3050.
Last chassis number:	P3488 Last car to be fitted, general production ended at chassis no. 3449.
Lowest Engine number:	TS 275ME (engines were not fitted in number sequence).
Highest engine no:	TS 8969ME

• PRODUCTION DATES

September 1954 June-1956.

• PRODUCTION BY BODY

Two-seater	245 + 1 Flat Radiator experimental car.
Four-seater	49
Drop Head Coupé Two-seater	21
Drop Head Coupé Four-seater	22
Sold in chassis form	1
Total cars & chassis	**338 + 1 Flat Radiator model.**

• PLUS 4 TR3-ENGINED SPECIFICATIONS

Engine:	Triumph TR3.
Type:	Four cylinders, in-line.
Bore and stroke:	83mm x 92mm.
Displacement:	1991cc.
Valves:	Overhead.
Brake horsepower:	100bhp at 5000rpm.
Compression ratio:	8.5 to 1.
Carburettors:	Twin SU H6.

(All other specifications are as TR2-engined cars).

• COWLED RADIATOR PRODUCTION BY ENGINE: TRIUMPH TR3

First chassis number:	3421. This is the first car fitted with a TR3 engine, but general production commenced from chassis no. 3450.
Last chassis number:	5574. Last car to be fitted, general production ended at chassis no. 5270.
Lowest engine number:	TS 9653. Engines are not fitted in number sequence.
Highest engine no:	TS 82524

• PRODUCTION DATES

March 1956-December 1962. (Then odd cars until last chassis number in February 1964).

• PRODUCTION BY BODY

Two-seater	1078
Four-seater	466
Drop Head Coupé	208
Super Sports *two-seater*	39
Super Sports Drop *Head Coupé*	1
Sold in chassis form	1
Experimental cars	2
Total number of cars	**1795**

• PLUS 4 TR4 & 4A-ENGINED SPECIFICATIONS

Engine:	Triumph TR4 & 4A.
Type:	Four cylinders, in-line.
Bore and stroke:	86mm x 92mm (both engines).
Displacement:	2138cc (both engines).
Valves:	Overhead.
Brake horsepower:	TR4: 100bhp at 4600rpm/TR4A: 104bhp at 4700rpm..
Compression ratio:	9.0:1 (both engines)
Carburettors:	Twin Stromberg CDs or SUs.

(All other specifications as TR2-engined cars).

• PRODUCTION BY ENGINE: TRIUMPH TR4

First chassis number:	5199. This is the first car fitted with a TR4 engine, but general production commenced from chassis no. 5271.
Last chassis number:	6853
Lowest engine number:	CT 6726 (engines were not fitted in number sequence).
Highest engine no:	CT 80108

• PRODUCTION DATES

October 1962-January 1969.

• PRODUCTION BY BODY

Two-seater	793
Four-seater	466
Drop Head Coupé	195
Plus 4 Plus Fixed Head coupé	26
Competition model	42 (including one works car)
Super Sports two-seater	55
Super Sports Drop Head Coupé	3
Super Sports four-seater	5 (Super Sports sold less engine. Engines supplied direct from Lawrencetune).
Chassis nos. not used	2 Experimental cars
Total number of cars	**1585**

Plus 4 Super Sports (works recognised) - factory record

No.	Chassis No.	Engine No.	Gearbox No.	Colour	Despatch date	Country	Notes	Drive
1	4749	81474	2475HR	White	10. 2.61	UK		RH
2	4826	82103	2202HR	Red	28. 6.61	USA		LH
3	4828	82104	2576HR	Black	21. 6.61	USA		LH
4	4837	82105	2585HR	Red	21. 6.61	USA		LH
5	4840	82061	2537HR	-	-	UK	To Lawrence	RH
6	4860	82114	2503HR	Blue	7. 6.61	UK		RH
7	4862	82101	2589HR	Green	31. 7.61	Switzerland		LH
8	4872	82110	2572HR	Black	12.10.61	USA	2.2-litre	LH
9	4884	82111	2575HR	Red	19. 9.61	France		LH
10	4919	82117	2628HR	Red	19. 9.61	France		LH
11	4958	82121	2656HR	Blue	15.11.61	USA		LH
12	4973	82161	2680HR	Green	13.11.61	Australia	UK Show Car	RH
13	4996	82113	2689HR	Flame/Black	22.11.61	Canada		LH
14	5004	82112	2688HR	White	5.12.61	USA		LH
15	5020	82252	2714HR	Black	20.12.61	USA		LH
16	5028	82109	2690HR	Green	3.1.62	USA		LH
17	5049	82233	2728HR	Blue	13. 2.62	USA		LH
18	5058	82325	2786HR	Black	12. 3.62	Switzerland	2.2-litre	LH
19	5063	82266	2767HR	Red	6. 3.62	USA		LH
20	5095	82304	2834HR	Green	23. 3.62	UK		RH
21	5101	82272	2831HR	Red	4. 4.62	USA		LH
22	5103	82303	2841HR	Red	13. 4.62	Australia		RH
23	5122	82376	2858HR	Blue	31. 5.62	USA		LH
24	5153	82214	2828HR	Green	28. 5.62		Works Test Car	RH
25	5159	82372	2864HR	Sun Beige	21.6.62	France	2.2-litre	LH
26	5161	82407	2899HR	Green	31. 7.62	Australia	2.2-litre	RH
27	5167	82406	2898HR	Express Blue	28.6 62	UK		LH
28	5193	82427	2921HR	Green	3. 9.62	USA		LH
29	5199	CT6726	2932HR	Gun Metal Grey	10.10.62	Cyprus	2.2-litre	LH
30	5204	82431	2919HR	White	12.9.62	USA		LH
31	5212	82455	2905HR	Red	26. 9.62	USA		LH
32	5220	82479	2937HR	White	10.10.62	USA		LH
33	5222	82478	2939HR	Green	5.10.62	USA	2.2-litre	RH
34	5237	CT14098	3015HR	Green	19.12.62	USA	2.2-litre	RH
35	5270	82515	3011HR	Green	13.12.62	Italy		LH
36	5276	CT14358	3000HR	White	1.12.62	UK	2.2-litre	RH
37	5287	CT14095	3045HR	Green	29.1.63	USA	2.2-litre	LH

No.	Chassis No.	Engine No.	Gearbox No.	Colour	Despatch date	Country	Notes	Drive
38	5297	CT14824	3051HR	Green	21.1.63	Denmark	All CT p/f 2.2-litre	LH
39	5305	TS82519	3050HR	Blue	19. 1.63	USA	Standard Engine 2-litre with Webers & exhaust of S/S by Lawrence	LH
40	5308	CT16775	3052HR	White	19. 1.63	USA	For Sebring Race	RH
41	5310	CT15605	3061HR	Black	6. 2.63	USA		LH
42	5322	CT16185	2977HR	Blue	27. 2.63	USA		LH
43	5326	CT14828	3020HR	Green	20. 3.63	USA		RH
44	5330	TS82518	3030HR	White/Black	21.2.63	Canada	2-litreTR3	LH
45	5339	CT14827	3019HR	Red	1. 5.63	USA		LH
46	5356	Less Engine	3079HR	Green	9.4.63	UK		RH
47	5376	CT16199	3058HR	Green	15. 5.63	USA		LH
48	5381	CT18954	3143HR	Red	8. 5.63	USA	Standard TR4 engine, with Webers & exhaust of Super Sports by Lawrence	LH
49	5398	CT17084	3099HR	Black	10. 5.63	Germany	S/Sports Coupé	LH
50	5425	CT16116	3165HR	Red	5. 7.63	Canada	Steel Body	LH
51	5434	CT16793	3171HR	White	6. 8.63	Cyprus		RH
52	5453	CT16228	3084HR	White	23. 8.63	Canada		LH
53	5553	TS82523	3296HR	Red	13.12.63	UK	2.0-litre TR3	RH
54	5567	TS82524	3301HR	White	14. 1.64	Holland		RH
55	5569	CT19088	3270HR	Marlborough Blue	11. 3.64	UK		RH
56	5611	CT26929		Black	10. 4.64	Holland	S/Sports Coupé	LH
57	5665	TS82521	4305HR	Blue	23. 6.64	USA	2-litreTR3	RH
58	5707	CT29738	3427HR	Green	23. 7.64	UK		RH
59	5711	CT29736	3434HR	Imperial Crimson	17.8.64	USA		LH
60	5721	CT31305	3462HR	Black	16. 9.64	USA		RH
61	5730	TS32513	3452HR	Green	29.8.64	USA	S/Sports Coupé	LH
62	5734	CT26727	3400HR	Jasmine Yellow	9.9.64	Canada	S/Sports Coupé	LH
63	5790	CT36788	3475HR	Green	28.10.64	USA		LH
64	5811	CT34297	3549HR	Imperial Crimson	20.11.64	UK		RH
65	5821	CT31419	3550HR	Green	24.11.64	UK		LH
66	5841	CT36785	3567HR	Green	24. 3.65	USA		LH
67	5936	CT36463	3620HR	Green	6.4.65	USA		LH
68	5927	TS82520	3621HR	Green	12.4.65	Sweden	2-litre TR3 Standard Plus 4 with Super Sports engine	LH
69	5954	CT36911	3638HR	Green	20.5.65	USA		LH
70	5973	CT40761	3697HR	Green	13.7.65	USA		RH
71	6030	CT39090	3752HR	Dark Blue	1.9.65	USA		LH
72	6032	CT40759	3752HR	Primrose	27.8.65	UK		RH
73	6116	CT40828	3869HR	Red	29.12.65	USA		LH
74	6126	CT57187	3848HR	White	6.1.66	France	Four-seater Super Sports	LH
75	6136	CT40835	3432HR	B.White	28.1.66	Germany		LH
76	6179	CT57191	3842HR	Red	31.1.66	Holland		LH
77	6220	CT58796	3959HR	Green	10.5.66	UK		RH
78	6235	CT57918	3937HR	Red	27.5.66	Italy	Four-seater Super Sports	LH

No.	Chassis No.	Engine No.	Gearbox No.	Colour	Despatch date	Country	Notes	Drive
79	6271	CT65976	3991HR	Green	30.6.66	UK	Four-seater Super Sports	RH
80	6290	CT65974	3979HR	Black	31.8.66	USA		RH
81	6328	TS82525	4092HR	Red	18.10.66	Australia	2-litre TR3 Limited Slip Rear Axle	RH
82	6359	CT68883	4089HR	Dauphine Red	10.11.66	UK		RH
83	6374	CT69181	4071HR	Green	29.11.66	UK	Four-seater Super Sports	RH
84	6377	CT70928	4128HR	Green	22.12.66	USA		LH
85	6388	CT69187	4144HR	Green	20.12.66	Holland		LH
86	6427	CT71245	4135HR	Red	8.2.67	Canada		LH
87	6452	CT73813	4200HR	Green	23.3.67	France	Four-seater Super Sports	LH
88	6464	CT73468	4172HR	Avion Blue	14.4.67	Spain		LH
89	6468	CT73465	4168HR	Blue	6.4.67	UK		RH
90	6472	CT73986	4208HR	Vivid Yellow	31.3.67	Spain		LH
91	6489	CT75197	4214HR	B.White	26.4.67	Canada		LH
92	6498	CT76028	4246HR	Green	17.5.67	Panama, USA		LH
93	6521	CT76030	4249HR	Red	19.7.67	Canada		LH
94	6537	CT76338	4268HR	Green	26.6.67	Canada		LH
95	6556	CT76339	4265HR	Red	19.7.67	UK		RH
96	6563	CT76973	4289HR	Vivid Yellow	31.7.67	Canada		LH
97	6596	CT75203	4216HR	Saphire Yellow	20.9.67	Spain		LH
98	6625	CT77135	4330HR	Indigo Blue	18.10.67	USA		LH
99	6653	CT77235	4366HR	B.White	7.11.67	UK		RH
100	6656	CT76755	4365HR	Green	28.12.67	UK		RH
101	6723	CT77137	4442HR	Black	24. 2.68	Canada		LH
102	6732	CT73618	4463HR	Kid Chrome	18.3.68	France		LH
103	6758	CT78927	4514HR	Green	17.5.68	UK		RH
104	6761	CT78936	4512HR	Green	15.5.68	UK		RH
						Jersey		RH

Plus 4 Lawrencetune-engined cars which, for various reasons, are not regarded by the Morgan Motor Company as true Super Sports - factory record

No.	Chassis No.	Engine No.	Gearbox No.	Colour	Despatch date	Country	Notes	Drive
1	4806	TS81379	2522HR	Blue	30.3.61	UK	4-seater	RH
2	4820	TS81954	2528HR	Green	26.4.61	UK	DHC	RH
3	5058	TS82325	2786HR	Black	12.3.62	Switzland	2.2-litre	LH
4	5084	TS82321	2836HR	Green	1.4.62	Australia	4-seater Westminster	RH
5	5328	CT17157	3034HR	Black	21.3.63	USA	DHC	LH
6	5352	TS81954	2940HR	P.O. Red	15.5.63	UK	RH	
7	5390	CT18955	3103HR	Black	24.5.63	U SA	DHC	LH
8	5568	CT25179	3269HR	Green	18.2.64	USA	DHC	LH
9	5611	CT26929	3367HR	Black	10.4.64	Holland	DHC 2.2-litre	LH
10	6334	CT69183	4045HR	Deep Bronze	20.10.66	USA	DHC	LH
11	6430	Supplied as rolling chassis		No engine or gearbox. In primer paint.	14. 2.67	USA		LH

Plus 4 Competition Model actual cars - factory record

Chassis No.	Engine No.	Colour	Date	Destination
6079	CT5662	Green	17.2.66	UK
6135	CT61082	Green	10.1.66	Belgium
6149	CT60334	White	11.2.66	Canada
6153	CT60588	Primer	2.3.66	USA
6156	CT60591	Light Maroon	17.2.66	France
6159	CT62081	Red	17.2.66	UK
6162	CT62082	Dark Blue	17.2.66	UK
6165	CT62079	Green	1.4.66	UK
6166	CT54337	Deep Crimson	Works car	-
6200	CT64771	Green	23.4.66	UK
6202	CT63199	Broken White	31.3.66	Sweden
6209	CT64774	Green	19.4.66	UK
6251	CT66247	Unpainted	29.6.66	USA
6255	CT66263	Blue	2.7.66	France
6263	CT67267	Aircraft Blue	21.6.66	UK
6264	CT66752	Green	12.7.66	Panama
6269	CT67552	Green	14.7.66	Canada
6288	CT68884	White	1.8.66	USA
6291	CT67256	not known	26.7.66	UK
6295	CT69185	Royal Maroon	31.8.66	USA
6312	CT69401	Deep Baize Green	12.9.66	UK
6313	CT68882	Broken White	26.8.66	Canada
6314	CT69069	Green	7.9.66	USA
6343	CT69990	Green	21.10.66	Germany
6346	CT71244	Grey	25.10.66	France
6351	CT71501	Green	20.10.66	Switzerland
6361	CT71243	Green	2.11.66	UK
6362	CT70660	Blue	18.4.67	USA
6363	CT72285	Green	21.3.67	USA
6369	CT70215	Green	17.11.66	UK
6375	CT72384	Dark Blue	28.11.66	UK
6384	CT71874	Dawn Grey	16.12.66	USA
6387	CT72388	Green	7.3.67	UK
6404	CT73814	Bright Red	3.3.67	UK
6405	CT73466	Green	2.2.67	Germany
6421	CT74347	Broken White	13.2.67	UK
6425	CT74345	Green	3.3.67	USA
6426	CT73979	Green	17.4.67	Australia
6428	CT74227	Black	14.2.67	Canada
6445	CT73615	Green	2.3.67	UK
6473	CT75883	Grey Body/Black Wings	20.4.67	USA
6484	CT76341	Broken White	28.4.67	Sweden

• PLUS 4 PLUS FIXED HEAD COUPÉ SPECIFICATIONS

Engine: Triumph TR4.
Type: Four cylinders, in-line.
Bore and stroke: 86mm x 92mm.
Displacement: 2138cc.
Valves: Overhead.
Brake horsepower: 105bhp at 4750rpm.
Carburettor: Twin Stromberg CD.
Chassis: 'Z' section steel pressings, underslung, with inverted 'U' section crossmembers.
Clutch: Borg & Beck 9in single dry plate, connected to gearbox by short shaft enclosed in tube.
Gearbox: Moss four-speed with synchromesh on second, third and top. Ratios 11.08, 6.51, 4.49, and 3.73 to 1. Reverse 11.08 to 1.
Final drive: Open Hardy Spicer 1300 series propeller shaft with needle bearing universals to hypoid bevel axle unit.
Front suspension: Independent with sliding stub axle assemblies on near vertical pillars, coil springs with Armstrong telescopic shock absorbers.

Rear suspension:	Half elliptic laminated springs, Armstrong lever arm shock absorbers.	*Steering:*	Cam gear, 2 turns from lock to lock.
Brakes:	Girling hydraulic 11in front discs and 9in rear drums with 1.75in wide shoes.	*Jacking system:*	Stephenson, simultaneously lifting two wheels on one side.
Wheels:	72 spoked.	*Dimensions:*	Wheelbase 8ft; Length 12ft 8in; Width 5ft 1in; Height 4ft 3in; Track 3ft 11 in front, 4ft 1in rear.
Tyres:	5.60 x 15in.	*Weight:*	1820lb.
Fuel:	10 gallon tank, rear mounted.		

Plus 4 Plus

No.	Chassis No.	Engine No.	Gearbox No.	Colour	Despatch date	Country	Drive
1	A5379	CT16164	3031HR	Red	Works Car	UK	RH
2	A5504	CT22723	3214HR	Red	30.11.64	UK	RH
3	A5530	CT21904	3159HR	Green	12.12.63	UK	RH
4	A5544	CT23482	3244HR	Blue	4.12.63	US	LH
5	A5558	CT23992	3239HR	Black	24.12.63	Canada	LH
6	A5575	CT25271	3274HR	Black	17.1 64	UK	RH
7	A5592	CT26233	3279HR	Red	18. 2.64	Canada	LH
8	A5606	CT26927	3330HR	Green	11. 3.64	US	LH
9	A5612	CT26167	3287HR	Red	2. 3.64	Switzerland	LH
10	A5619	CT27437	3345HR	White	18. 3.64	US	LH
11	A5625	CT25298	3272HR	Grey	19.12.64	UK	RH
12	A5650	CT25448	3283HR	Red	1.5.64	UK	RH
13	A5657	CT26511	3313HR	White	6. 5.64	US	LH
14	A5678	CT27664	3344HR	Red	3. 6.64	US	LH
15	A5686	CT28327	3351HR	Blue	17.6.64	US	LH
16	A5702	CT28328	3369HR	Green	8.7.64	US	LH
17	A5722	CT26991	3311HR	Blue	17.7.64	UK	RH
18	A5723	CT29259	3408HR	Black	29.7.64	US	LH
19	A5737	CT29795	3407HR	Green	28.8.64	Belgium	LH
20	A5758	CT29261	3401HR	Red	12.9.64	US	LH
21	A5794	CT30637	3453HR	Black	31.3.65	US	LH
22	A5908	CT26117	3357HR	Blue	12.3.65	UK	RH
23	A5989	CT26449	3314HR	White	6.12.65	Japan	LH
24	A6124	CT26210	3304HR	Crimson	5.1.65	UK	RH
25	A6373	CT64673	Unknown	Green	9.12.66	Canada	LH
26	A6436	CT75058	Unknown	Red	7.2.67	Holland	LH

• SPECIFICATION CHANGES PLUS 4 INTERIM AND COWLED RADIATOR MODELS

3000	First interim radiator car (circa October 1953).
3043	Shock absorbers raised 0.5in (January 1954).
3050	TR2 engine fitted as standard (April 1954).
3136	Tie rods to radiator. Raised lamps, rounded stone guards (cowls) and cam steering fitted as standard (July 1954).
3146	New pattern bonnet clips introduced (August 1954).
3242	Chassis used for experimental 4/4 Series II. Engine no. 3601 (March 1955).
3313	First Plus 4 fitted with Armstrong shock absorbers (July 1955).
3362	Single spare wheel and 9in x 1.75in front brakes on and after this chassis (November 1955).
3421	First TR3-engined car, commencing no. TS 9653 (February 1956).
3428	Final four-seater Drop Head Coupé (February 1956).
3450	TR3 engine standard (March 1956).
3485	All two-seaters now fixed screen (May 1956).
3493	Small warning lights introduced (June 1956).
3500	(circa) Salisbury rear axles 3HA crownwheel drive gear bolts increased from 0.37in to 0.44in diameter, as from axle serial number G56 37 (June 1956).
3536	Gearbox no. 1217. All gearboxes on and after this number fitted with high ratio (September 1956).
3540	9in x 1.25in rear brakes fitted (September 1956).
3565	Stronger and shorter stub axle arms adopted (November 1956).
3779	Works car (TUY 275) (October 1957).

3790 Lower front cowl introduced to special order (November 1957).
3903 Larger stop lamps adopted (April 1958).
3922 Last Vanguard-engined car (May 1958).
3923 Lower front cowl withdrawn (June 1958).
4023 Altered rear end on all models with single spare wheel recessed into and following the line of a sloping panel. 12 gallon fuel tank in place of 11 gallon mounted on chassis. Narrower running boards. Amber flashers now standard (October 1958).
4026 Engine no. TS 34038E, long steering arms fitted. Also fitted to cars with engine nos. 28931, 28934, 33530, 33531, 33532, 33533, 33536, 33846 and 33849 (circa October/November 1958).
4033 4in wider body introduced (November 1958).
4096 New-type Girling front and rear brakes fitted. A few cars after this were fitted with old type for a short time, until old stocks cleared (December 1958).
4127 Engine no. TS 39202. 1959 Sebring race car (February 1959).
4203 Front wheel disc brakes available as optional extra, but only with wire wheels (May 1959).
4277 Cranked windscreen brackets on all two-seaters (August 1959).
4285 From engine no. TS 50001E onwards, all engines were fitted with modified flywheel and starter motor (August 1959).
4335 New type starter motor introduced (November 1959).
4368 Front wheel disc brakes available as optional extra with pressed steel wheels. Pedals on all left-hand drive cars moved in. Hypoid oil in all gearboxes (December 1960).
4394 Drop Head Coupé with hard top. Monte Carlo Rally and Tulip Rally car (January 1960).
4445 Engine no. TS 63946. 1960 Sebring race car (February 1960).
4472 (circa) New type steering columns and shorter drop arms introduced (April 1960).
4647 Improved steering introduced (October 1960).
4650 Revised instrument panel with toggle switches introduced (October 1960).
4671 Front wheel disc brakes fitted as standard. These had been on offer as an optional extra for several months (November 1960).
4749 First Super Sports (February 1961).
4749 Rear axles changed to type 7HA. Longer Hardy Spicer propshafts used. A few cars after this number were fitted with old type 3HA axles, until old stocks cleared (February 1961).
4872 First 2.2-litre Super Sports (June 1961).
4923 All bellhousings 0.09in longer (September 1961).
4974 Deeper windscreen and lower door sills introduced on Drop Head Coupé (November 1961).
5063 Five leaf springs now fitted (March 1962).
5101 (circa) All Plus 4s built after January 26, 1962 fitted with EN12 steel stub axles.
5136 First 2.2-litre engined Plus 4. Engine no. CT 511535 (June 1962).
5153 Fitted with 4/4 front frame and body. Works test car (June 1962).
5199 TR4 2138cc engine adopted as standard but 1991cc engine still available to order. Instruments changed to white figures on black dials (September 1962).
5237 Super Sports now fitted with 2138cc TR4 engine as standard but 1991cc engine still available to order (October 1962).
5356 Super Sports supplied less engine. Aluminium body with hard top. Supplied to Pip Arn old (XRX 1) (April 1963). 5379 Prototype Plus 4 Plus fixed head coupé (May 1963).
5408 New-type propshafts fitted (June 1963).
6060 Engine no. CT 56217. From this chassis number all cars with negative earth electrical system fitted negative earth (October 1965).
6079 Competition model introduced with 'Low Line' body as on Super Sports. With 72 spoke wire wheels as standard (October 1965).
6204 Engine no. CT 64769. From this engine number, SU carburettors fitted as standard (May 1966).
6331 First 'Low Line' body on standard two-seater (September 1966).
6393 'Low Line' body now standard on all two-seaters (December 1966).
6431 Works car, NWP 213 (February 1967).
6480 Used for prototype Plus 8. Now chassis no. R7000 (April 1967).
6484 Competition model discontinued (April 1967).
6496 Engine no. CT 75738. First TR4A engine (May 1967).
6583 Engine no. CT 78483ME. New-type brake discs, hoses and pads fitted (September 1967).
6673 Fitted with Series V 4/4 body. For J H Sheally II (USA) (January 1968).
6761 Last Super Sports (May 1968).
6818 New-type handbrake cable introduced (August 1968).
6842 Last two-seater (October 1968).
6850 Last four-seater (November 1968).
6853 Last Drop Head Coupé and last Plus 4 (September 1969).

Note: actual production ended in January 1969 with chassis no. 6852. This car was produced to special order for export to Australia.

• CHASSIS NUMBER SEQUENCE
3000-6835

• PRODUCTION DATES
September 1954-January 1969

• PRODUCTION BY BODY

Two-seaters	2176
Four-seaters	1011

Drop Head Coupés two-seaters	435
Drop Head Coupés four-seaters	49
Plus 4 Plus Fixed Head Coupé	26
Competition Model	42
Super Sports two-seater	94
Super Sports Drop Head Coupé	4
Super Sports four-seater	5
Super Sports sold less engine	2 (engines supplied direct from Lawrencetune)
Chassis nos. not used	4
Experimental cars	4
Sold in chassis form	2
Total number of cars & chassis	**3850**
Total number of cars exported	**2806**
Total no.cars exported to USA	**2206***

** 78.6% of exports or 57.2% of total production*
Note: one chassis used twice.

Plus 4: other useful information

• DUNLOP WHEELS FOR PLUS 4S

First type Dunlop code no. LP 643 HAS. No ventilating slots. Next Dunlop code no. LP 643A. Disc wheels with ventilating slots.

Next Dunlop code no. LP 658. Disc wheels with ventilating slots and 15 drilled lightening holes.

• SMITHS SPEEDOMETERS

For Plus 4s with 3.77 to 1 rear axle ratio are reference no. 58956/67 and are geared to 1350rpm for 16 x 5.25 tyres. Speedometer gears 2.2 to 1.

For Plus 4s with standard axle ratio reference nos. X5 8956/45: Speedometer/clock rated in mph. X58956/44: Speedometer/clock rated in kph.

• GEARBOX BEARINGS

Primary shaft:
1 off FE single row 30mm x 72mm x 19mm SKF 6306 or Hoffmann 330KP (front).
1 off single row 40mm x 90mm x 23mm SKF 6308NR with circlip (rear).
Constant mesh pinion: 1 off Salters SC 104G.
Mainshaft:
1 off single row with circlip 1.37in x 3.5in x 0.87in SKF RMS11 NR.
6 off 0.31in steel balls.
58 off 3mm x 23.8mm needle rollers.
82 off 3.5mm x 29.8mm needle rollers.

• SALISBURY REAR AXLE BEARINGS

1 off 128 72/02830 pinion bearing front.
1 off 31593 4/31520 pinion bearing rear.
2 off 24780 4/24721 differential.
2 off 14132/14276 hubs.

• MOSS GEARBOX RATIOS

High ratio gearboxes have 28 teeth constant mesh pinions (fitted from chassis no. 3536) and low ratio gearboxes have 26 teeth constant mesh pinions. High ratio layshaft cluster reference no. 60559.

• MOSS GEARBOX RATINGS

All Moss gearboxes marked with three stars were the quietest when tested at the Moss factory and were reserved for fitting in Drop Head Coupé models. However, some were used in two- and four-seater models.

• TRIM AMOUNTS

The amount of leather or leathercloth (Dulon) required to trim a Plus 4 is:

Two-seater	40 sq ft
Four-seater	68sq ft
Drop Head Coupé	70sq ft

• TRS AND TR3 ENGINE CHANGES

The changes that follow were introduced at the Triumph factory. As far as dates are concerned, it would be approximately three months to a year before the engines fitted in Morgans were affected. However, changes are relevant to engine numbers after those listed.

TS8213	Distributor altered (August 1955).
TS8636	Last production TR2 (November 1955).
TS8637	First production TR3 (September 1955).
TS8998	Change to 4 bolt carburettors (October 1955).
TS9095	Replacement cam bearings (November 1955)
TS13502	High port cylinder head (September 1956)
TS18478	Change to racing clutch plate (May 1957).
TS20780	Changed rod bearings for increased running clearance and greater oil film strength (July 1957).
TS22013	Last production TR3 (August 1957).
TS22014	First production TR3A (August 1957).
TS26698	Pistons with strengthened crowns introduced (May 1958).
TS50001	Starter changed from pusher to puller (May 1959).
RS59289	Larger capacity generator (October 1959).
TS83572	Last production TR3A (October 1961).

TR engines: dimensions and tolerances

This data applies as far as is known to the Triumph TR2, 3, 3A, 4 and 4A engines as used in the Plus 4 Morgan. In addition, most of the information can be applied in general terms to the Vanguard engined Plus 4s as well.

Part & description	Dimensions (new)	Clearance (new)	Comments
Crankshaft			
Journal diameter	2.4795in to 2.4790in	0.0010in to 0.0025in	
Bearing internal diameter	2.4815in to 2.4805in		
Bearing Housing internal diameter	2.6255in to 2.6250in		
Undersize bearings are available in sizes -0.010", -0.020", -0.030" and -0.040"			
Crankshaft end float			
Intermediate journal length	1.7507in to 1.7498in	0.0048in to 0.0117in	Clearance of 0.004in-0.006in is specified and obtained by selective assembly of thrust washers
Intermediate bearing cap width (plus thickness of two thrust washers)	1.7450in to 1.7390in		
Main bearing cap width	1.5050in to 1.4950in		
Big end			
Crank pin diameter	2.0866in to 2.0860in	0.0016in to 0.0035in	
Bearing internal diameter	2.0895in to 2.0882in		
Internal diameter of bearing housing	2.2335in to 2.2327in		
Bearing width	0.9670in to 0.9650in		
Undersize bearings are available in sizes -.010", -.020", -.030" and -.040"			
Big end float			
Crankpin width	1.1915in to 1.1865in	0.007in to 0.014in	
Con rod width	1.1795in to 1.1775in		
Small end			
Bore for bush	1.0000in to 0.9950in	Press fit in con rod	
Bush external diameter	1.0005in to 0995in		
Internal diameter of bush	0.8752in to 0.8748in		
Gudgeon pin diameter	0.87510in to 0.87485in	0.002in at 68°F	
Piston rings			
Compression ring width	0.062in to 0.061in		
Groove width	0.0645in to .0635in	0.0015in to 0.0035in	

Part & description	Dimensions (new)	Clearance (new)	Comments
Scraper ring width	0.156in to 0.155in		
Groove width	0.158in to 0.157in	0.001in to 0.003in	
Ring gap in cylinder sleeves	0.003in to 0.010in		

Piston rings are available in sizes +0.010", +0.020", +0.030" and +0.040"

Pistons and cylinder sleeves	F Pistons	G Pistons	H Pistons
Bore diameter	3.2676in to 3.2673in	3.2680in to 3.2677in	3.2684in to 3.2681in
Top diameter of piston skirt	3.2626in to 3.2622in	3.2630in to 3.2626in	3.2634in to 3.2630in
Bottom diameter of piston skirt	3.2641in to 3.2627in	3.2645in to 3.2641in	3.2649in to 3.2645in

Top - skirt clearance	0.0054in to 0.0047in }	} applicable to 'F', 'G' & 'H' pistons	
Bottom - skirt clearance	0.0039in to 0.0032in }		
Height of cylinder sleeves above face of cylinder block	0.003in to 0.0055in		

Pistons are available in the following oversizes: +0.020", +0.030", +0.040"

Camshaft			
Front journal diameter	1.8720in to 1.8710in	0.0028in to 0.0047in	
Front journal bearing bore	1.8757in to 1.8748in		
External diameter of front bearing	2.2498in to 2.2493	Push fit in cylinder block	
Bore in block for front bearing	2.2507in to 2.2498in		
Diameter of 2nd, 3rd and rear camshaft journal	1.7157in to 1.7152in	0.0026in to 0.0046in	
Bore in cylinder block for 2nd 3rd and rear journals	1.7198in to 1.7183in		
End float	0.003in to 0.0075in		
Valves and valve guides			
Inlet stem diameter	0.3110in to 0.3100in	0.001in to 0.003in	
Inlet guide diameter	0.3130in to 0.3120in		
Exhaust stem diameter	0.3715in to 0.3705in	0.003in to 0.005in	
Exhaust guide diameter	0.3755in to 0.3745in		
Included angle of valve faces	90°		
Inlet valve head diameter	1.5620in to 1.5580in		
Width of inlet valve seating	0.0469" approx		

TR ENGINES TIGHTENING TORQUES

Component	Description	Specified torque range lbf/ft	Comments
ENGINE			
Cylinder head	1/2in UNF & UNC stud	100-105	Tighten nuts with engine cold
Connecting rod caps	7/16in UNF bolt	55-60	
Main bearing caps	1/2in x 13 NC setscrew	85-90	
Flywheel attachment to crankshaft	5/8in x 24 NF setscrew	42-46	
Timing chain wheel to camshaft	5/16in x 18 NC setscrew	24-26	
Manifold attachment	3/8in NC stud	22-24	
Oil pump attachments	5/16in x 24 UNF stud	12-14	
Rear oil seal attachment	1/4in x 20 UNC setscrew	8-10	
Clutch attachment	5/16in x 18 UNC setscrew	20	
Attachment of end plates	5/16in x 18 UNC bolt	14-16	Tapped into aluminium
Attachment of oil filters	5/16in x 18 x 24 UNC bolts Cap nut Bolt	18-20	
Timing cover	5/16in x 18 & 24 NC setscrew	18-20	
Sump attachment	5/16in x 18 NC setscrew	16-18	
Pulley to water pump spindle	5/16in x 24 UNF Simmonds Nyloc nut	16-18	
Dynamo bracket to block	5/16in x 18 UNC setscrew	16-18	
Dynamo to bracket & pedestal	5/16in x 24 UNF setscrew & bolt	16-18	
Rocker pedestal	3/8in NF & NC stud	24-26	
Oil gallery plugs	3/16in x 14 UNC 3/8in x 16 UNC	32-36 24-26	Tighten on to copper washer
Attachment of starter motor	3/8in x 24 NF bolt	26-26	
Water pump attachment	3/8in x 16 UNC bolt	26-28 26-26	
Petrol pump attachment	5/16in NF & NC stud	12-14	
Thermostat assembly to cylinder head	5/16in x 18 UNC bolt 5/16in x 18 UNC bolt	16-18 16-18	
Inlet to exhaust manifold	5/16in x 24 UNF stud	12-14	
Dynamo to pedestal front	5/16in x 24 UNF bolt	16-18	

Morgan

APPENDIX II

TECHNICAL DATA SHEETS, WIRING DIAGRAMS, LUBRICATION CHARTS & LUCAS ELECTRICAL COMPONENT DATA SHEETS

With the exception of those sheets applicable to the 4/4 Series I, which were compiled over twenty years by the author, the following technical data sheets are reproduced courtesy of Glass's Technical Services Limited.

Certain discrepancies occur, in some instances, between the information contained in the sheets and the model specifications. In the case of the latter, the author obtained the information from magazine road tests and articles, published at the time the models were current. In many instances, these tests were carried out using a prototype or very early example of the production vehicle. It therefore follows that the Morgan Motor Company and/or manufacturers of major components used in the cars might subsequently alter the specifications to produce a more efficient product. Details of these changes were passed to the data sheet compilers by the Morgan Motor Company but, of course, would not have appeared in print until the following year.

ENGINE

No.	Item	Value
1.	No. of cylinders	4
2.	Treasury rating	9.8hp
3.	C.C.	1,122.
4.	B.H.P. Q.R.P.M.	34/4,500
5.	Bore (mm)	63
6.	Stroke (mm)	90
7.	Compression Ratio.	6.85
8.	System of Cooling	Thermo-syphon
9.	1st O.S. Bore	–
10.	Max. O.S. for boring.	–
11.	Firing Order.	1, 3, 4, 2.
12.	Piston clearance top.	0.002–0.003in
13.	Piston clearance bottom.	Zero–0.0005in
14.	Ring Gap.	0.015in
15.	No. of Compression rings.	2
16.	Width of Compression rings.	4.5mm
17.	No. of oil rings.	1
18.	Width of oil rings.	4.5mm
19.	Oil pressure (lb sq in)	40
20.	Gudgeon pin type.	Fully-floating.
21.	Fit in piston.	Thumb push fit. Hot.
22.	Fit in Con. rod.	Push fit.
23.	Crankpin Diam. (Standard).	1⅝in
24.	Min. diam. for regrind Crankpin.	–
25.	Con Rod – Length between centres	–
26.	Type of bearing – Con. Rod.	White Metal
27.	Side clearance – Con. Rod.	0.002–0.0055in
28.	Diam. Clearance – Con. Rod.	0.0015–0.002in
29.	No. Crankshaft bearings.	3
30.	Type of main bearing.	Bronze backed white metal shells.
31.	Standard main journal diam.	2in
32.	Min. diam. for regrind. Main journals.	–
33.	Main bearing length	1⅝in
34.	End clearance. Main bearing.	0.003–0.007in
35.	Diam. Clearance. Main bearing	0.0015–0.002in
36.	End thrust taken on – Crankshaft.	Front main bearing thrust washer, behind timing gear.
37.	No. of Crankshaft bearings	3
38.	Type of bearing – Camshaft.	Front & rear bushed. Centre direct in crankcase.
39.	Bearing clearance – Camshaft.	–
40.	End thrust taken on – Camshaft.	Front bush bolted on crankcase.
41.	Camshaft drive (Type).	Duplex roller endless chain.
42.	Valve timing markings.	Mark on flywheel.
43.	Exhaust valve diam. – Head & Stem.	1.224in 0.309in
44.	Inlet valve diam. – Head & Stem.	1.433in 0.311in
45.	Valve seat angle.	45
46.	Tappet type.	Mushroom. In two detachable blocks of four.
47.	Inlet Valve Clearance for timing.	0.006 (cold)
48.	Inlet valve opens – Degrees.	10 before T.D.C.
49.	Inlet valve opens. – No. teeth on flywheel.	–
50.	Inlet valve opens – piston transverse.	50 after B.D.C.
51.	Inlet valve working clearance.	0.006 (cold)
52.	Exhaust valve working clearance.	0.008 (cold) 0.010 Later engines
53.	Are valve guides removable?	Yes
54.	Carburettor make & type.	Solex 30 HBFG
55.	Choke setting.	30–32
56.	Main jet setting.	115
57.	Compensating jet.	210
58.	Pilot jet.	055
59.	Starter jet. Air/petrol.	G.A. 3.5/G.S. 110

CLUTCH & GEARBOX

No.	Item	Value
60.	Clutch type	Borg & Beck
61.	Type of facing.	Single dry plate
62.	Gear Box type.	Meadows early cars Moss later & all post war cars.

No.	Item	MEADOWS	MOSS
63.	Gear Ratio. Top.	5	5
64.	Gear Ratio. Third.	7	6.7
65.	Gear Ratio. Second.	12.1	11.95
66.	Gear Ratio. First.	17.1	19.3
67.	Gear Ratio. Reverse.	22.6	22.35

FRONT AXLE & STEERING, ETC.

No.	Item	Value	
68.	Camber.	3	
69.	Caster angle.	3	
70.	Toe-in.	3	
71.	King pin inclination.	⅛in max.	
72.	Track	3ft 9in	
73.	Turning Circle.	33ft	
74.	Wheelbase.	7ft 8in	
75.	Tyre size & pressures.	50 x 17	5.00 x 16
	Two-seater	2lb/sq in	18lb/sq in
	Four seater & Drop Head Coupes	21lb/sq in	20lb/sq in

REAR AXLE

No.	Item	Value
76.	Type of axle.	Semi-floating
77.	Type of drive	Spiral bevel
78.	Ratio/Crown wheel/Bevel teeth.	5.0/40/8
79.	Adjustment.	Crown wheel & pinion shims.
80.	End play-ins.	Nil
81.	Lash- ins	0.003

BRAKES

No.	Item	Value
82.	Type	Girling. 8 in drum
83.	Type of linings.	Girling
84.	Linings size – front	1½in x 5/16in early cars 1¾in x 5/16in later cars
85	Linings size – rear	1½in x 5/16in early cars 1¾in x 5/16in later cars

SPRINGS

No.	Item	Value
86.	Type.	Front, independent coil; Rear, Semi-elliptic.
87.	Length front	5 9/16 free
88.	Width front	–
89.	No. leaves – front	–
90.	Thickness of leaves. Front.	–
91.	Camber front.	3
92.	Length – Rear.	40in
93.	Width – Rear.	1¾in
94.	No. leaves – Rear.	Two-seater = 5 Four-seater & D.H.C.=6
95.	Thickness of leaves – Rear	¼in
96.	Camber. – Rear.	Nil

ELECTRICAL

No.	Item	Value
97.	Distributor rotation.	Anti-clockwise
98.	Manual advance.	No
99.	Automatic advance.	Yes
100.	Contact breaker gap.	0.012–0.015in
101.	Plug make & type.	KLG F50X
102.	Plug gap.	0.025in
103.	Firing order.	1, 3, 4, 2
104.	Ignition timing – degrees.	T.D.C.
105.	Ignition timing – No. flywheel teeth.	T.D.C.
106.	Charging system.	Voltage control.
107.	Battery – capacity/make.	57amp hrs/Lucas.
108.	Battery – voltage.	12 (On early cars two 6 vols batteries wired in series were used).

CAPACITIES

No.	Item	Value
110.	Sump – pints	8
111.	Gear box – pints	1¾
112.	Rear Axle – pints	1¾
113.	Cooling system – pints.	16
114.	Radiator hose top. Length/Intern. Diam.	5½in 48mm
115.	Radiator hose bottom. Length/Intern. Diam.	6in 38mm

4-4 1938-1939

ENGINE.

1. No. of Cylinders ...	... 4
2. Treasury Rating ...	... 9.99 h.p. o.h.v.
3. C.C.	... 1,267
4. B.H.P.	... 40 at 4,300 r.p.m.
5. Bore	... 63.5 mm.
6. Stroke	... 100 mm.
7. Compression Ratio	... 7 to 1
8. System of Cooling	... Thermo-Syphon, Fan
9. 1st O.S. Bore ...	... .020in.
10. Max. O.S. for boring	... Sleeve after first rebore
11. Firing order ...	... 1—3—4—2
12. Piston clearance top	... .0022 to .0035in.
13. Piston clearance bottom	... .0007 to .002in.
14. Ring Gap	... .003 to .007in.
15. No. of compression rings	2
16. Width of compression rings	... .0937in.
17. No. of oil rings ...	... 1
18. Width of oil rings	... .177
19. Oil pressure, lb. sq. in.	... 40—60
20. Gudgeon pin type	... Fully-floating
21. Fit in piston ...	... Push in
22. Fit in con. rod ...	... Push in
23. Crankpin diam. (standard)	44.451 mm. (1¾in.)
24. Min. diam. for regrind—Crankpin	.040in. undersize
25. Con. Rod—Length between centres	... 7¼in. plus or minus .002in.
26. Type of bearing—Con. Rod	... Shell, White Metal lined
27. Side clearance—Con. Rod	.0075 to .010in.
28. Diam. clearance—Con. Rod	... .001 to .002in.
29. No. of crankshaft bearings	3
30. Type of bearing—Main	... Shell, White Metal lined
31. Standard main journal diam.	... 50.8001 mm. (2in.)
32. Min. diam. for regrind—Main journals	49.784 mm. (.040in. undersize)
33. Main Bearings—Length	... 1.380 to 1.370in.
34. End clearance—Main Bearing	... .005 to .010in.
35. Diam. clearance—Main Bearing	... .0015 to .003in.
36. End thrust taken on—Crankshaft	... Washers each side of Rear Main Bearing
37. No. of Camshaft bearings	4
38. Type of bearing—Camshaft	... Direct in Cylinder Block
39. Bearing clearance—Camshaft	... .0027 to .0047in.
40. End thrust taken on—Camshaft	... Collar at front end
41. Camshaft drive (type)	... Chain
42. Valve timing markings	... Marked on flywheel
43. Exhaust Valve diam.—Head & Stem	... 1.100 × .309in.
44. Inlet Valve diam.—Head and Stem	1.220 × .309in.
45. Valve seat angle ...	... 45°
46. Tappet type ...	... Mushroom
47. Inlet Valve Clearance for timing	.020in.
48. Inlet Valve opens—degrees	... 10° before T.D.C.
49. Inlet Valve opens—No. of flywheel teeth	2½
50. Inlet Valve opens—piston traverse	3/64in.
51. Inlet Valve working clearance	... .020in. cold
52. Exhaust Valve working clearance	... .020in. cold
53. Are guides removable?	... Yes
54. Carburettor make & type	Solex Down Draught
55. Choke setting ...	... 100
56. Main setting ...	... 125
57. Compensator setting	... 170
58. Idling setting ...	... 45
59. Progression setting	... —

CLUTCH AND GEARBOX.

60. Clutch type ...	... Borg & Beck
61. Type of facing ...	... Borg & Beck
62. Gear ratio, top ...	... 4.72
63. Gear ratio, third ...	... 6.70
64. Gear ratio, second	... 11.42
65. Gear ratio, first ...	... 16.14
66. Gear ratio, reverse	... 21.33

FRONT AXLE AND STEERING, ETC.

67. Camber	... 2°
68. Caster angle ...	... 4¾in.
69. Toe-in	... ⅛in.
70. King pin inclination	... —
71. Track	... 3ft. 9in.
72. Turning circle ...	... 33ft.
73. Wheelbase	... 7ft. 8in
74. Tyre size and pressures	2 & 4 Str., 4.50—17; Coupé, 5.00—16; 18—20 Front & Rear

REAR AXLE.

75. Type of axle ...	... ¾ floating
76. Type of drive ...	... Spiral Bevel
77. Ratio or No. of teeth	... 4.72
78. Adjustment ...	... Crown Wheel & Pinion—Adjusting Rings
79. End play—ins. ...	... Nil
80. Lash—ins. ...	... .007—.010in.

BRAKES.

81. Type	... Girling
82. Type of linings ...	... Girling
83. Lining size—Front	... 1½ × 3/16in.
84. Lining size—Rear	... 1½ × 3/16in.

SPRINGS.

85. Type	... Front, Independent—Coil; Rear, Semi-elliptic
86. Length—Front ...	... Main, 6½in.; Rebound, 8in.
87. Width—Front ...	... —
88. No. of leaves—Front	... —
89. Thickness of leaves—Front	... —
90. Camber—Front ...	... 2°
91. Length—Rear ...	... 3ft. 6¾in.
92. Width—Rear ...	... 1⅜in.
93. No. of leaves—Rear	... 6
94. Thickness of leaves—Rear	¼in.
95. Camber—Rear ...	... Nil

ELECTRICAL.

96. Distributor rotation	... Anti-clockwise
97. Manual advance ...	... No
98. Automatic advance	... Yes
99. Breaker gap ...	... .012in.
100. Plug make and type	... Champion N8
101. Plug gap	... .025in.
102. Firing order ...	... 1—3—4—2
103. Ignition timing—degrees	T.D.C.
104. Ignition timing — No. of flywheel teeth ...	T.D.C. ...
105. Charging system ...	... V.C.
106. Battery—Capacity and make, type	... 57 amp. hrs. Lucas 12v
107. Battery earth—pos. or neg.	... Positive

CAPACITIES.

108. Sump—pints ...	... 11
109. Gearbox—pints ...	... 2
110. Rear Axle—pints ...	... 2
111. Cooling system—pints	... 16
112. Radiator hose (top) length. Int. diam.	... 8 × 1¾in.
113. Radiator hose (bottom) length. Int. diam.	... 16½ × 1¾in.
114. Petrol galls. ...	... 9

GENERAL DIMENSIONS.

	2-Str.	4-Str.	Coupé
115. Overall Length ...	... 11ft. 4in.	11ft. 7in.	11ft. 7¼in.
116. Overall Height ...	... 4ft. 4¼in. with hood erected	4ft. 6½in.	4ft. 1¾in.
117. Overall Width ...	... 4ft. 6in.	4ft. 6in.	4ft. 6in.
118. Ground clearance ...	... 6in.	6in.	6in.
119. Total weight ...	... 14½ cwt.	14½ cwt.	15 cwt.

4/4 1949-1950

ENGINE.

1.	No. of Cylinders	4 (Modified Standard Vanguard)
2.	Treasury Rating	18 h.p.
3.	C.C.	2088 c.c.
4.	B.H.P.	68 at 4,300 r.p.m.
5.	Bore	85 mm. (3·347in.)
6.	Stroke	92 mm. (3·622in.)
7.	Compression Ratio	6·7 to 1 & 1954, 7 to 1
8.	System of Cooling	Pump and Fan
9.	1st O.S. Bore	Not recommended—fit new sleeves and and standard pistons when worn
10.	Max. O.S. for boring	
11.	Firing order	1—3—4—2
12.	Piston clearance top	·003 to ·00375in.
13.	Piston clearance bottom	·0015 to ·002in.
14.	Ring Gap	·003 to ·007in.
15.	No. of compression rings	2
16.	Width of compression rings	·0777 to ·0787in.
17.	No. of oil rings	2
18.	Width of oil rings	·155 to ·156in.
19.	Oil pressure, lb. sq. in.	40—70 hot
20.	Gudgeon pin type	Fully floating, circlip retained
21.	Fit in piston	Floating
22.	Fit in con. rod	·0002in. clearance at 68° F.
23.	Crankpin dia. (standard)	2·0861 to 2·0866in.
24.	Min. diam. for regrind—Crankpin	·040in. undersize
25.	Con. Rod—Length between centres	$6\frac{1}{4}$in. ± ·002in.
26.	Type of bearing—Con. Rod	Precision, steel-backed, w.m. lined
27.	Side clearance—Con.Rod	·0075 to ·0105in.
28.	Diam. clearance—Con. Rod	·0006 to ·0025in.
29.	No. of crankshaft bearings	3
30.	Type of bearings—Main	Precision, steel-backed, w.m. lined
31.	Standard main journal diam.	2·479 to 2·4795in.
32.	Min. diam. for regrind—Main journals	·040in. undersize
33.	Main Bearings—Length	$1\frac{3}{4}$in.—·0007 or + ·0002in.
34.	End clearance—Main Bearing	·004 to ·006in.
35.	Diam. clearance—Main Bearing	·001 to ·002in.
36.	End thrust taken on—Crankshaft	W.M. covered steel washers on each side of Centre Main Bearing
37.	No. of Camshaft bearings	4—End float ·004 to ·012in.
38.	Type of bearing—Camshaft	Front—phos. bronze ; Inter. and Rear—direct in block
39.	Bearing clearance—Camshaft	·0025in.
40.	End thrust taken on—camshaft	Flange on Front bearing—shoulder on camshaft
41.	Camshaft drive (type)	Duplex silent roller chain
42.	Valve timing markings	Punched and scribed on timing gears
43.	Exhaust Valve diam.—Head and Stem	$1\frac{9}{32}$in. ± ·002in. and $\frac{5}{16}$in. stem
44.	Inlet Valve diam.—Head and Stem	$1\frac{1}{2}$in. ± ·002in. and $\frac{5}{16}$in. stem
45.	Valve seat angle	45°. Width approx. $\frac{1}{16}$in.
46.	Tappet type	Hollow cylindrical, flat base, chilled face
47.	Inlet Valve Clearance for timing	·014in.
48.	Inlet Valve opens—degrees	10° before T.D.C. Closes 50° after B.D.C.
49.	In. Valve opens—Flywheel	1in. on flywheel rim (4 teeth)
50.	Inlet Valve opens—piston traverse	·035in.
51.	Inlet Valve working clearance	·010in. cold
52.	Exhaust Valve working clearance	·012in. cold
53.	Are guides removable ?	Yes
54.	Carburettor make and Type	Solex D/Draught

		Standard	*Economy*	*Power*
55.	Choke setting	25	23	25
56.	Main setting	135	120	140
57.	Air correction	190	180	190
58.	Idling setting	55	55	55
59.	Starter setting	130	130	130

CLUTCH AND GEARBOX.

60.	Clutch type	Borg & Beck, single dry plate	
61.	Type of facing	B. & B. composition	
62.	Gear ratio, top	4·1	1954, 3·73
63.	Gear ratio, third	5·4	5·1
64.	Gear ratio, second	8·0	7·3
65.	Gear ratio, first	13·5	12·5
66.	Gear ratio, reverse	13·5	12·5

FRONT AXLE AND STEERING, ETC.

67.	Camber	2°
68.	Caster angle	4°
69.	Toe-in	$\frac{1}{8}$in.
70.	King pin inclination	2°
71.	Track	3ft. 11in.
72.	Turning circle	33ft. 0in.
73.	Wheelbase	8ft. 0in.
74.	Tyre size and pressures	5·25 × 16in. 16/18 lb. F. and R.

REAR AXLE.

75.	Type of axle	Hypoid	Make—Salisbury—Follow maker's Service Instructions
76.	Type of drive		
77.	Ratio or No. of teeth	4·1 (1954, 3·73	
78.	Adjustment	Shim	
79.	End play—ins.	—	
80.	Lash—ins.	·003 to ·006in.	

BRAKES.

81.	Type	Girling, hydraulic, two-leading shoe
82.	Type of linings	—
83.	Lining size—Front	9in. dia. × $1\frac{1}{4}$in. wide
84.	Lining size—Rear	9in. dia. × $1\frac{1}{4}$in. wide

SPRINGS.

85.	Type	Front—independent coil ; semi-elliptic
86.	Length—Front	$12\frac{1}{4}$in.
87.	Width—Front	—
88.	No. of leaves—Front	—
89.	Thickness of leaves—Front	—
90.	Camber—Front	—
91.	Length—Rear	$42\frac{3}{4}$in.
92.	Width—Rear	$1\frac{3}{8}$in.
93.	No. of leaves—Rear	6
94.	Thickness of leaves—Rear	$\frac{1}{4}$in.
95.	Camber—Rear	$3\frac{1}{2}$in.

ELECTRICAL.

96.	Distributor rotation	Anti-clockwise
97.	Manual advance	No
98.	Automatic advance	Yes
99.	Breaker gap	·012in.
100.	Plug make and type	Champion L10. 14 mm. $\frac{1}{2}$in. reach
101.	Plug gap	·032in.
102.	Firing order	1—3—4—2
103.	Ignition timing—degrees	4° before T.D.C., fully retarded (measured at crankshaft)
104.	Ignition timing—flywheel	0·4in. on flywheel rim, before T.D.C.
105.	Charging sytem	A.V.C.
106.	Battery—Capacity and make, type	12v. 54 a.h. Lucas
107.	Battery earth—pos. or neg.	Positive

CAPACITIES (Imperial)

108.	Sump-pints	11	
109.	Gearbox—pints	$2\frac{1}{2}$	
110.	Rear Axle—pints	$2\frac{1}{2}$	
111.	Cooling system—pints	16	
112.	Radiator hose (top) length	$6\frac{1}{2}$in. × $1\frac{3}{16}$in.	Moulded hose
113.	Radiator hose (bottom) length. Int. diam.	13in. × $1\frac{1}{2}$in.	Moulded hose
114.	Petrol, galls.	11	

GENERAL DIMENSIONS.

		2 Str.	Coupé	4 Str.
115.	Overall Length	11ft. 8in.	11ft. 10in.	11ft. 8in.
116.	Overall Height (Hood up)	4ft. $4\frac{1}{2}$in.	4ft. 2in.	4ft. $6\frac{1}{2}$in.
117.	Overall Width	4ft. 8in.	4ft. 8in.	4ft. 8in.
118.	Ground clearance	7in.		
119.	Total Weight	16 cwt. 3 qr.	17 cwt. 2 qr.	17 cwt. 0 qr.
		with Tools and full Tank		

Plus 4 1953-1954

ENGINE		
1.	No. of Cylinders	4
2.	Treasury Rating	17·08 h.p.
3.	C.C.	1,991 c.c. (121·5 cu. in.)
4.	B.H.P.	90 at 4,800 r.p.m.
5.	Bore	83mm. (3·268in.)
6.	Stroke	92mm. (3·622in.)
7.	Compression Ratio	8·5 to 1
8.	System of Cooling	Pump and fan, thermostat controlled, pressurised
9.	1st O.S. Bore	+·020in.
10.	Max. O.S. for boring	+·040in.
11.	Firing order	1—3—4—2
12.	Piston clearance top	·0230 to ·0207in.
13.	Piston clearance bottom	·0039 to ·0032in.
14.	Ring gap	·008in.
15.	No. of compression rings	2
16.	Width of compression rings	·0625in.
17.	No. of oil rings	1
18.	Width of oil rings	·1562in.
19.	Oil pressure, lb. sq. in.	40 to 60
20.	Gudgeon pin type	Fully floating, circlip located
21.	Fit in piston	Floating
22.	Fit in con. rod	Floating
23.	Crankpin diam. (standard)	2·0861 to 2·0866in.
24.	Min. diam. for regrind—Crankpin	−·040in.
25.	Con. Rod—Length between centres	6·25in.
26.	Type of bearing—Con. Rod	Shell, lead-indium, steel backed
27.	Side clearance—Con. Rod	·014 to ·007in.
28.	Diam. clearance—Con. Rod	·0011 to ·0035in.
29.	No. of crankshaft bearings	3
30.	Type of bearings—Main	Shell, white metal lined, steel backed
31.	Standard main journal diam.	2·479 to 2·4795in.
32.	Min. diam. for regrind—Main journals	−·040in.
33.	Main Bearings—Length	1·50in.
34.	End clearance—Main Bearing	·004 to ·006in.
35.	Diam. clearance—Main Bearing	·001 to ·002in.
36.	End thrust taken on—Crankshaft	White metal lined, steel washers each side centre main
37.	No. of Camshaft bearings	4
38.	Type of bearing—Camshaft	Cast iron sleeve front; three others direct in block
39.	Bearing clearance—Camshaft	·00275 to ·00475in.
40.	End thrust taken on—Camshaft	Between flange on front bearing and shoulder on camshaft
41.	Camshaft drive (type)	Double roller silent chain
42.	Valve timing markings	Punch and scribed markings on timing wheels
43.	Exhaust Valve diam.—Head and Stem	1·303/1·299in. & ·3715/·3705in.
44.	Inlet Valve diam.—Head and Stem	1·562/1·558in. & ·311/·310in.
45.	Valve seat angle	45°
46.	Tappet type	Hollow, flat based, chilled face
47.	Inlet Valve clearance for timing	·015in.
48.	Inlet Valve opens—degrees	15° B.T.D.C.
49.	Inlet Valve opens—No. of flywheel teeth	3·9
50.	Inlet Valve opens—piston traverse	·081in.
51.	Inlet Valve working clearance	·012in. cold
52.	Exhaust Valve working clearance	·012in. cold
53.	Are guides removable?	Yes
54.	Carburettor make and type	Twin S.U. Type H4
55.	Choke setting	Single Jet—Standard Needle FV, Competition Needle GC
56.	Main setting	
57.	Compensator setting	
58.	Idling setting	
59.	Progression setting	

CLUTCH AND GEARBOX			
60.	Clutch type	9in. Single dry plate, Borg & Beck	
61.	Type of facing	Borglite	
62.	Gear ratio, top	1 to 1	Overall 3·7
63.	Gear ratio, third	1·325	„ 4·9
64.	Gear ratio, second	2·00	„ 7·4
65.	Gear ratio, first	3·38	„ 12·5
66.	Gear ratio, reverse	4·28	„ 15·8

FRONT AXLE AND STEERING, Etc.		
67.	Camber	2° (static laden)
68.	Caster angle	0°
69.	Toe-in	·125in.
70.	King pin inclination	7°
71.	Track	Front 3ft. 9in.; Rear 3ft. 9½in.
72.	Turning circle	32ft.
73.	Wheelbase	7ft. 4in.
74.	Tyre size and pressures	(550×15in.); Front 22 lb., Rear, 24 lb.

REAR AXLE		
75.	Type of axle	Semi-floating
76.	Type of drive	Hypoid
77.	Ratio or No. of teeth	3·7 to 1
78.	Adjustment	Shims for Pinion and Diff.
79.	End play—ins.	None
80.	Lash—ins.	·004 to ·006in.

BRAKES			
81.	Type	Lockheed hydraulic	
82.	Type of linings	Ferodo DM7	
83.	Lining size—Front	10×2¼in.	148 sq. in. area
84.	Lining size—Rear	9×1¾in.	

SPRINGS		
85.	Type	Front, coil; Rear, semi-elliptic (laminated)
86.	Length—Front	Coil springs
87.	Width—Front	
88.	No. of leaves—Front	
89.	Thickness of leaves—Front	
90.	Camber—Front	
91.	Length—Rear	41·0±·13in. (static laden)
92.	Width—Rear	2in.
93.	No. of leaves—Rear	6
94.	Thickness of leaves—Rear	Main and No. 2, ·2187in.; Nos. 3 to 5, ·1875in.; No. 6, ·1719in.
95.	Camber—Rear	⅞ to 1⅛in.

ELECTRICAL		
96.	Distributor rotation	Anti-clockwise
97.	Manual advance	Micro adjuster
98.	Automatic advance	Centrifugal and suction
99.	Breaker gap	·015in.
100.	Plug make and type	L10S Champion (Competition, L11S)
101.	Plug gap	·032in.
102.	Firing order	1—3—4—2
103.	Ignition timing—degrees	4° B.T.D.C.
104.	Ignition timing—No. of flywheel teeth	1
105.	Charging system	C.V.C.
106.	Battery—Capacity and make, type	Lucas 12v. 51 a.h.
107.	Battery earth-pos. or neg.	Positive

CAPACITIES (Imperial)		
108.	Sump—pints	11
109.	Gearbox—pints	1½ (3½ with O/drive)
110.	Rear Axle—pints	1½
111.	Cooling system—pints	13 (14 with heater)
112.	Radiator hose (top) length. Int. diam.	4×1·38in.
113.	Radiator hose (bottom) length. Int. diam.	1·38in. diam. elbow
114.	Petrol—galls.	12½in.

GENERAL DIMENSIONS		
115.	Overall Length	12ft. 7in.
116.	Overall Height	4ft. 2in., hood erect
117.	Overall Width	4ft. 7½in.
118.	Ground clearance	6in.
119.	Total weight	17 cwt. 2 qr. 21 lb.

Triumph TR2 1953-1954

ENGINE (Ford Anglia)

1.	No. of Cylinders	4, in line, side valve
2.	Treasury Rating	10 h.p.
3.	C.C.	1172 c.c. (71·55 cu. in.)
4.	B.H.P.	36 at 4,400 r.p.m.
5.	Bore	2½in. (63·5mm.)
6.	Stroke	3·64in. (92·5mm.)
7.	Compression Ratio	7 to 1
8.	System of Cooling	Water Pump and Thermostat
9.	1st O.S. Bore	·010in. oversize
10.	Max. O.S. for boring	·030in. oversize
11.	Firing order	1—2—4—3
12.	Piston clearance top	6 to 12 lb. pull using ·0015in. feeler
13.	Piston clearance bottom	—
14.	Ring Gap	Top, ·007 to ·012in.; Centre, ·008 to ·014in.; Oil, ·007 to ·012in.
15.	No. of compression rings	2
16.	Width of comp. rings	Top, ·076 to ·0765in.; Centre, ·093 to ·0935in.
17.	No. of oil rings	1
18.	Width of oil rings	·1545 to ·1550in.
19.	Oil pressure, lb. sq. in.	30
20.	Gudgeon pin type	Fully floating, circlip retention
21.	Fit in piston	Up to ·0002in. clearance
22.	Fit in con. rod	Up to ·0005in. clearance
23.	Crankpin diam. (standard)	1·698 to 1·699in.
24.	Min. diam. for regrind—Crankpin	·030in. undersize
25.	Con. Rod—Length between centres	6·123 to 6·127in.
26.	Type of bearing—Con. Rod	Integral babbit
27.	Side clearance—Con. Rod	·004 to ·010in.
28.	Diam. clearance—Con. Rod	·0005 to ·002in.
29.	No. of crankshaft beargs.	3
30.	Type of bearings—Main	Detachable shells
31.	Standard main journal diam.	2·0010 to 2·0015in.
32.	Min. diam. for regrind—Main journals	·030in. undersize
33.	Main Bearings—Length	Front, 1·615 to 1·635in.; Centre, 1·310 to 1·314in.; Rear 2·003 to 2·006in.
34.	End clearance—Main Bearing	·002 to ·011in.
35.	Diam. clearance—Main Bearing	·000 to ·0015in.
36.	End thrust taken on—Crankshaft	Split thrust washers each side of rear main bearing
37.	No. of Camshaft bearings	3
38.	Type of bearing—Camshaft	Steel-backed, babbit liners
39.	Bearing clearance—Camshaft	·001 to ·002in.
40.	End thrust taken on—Camshaft	Thrust washer and spring plunger at front end
41.	Camshaft drive (type)	Duplex roller chain
42.	Valve timing markings	Lines on timing gears
43.	Exhaust Valve diam.—Head and Stem	1·05 to 1·06in. and ·3086 to ·3096in.
44.	Inlet Valve diam.—Head and Stem	1·15 to 1·16in. and ·3095 to ·3105in.
45.	Valve seat angle	45¼°
46.	Tappet type	Mushroom
47.	Inlet Valve clearance for timing	·015in.
48.	Inlet Valve opens—degs.	3° 30′ B.T.D.C.
49.	Inlet Valve opens—No. of flywheel teeth	—
50.	Inlet Valve opens—piston traverse	—
51.	Inlet Valve working clearance	·0115 to ·0135in. (cold)
52.	Exhaust Valve working clearance	·0115 to ·0135in. (cold)
53.	Are guides removable?	Yes
54.	Carburettor make & type	Downdraught
55.	Choke tube	21mm.
56.	Main jet	110
57.	Main air correction	160
58.	Idling jet	50; air bleed, 1·2mm.
59.	Starter jet	130; starter air jets, 5mm.

CLUTCH AND GEARBOX

60.	Clutch type	Single dry plate
61.	Type of facing	Asbestos composition
62.	Gear ratio, top	4·4
63.	Gear ratio, third	Not fitted
64.	Gear ratio, second	8·25
65.	Gear ratio, first	15·07
66.	Gear ratio, reverse	19·71

FRONT AXLE AND STEERING, Etc.

67.	Camber	2°
68.	Caster angle	4°
69.	Toe-in	⅛in.
70.	King pin inclination	2°
71.	Track	3ft. 11in.
72.	Turning circle	33ft. 0in.
73.	Wheelbase	8ft.
74.	Tyre size and pressures	5·00 × 16; 16 to 18 lb., front and rear

REAR AXLE

75.	Type of axle	Hypoid. Make: Salisbury HA; follow maker's service instructions
76.	Type of drive	Bevel
77.	Ratio or No. of teeth	4·4 to 1
78.	Adjustment	Shim
79.	End play—ins.	·006 to ·008in.
80.	Lash—ins.	Min., ·004in. (etched on drive gear)

BRAKES

81.	Type	Girling hydraulic
82.	Type of linings	Moulded
83.	Lining size—Front	9 × 1¾in.
84.	Lining size—Rear	9 × 1¾in.

SPRINGS

85.	Type	Front, independent coil; Rear, semi-elliptic
86.	Length—Front (coil)	12¼in.
87.	Width—Front	—
88.	No. of leaves—Front	—
89.	Thickness of leaves—Front	—
90.	Camber—Front	—
91.	Length—Rear	42¾in.
92.	Width—Rear	1¾in.
93.	No. of leaves—Rear	6
94.	Thickness of leaves—Rear	¼in.
95.	Camber—Rear	3¼in.

ELECTRICAL

96.	Distributor rotation	Anti-clockwise
97.	Manual advance	None
98.	Automatic advance	Centrifugal, 14°
99.	Breaker gap	·014 to ·016in.
100.	Plug make and type	Champion L10, 14mm.
101.	Plug gap	·025in.
102.	Firing order	1—2—4—3
103.	Ignition timing—degrees	5° B.T.D.C.
104.	Ignition timing—No. of flywheel teeth	—
105.	Charging system	Voltage control
106.	Battery—Capacity and make, type	12-volt, 40 amp. hr.
107.	Battery earth-pos. or neg.	Positive

CAPACITIES (Imperial)

108.	Sump—pints	4½ plus ¾ for filter
109.	Gearbox—pints	1¾
110.	Rear Axle—pints	2½
111.	Cooling system—pints	12
112.	Radiator hose (top) length. Int. diam.	11·1, 1·25 and 1·84in. (moulded)
113.	Radiator hose (bottom) length. Int. diam.	5·5 and 1·25in. (moulded)
114.	Petrol—galls.	8

GENERAL DIMENSIONS

115.	Overall Length	12ft.
116.	Overall Height	4ft. 4in.
117.	Overall Width	4ft. 8in.
118.	Ground clearance	7in. (approx.)
119.	Total weight	12¾ cwt.

4/4 Series II 1955-1956

ENGINE

1. No. of Cylinders	4
2. Treasury Rating	17·14 h.p.
3. C.C.	1,991 c.c.
4. B.H.P.	100 at 5,000 r.p.m.
5. Bore	83mm. (3·268in.)
6. Stroke	92mm. (3·622in.)
7. Compression Ratio	8·5 :1
8. System of Cooling	Pump and fan, thermostat controlled, pressurised
9. 1st O.S. Bore	+·020in.
10. Max. O.S. for boring	+·040in.
11. Firing order	1—3—4—2
12. Piston clearance (top)	0·023in. to 0·0207in.
13. Piston clearnce (bottom)	0·0039in. to 0·0032in.
14. Ring Gap	0·008in.
15. No. of compression rings	2
16. Width of compression rings	·0625in.
17. No. of oil rings	1
18. Width of oil rings	·1562in.
19. Oil pressure, lb. sq. in.	70 lb. sq. in. at 70° C. and 2,000 r.p.m.
20. Gudgeon pin type	Fully floating (located by circlips)
21. Fit in piston	Floating
22. Fit in con. rod	Floating
23. Crankpin diam. (std.)	2·0860in. to 2·0866in.
24. Min. diam. for regrind—Crankpin	—·040in.
25. Con. Rod—Length between centres	6·252in. to 6·248in.
26. Type of bearing—Con. Rod	Shell, lead indium, steel backed
27. Side clearance—Con. Rod	0·007in. to 0·014in.
28. Diam. clearance—Con. Rod	0·0016in. to 0·0035in.
29. No. of crankshaft bearings	3
30. Type of bearings—Main	Steel shell, white metal lined
31. Standard main journal diam.	2·4790in. to 2·4795in.
32. Min. diam. for regrind—Main journals	—·040in.
33. Main Bearings—Length	1·5050in. to 1·4950in.
34. End clearance—Main Bearing	·004in. to ·006in.
35. Diam. clearance—Main Bearing	0·001in. to 0·0025in.
36. End thrust taken on—Crankshaft	Between flanks of centre main bearing by white metal covered steel washers
37. No. of Camshaft bearings	4
38. Type of bearing—Camshaft	Front: cast iron; Remainder: steel shell, white metal lined (replaceable)
39. Bearing clearance—Camshaft	0·0029in. to 0·0044in.
40. End thrust taken on—Camshaft	Between the flanged front bearing and a shoulder on the shaft
41. Camshaft drive (type)	Double roller silent chain
42. Valve timing markings	Centre punched and scribed lines on timing sprockets
43. Exhaust Valve diam.—Head and Stem	Head: 1·303in. to 1·299in.; Stem: 0·3715in. to 0·3705in.
44. Inlet Valve diam.—Head and Stem	Head: 1·562in. to 1·558in.; Stem: 0·310in. to 0·311in.
45. Valve seat angle	45°
46. Tappet type	Hollow, cylindrical, flat based, chilled face
47. Inlet Valve clearance for timing	·015in.
48. Inlet Valve opens—degrees	15° B.T.D.C.
49. Inlet Valve opens—No. of flywheel teeth	3·8
50. Inlet Valve opens—piston traverse	·081in.
51. Inlet Valve working clearance	·012in. cold
52. Exhaust valve working clearance	·012in. cold
53. Are guides removable?	Yes
54. Carburettor make and type	Twin S.U. Type H6
55. Choke setting 56. Main setting 57. Compensator setting 58. Idling setting 59. Progression setting	Single jet: TD needle fitted from Eng. No. TS 9091E to TS 9350E, TE needle from TS 9350E to TS 10073E; SM needle from TS 10073E and future production

CLUTCH AND GEARBOX

60. Clutch type	Borg and Beck single dry plate, Model 9A.6G
61. Type of facing	Borg and Beck composition
62. Gear ratio, top	1 :1 (overall 3·7)
63. Gear ratio, third	1·325 :1 (overall 4·9)
64. Gear ratio, second	2·00 :1 (overall 7·4)
65. Gear ratio, first	3·38 :1 (overall 12·5)
66. Gear ratio, reverse	4·28 :1 (overall 15·8)

FRONT AXLE AND STEERING, Etc.

67. Camber	2° pos. (static laden)
68. Caster angle	Nil
69. Toe-in	$\frac{1}{8}$in.
70. King pin inclination	7°
71. Track	Front: 3ft. 9in.; Rear: 3ft. 9$\frac{1}{2}$in.
72. Turning circle	34ft. 0in.
73. Wheelbase	7ft. 4in.
74. Tyre size and pressures	550 × 15in.: Front, 22 lb. sq. in.; Rear: 24 lb. sq. in.

REAR AXLE

75. Type of axle	Semi-floating
76. Type of drive	Hypoid
77. Ratio or No. of teeth	3·7 :1
78. Adjustment	Shims for pinion and differential
79. End play—ins.	None
80. Lash—ins.	0·004in. to 0·006in.

BRAKES

81. Type	Lockheed hydraulic
82. Type of linings	Mintex M20
83. Lining size—Front	10in. × 2$\frac{1}{4}$in.
84. Lining size—Rear	10in. × 2$\frac{1}{4}$in.

SPRINGS

85. Type	Front: coil; Rear: semi-elliptic (laminated)
86. Length—Front	9·75in. (free)
87. Dia. of coils—Front	3·5in. + ·020in.; − ·010in.; coil springs fitted to front in all cases
88. No. of working coils—Front	6$\frac{3}{4}$
89. Thickness (section) of coils—Front	0·50in. ± 0·002in.
90. Fitted load—Front	925 lb.
91. Length—Rear	41·0in. ± 0·13 (static laden)
92. Width—Rear	2·00in.
93. No. of leaves—Rear	6
94. Thickness of leaves—Rear	Blade 1, $\frac{7}{32}$in.; Remainder, $\frac{3}{16}$in.
95. Camber—Rear	$\frac{7}{8}$in. to 1$\frac{1}{8}$in. neg.; identified by red paint mark

ELECTRICAL

96. Distributor rotation	Anti-clockwise
97. Manual advance	Micro adjuster
98. Automatic advance	Centrifugal and vacuum
99. Breaker gap	·015in.
100. Plug make and type	Normal use, Champion L10S
101. Plug gap	·025in.
102. Firing order	1—3—4—2
103. Ignition timing—degrees	4° B.T.D.C.
104. Ignition timing—No. of flywheel teeth	1
105. Charging system	C.V.C.
106. Battery—capacity and make, type	Lucas 12v. 51 a.h. Model G.T.W. 9A/2
107. Battery earth—pos. or neg.	Positive

CAPACITIES

108. Sump—pints	11
109. Gearbox—pints	1$\frac{1}{2}$ (3$\frac{1}{2}$ with overdrive)
110. Rear Axle—pints	1$\frac{1}{2}$
111. Cooling system—pints	13 (14 with heater)
112. Radiator hose (top) length. Int. diam.	4in. × 1$\frac{3}{8}$in. dia.
113. Radiator hose (bottom) length. Int. diam.	1·38in. dia. elbow
114. Petrol galls.	12

GENERAL DIMENSIONS

115. Overall length	12ft. 7in.
116. Overall Height	4ft. 2in. hood erect
117. Overall Width	4ft. 7$\frac{1}{2}$in.
118. Ground clearance	6in.
119. Total weight	18-cwt. (without tools, fuel or water)

Triumph TR3 1955-1956

NOTE—As an alternative some 1954 2-seaters were fitted with Triumph TR2 engine and chassis number prefixed with letter T. The alternative was available for all models in 1956 (see page 106 of 1954 edition for engine details). Current models have the TR3 engine (described in this edition) but the Vanguard engine is still available if ordered. TR3 engines commenced at TS9653ME. Chassis numbers do not now carry a prefix letter.

ENGINE

1. No. of Cylinders ... ... 4 (Modified Standard Vanguard)
2. R.A.C. Rating ... ... 18 h.p.
3. Cubic capacity ... ... 2,088 c.c.
4. B.H.P. ... ... 68 at 4,300 r.p.m.
5. Bore ... ... 85mm. (3·347in.)
6. Stroke ... ... 92mm. (3·622in.)
7. Compression Ratio ... 6·7 to 1 & 1954, 7 to 1
8. System of Cooling ... Pump and Fan
9. 1st O.S. Bore ... ... } Not recommended—fit new sleeves and
10. Max. O.S. for boring ... } standard pistons when worn
11. Firing order ... ... 1—3—4—2
12. Piston clearance top ... ·003 to ·00375in.
13. Piston clearance bottom ·0015 to ·002in.
14. Ring Gap ... ... ·003 to ·007in.
15. No. of compression rings 2
16. Width of compression rings ·0777 to ·0787in.
17. No. of oil rings 2
18. Width of oil rings ... ·155 to ·156in.
19. Oil pressure, lb. sq. in. ... 40—70 hot
20. Gudgeon pin type ... Fully floating, circlip retained
21. Fit in piston ... ... Floating
22. Fit in con. rod ... ... ·0002in. clearance at 68° F.
23. Crankpin dia. (standard) 2·0861 to 2·0866in.
24. Min. diam. for regrind—Crankpin ·040in. undersize
25. Con. Rod—Length between centres 6¼in. ± ·002in.
26. Type of bearing—Con. Rod Precision, steel-backed, w.m. lined
27. Side clearance—Con. Rod ·0075 to ·0105in.
28. Diam. clearance—Con. Rod ·0006 to ·0025in.
29. No. of crankshaft bearings 3
30. Type of bearings—Main Precision, steel-backed, w.m. lined
31. Standard main journal diam. 2·479 to 2·4795in.
32. Min. diam. for regrind—Main journals ·040in. undersize
33. Main Bearings—Length 1¾in.—·0007 or + ·0002in.
34. End clearance—Main Bearing ·004 to ·006in.
35. Diam. clearance—Main Bearing ·001 to ·002in.
36. End thrust taken on—Crankshaft W.M. covered steel washers on each side of Centre Main Bearing
37. No. of Camshaft bearings 4—End float ·004 to ·012in.
38. Type of bearing—Camshaft ... ... Front—phos. bronze; Inter. and Rear—direct in block
39. Bearing clearance—Camshaft ... ·0025in.
40. End thrust taken on—Camshaft Flange on Front bearing—shoulder on camshaft
41. Camshaft drive (type) ... Duplex silent roller chain
42. Valve timing markings ... Punched and scribed on timing gears
43. Exhaust Valve diam.—Head and Stem $1\frac{9}{32}$in ± ·002in. and $\frac{5}{16}$in. stem
44. Inlet Valve diam.—Head and Stem 1½in. ± ·002in. and $\frac{5}{16}$in. stem
45. Valve seat angle ... ... 45°. Width approx. $\frac{1}{16}$in.
46. Tappet type ... ... Hollow cylindrical, flat base, chilled face
47. Inlet Valve Clearance for timing ·014in.
48. Inlet Valve opens—degrees 10° before T.D.C. Closes 50° after B.D.C.
49. Inlet Valve opens—Flywheel 1in. on flywheel rim (4 teeth)
50. Inlet Valve opens—piston traverse ·035in.
51. Inlet Valve working clearance ·010in. cold
52. Exhaust Valve working clearance ·012in. cold
53. Are guides removable? ... Yes
54. Carburettor make and Type Solex D/Draught

	Standard	*Economy*	*Power*
55. Choke setting	25	23	25
56. Main setting	135	120	140
57. Air correction	190	180	190
58. Idling setting	55	55	55
59. Starter setting	130	130	130

CLUTCH AND GEARBOX

60. Clutch type ... ... Borg & Beck, single dry plate
61. Type of facing ... ... B. & B. composition

		1954,
62. Gear ratio, top	4·1	3·73
63. Gear ratio, third	5·4	5·1
64. Gear ratio, second ...	8·0	7·3
65. Gear ratio, first	13·5	12·5
66. Gear ratio, reverse ...	13·5	12·5

FRONT AXLE AND STEERING, Etc.

67. Camber ... ... 2°
68. Caster angle ... ... 4°
69. Toe-in ... ... ⅛in.
70. King pin inclination ... 2°
71. Track ... ... 3ft. 11in.
72. Turning circle ... ... 33ft. 0in.
73. Wheelbase ... ... 8ft. 0in.
74. Tyre size and pressures ... 5·25 × 16in. 16/18 lb. F. and R.

REAR AXLE

75. Type of axle ... ... Hypoid
76. Type of drive ... ...
77. Ratio or No. of teeth ... 4.1 (1954, 3.73)
78. Adjustment ... ... Shim
79. End play—ins. ... ... —

(75–79: Make—Salisbury—Follow maker's Service Instructions)

80. Lash—ins. ... ... ·003 to ·006in.

BRAKES

81. Type ... ... Girling, hydraulic, two-leading shoe
82. Type of linings ... ... —
83. Lining size—Front ... 9in. dia. × 1¼in. wide
84. Lining size—Rear ... 9in. dia. × 1¼in. wide

SPRINGS

85. Type ... ... Front—independent coil; semi-elliptic [Rear
86. Length—Front ... ... 12¼in.
87. Width—Front ... ... —
88. No. of leaves—Front —
89. Thickness of leaves—Front —
90. Camber—Front ... ... —
91. Length—Rear ... ... 42¾in.
92. Width—Rear ... ... 1⅜in.
93. No. of leaves—Rear ... 6 (1957/8—4-seater and Coupe, 7)
94. Thickness of leaves—Rear ¼in.
95. Camber—Rear ... ... 3½in.

ELECTRICAL

96. Distributor rotation ... Anti-clockwise
97. Manual advance ... ... No
98. Automatic advance ... Yes
99. Breaker gap ... ... ·012in.
100. Plug make and type ... Champion L10. 14mm. ½in. reach
101. Plug gap ... ... ·032in.
102. Firing order ... ... 1—3—4—2
103. Ignition timing—degrees 4° before T.D.C., fully retarded (measured at crankshaft)
104. Ignition timing—flywheel ... 0·4in. on flywheel rim, before T.D.C.
105. Charging system ... ... A.V.C.
106. Battery—Capacity and make, type 12v. 54 a.h. Lucas
107. Battery earth—pos. or neg. Positive

CAPACITIES (Imperial)

108. Sump—pints 11
109. Gearbox—pints 2½
110. Rear Axle—pints 2½
111. Cooling system—pints ... 16
112. Radiator hose (top) length 6½in. × $\frac{13}{16}$in. Moulded hose
113. Radiator hose (bottom) length. Int. diam. 13in. × 1½in. Moulded hose
114. Petrol, galls. 11

GENERAL DIMENSIONS

	2 Str.	Coupé	4 Str.
115. Overall Length	11ft. 8in.	11ft. 10in.	11ft. 8in.
116. Overall Height (Hood up)	4ft. 4½in.	4ft. 2in.	4ft. 6½in.
117. Overall Width	4ft. 8in.	4ft. 8in.	4ft. 8in.
118. Ground clearance ...	7in.		
119. Total Weight	16 cwt. 3 qr.	17 cwt. 2 qr.	17 cwt. 0 qr.

with Tools and full Tank

Plus 4 1957-1958

ENGINE

1.	No. of Cylinders	4
2.	R.A.C. Rating	17·14 h.p.
3.	Cubic capacity	1,991 c.c.
4.	B.H.P.	100 at 5,000 r.p.m.
5.	Bore	83mm. (3·268in.)
6.	Stroke	92mm. (3·622in.)
7.	Compression Ratio	8·5 : 1
8.	System of Cooling	Pump and fan, thermostat controlled, pressurised
9.	1st O.S. Bore	+ ·020in.
10.	Max. O.S. for boring	+ ·040in.
11.	Firing order	1—3—4—2
12.	Piston clearance (top)	0·023in. to 0·0207in.
13.	Piston clearance (bottom)	0·0039in. to 0·0032in.
14.	Ring Gap	·003in. to ·001in.
15.	No. of compression rings	2
16.	Width of compression rings	·0625in.
17.	No. of oil rings	1
18.	Width of oil rings	·156 to ·155in.
19.	Oil pressure, lb. sq. in.	70 lb. sq. in. at 70° C. and 2,000 r.p.m.
20.	Gudgeon pin type	Fully floating (located by circlips)
21.	Fit in piston	Floating
22.	Fit in con. rod	Floating
23.	Crankpin diam. (std.)	2·0860in. to 2·0866in.
24.	Min. diam. for regrind—Crankpin	—·040in.
25.	Con. Rod—Length between centres	6·252in. to 6·248in.
26.	Type of bearing—Con. Rod	Shell, lead indium, bronze bearings
27.	Side clearance—Con. Rod	0·007in. to 0·014in.
28.	Diam. clearance—Con. Rod	0·0016in. to 0·0035in.
29.	No. of crankshaft brgs.	3
30.	Type of bearings—Main	Steel shell, white metal lined
31.	Standard main journal diam.	2·4790in. to 2·4795in.
32.	Min. diam. for regrind—Main journals	—·040in.
33.	Main Bearings—Length	1·5050in. to 1·4950in.
34.	End clearance—Main Brg.	·004in. to ·006in.
35.	Diam. cl'rance—Main Brg.	0·001in. to 0·0025in.
36.	End thrust taken on—Crankshaft	Between flanks of centre main bearing by white metal covered steel washers
37.	No. of Camshaft bearings	4
38.	Type of brg.—Camshaft	Front: cast iron; Remainder: steel shell, white metal lined (replaceable)
39.	Brg. clearance—Camshaft	0·0029in. to 0·0044in.
40.	End thrust taken on—Camshaft	Between the flanged front bearing and a shoulder on the shaft
41.	Camshaft drive (type)	Double roller silent chain
42.	Valve timing markings	Centre punched and scribed lines on timing sprockets
43.	Exhaust Valve diam.—Head and Stem	Head: 1·303in. to 1·299in.; Stem: 0·3715in. to 0·3705in.
44.	Inlet Valve diam.—Head and Stem	Head: 1·562in. to 1·558in.; Stem: 0·310in. to 0·311in.
45.	Valve seat angle	45°
46.	Tappet type	Hollow, cylindrical, flat based, chilled face
47.	Inlet Valve clearance for timing	·015in.
48.	Inlet Valve opens—degrees	15° B.T.D.C.
49.	Inlet Valve opens—No. of flywheel teeth	3·8
50.	Inlet Valve opens—piston traverse	·081in.
51.	Inlet Valve working clearance	·010in. cold
52.	Exhaust Valve working clearance	·010in. cold
53.	Are guides removable?	Yes
54.	Carburettor make & type	Twin S.U. Type H6
55.	Choke setting	Single jet: SM needle from Eng. TS 10073E (needle CIW used with oil bath air cleaner)
56.	Main setting	
57.	Compensator setting	
58.	Idling setting	
59.	Progression setting	

CLUTCH AND GEARBOX

60.	Clutch type	Borg & Beck single dry plate, Model 9A.6G
61.	Type of facing	Borg & Beck composition
62.	Gear ratio, top	1 : 1 (overall 3·7)
63.	Gear ratio, third	1·325 : 1 (overall 4·9)
64.	Gear ratio, second	2·00 : 1 (overall 7·4)
65.	Gear ratio, first	3·38 : 1 (overall 12·5)
66.	Gear ratio, reverse	4·28 : 1 (overall 15·8)

FRONT AXLE AND STEERING, Etc.

67.	Camber	2° pos. (static laden)
68.	Caster angle	Nil
69.	Toe-in	⅛in.
70.	Kin pin inclination	7°
71.	Track	Front; 3ft. 9in.; Rear, 3ft. 9½in.
72.	Turning circle	34ft. 0in.
73.	Wheelbase	7ft. 4in.
74.	Tyre size and pressures	550 × 15in. Front, 22 lb. sq. in.; Rear, 24 lb. sq. in.

REAR AXLE

75.	Type of axle	Semi-floating
76.	Type of drive	Hypoid
77.	Ratio or No. of teeth	3·7 : 1
78.	Adjustment	Shims for pinion and differential
79.	End play—ins.	None
80.	Lash—ins.	0·004in. to 0·006in.

BRAKES

81.	Type	Girling. Hydraulic
82.	Type of linings	Ferodo. DM8. (Front Disc. Pad DS1)
83.	Lining size—Front	Calliper Disc type
84.	Lining size—Rear	10in. × 2¼in.

SPRINGS

85.	Type	Front, coil; Rear, semi-elliptic (laminated)
86.	Length—Front	9·75in. (free)
87.	Dia. of coils—Front	3·5in. + ·020in.; — ·010in.; coil springs fitted to front in all cases
88.	No. of working coils—Front	6¾
89.	Thickness (section) of coils—Front	0·50in. ± 0·002in.
90.	Fitted load—Front	925 lb.
91.	Length—Rear	41·0in. ± 0·13 (static laden)
92.	Width—Rear	2·00in.
93.	No. of leaves—Rear	6
94.	Thickness of leaves—Rear	Blade 1, 7/32in.; Remainder, 3/16 in.
95.	Camber—Rear	⅞in. ± ·25in. neg.

ELECTRICAL

96.	Distributor rotation	Anti-clockwise
97.	Manual advance	Micro adjuster
98.	Automatic advance	Centrifugal and vacuum
99.	Breaker gap	·015in.
100.	Plug make and type	Normal use, Champion L10S
101.	Plug gap	·025in.
102.	Firing order	1—3—4—2
103.	Ignition timing—degrees	4° B.T.D.C.
104.	Ignition timing—No. of flywheel teeth	1
105.	Charging system	C.V.C.
106.	Battery—capacity and make, type	Lucas 12v. 51a.h. Model BT.9A.
107.	Battery earth—pos. or neg.	Positive

CAPACITIES (Imperial)

108.	Sump—pints	11
109.	Gearbox—pints	1½ (3½ with overdrive)
110.	Rear Axle—pints	1½
111.	Cooling System—pints	13 (14 with heater)
112.	Radiator hose (top) length. Int. diam.	4in. × 1⅜in. dia.
113.	Radiator hose (bottom) length. Int. diam.	1·38in. dia. elbow
114.	Petrol, galls.	12

GENERAL DIMENSIONS

115.	Overall length	12ft. 7in.
116.	Overall Height	4ft. 2in. hood erect
117.	Overall Width	4ft. 7½in.
118.	Ground clearance	6in.
119.	Total weight	18-cwt. (without tools, fuel or water)

ENGINE (Ford, side valve)

1.	No. of Cylinders ...	... 4
2.	R.A.C. Rating ...	... 10 h.p.
3.	Cubic capacity ...	... 1172 c.c. (71·55 cu. in.)
4.	B.H.P.	... 36 at 4,500 r.p.m.
5.	Bore	... 2½in. (63·5mm.)
6.	Stroke	... 3·64in. (92·5mm.)
7.	Compression Ratio	... 7 to 1
8.	System of Cooling	... Water Pump and thermostat
9.	1st O.S. Bore ...	... ·010in.
10.	Max. O.S. for boring	... ·030in.
11.	Firing order ...	... 1—2—4—3
12.	Piston clearance top	... 6 to 12 lb. pull using ·0015in. feeler
13.	Piston clearance bottom	—
14	Ring Gap ...	... Top ·007 to ·012in.; Centre, ·008 to ·014in.; Oil, ·007 to ·012in.
15	No. of compression rings	2
16.	Width of comp. rings	... Top, ·076 to ·0765in.; Centre, ·093 to ·0935in.
17.	No. of oil rings ...	... 1
18.	Width of oil rings	... ·01545 to ·01550in.
19.	Oil pressure, lb. sq. in.	... 30
20.	Gudgeon pin type	... Fully floating, circlip retention
21.	Fit in piston ...	... Up to ·0002in. clearance
22.	Fit in con. rod ...	... Up to ·0005in. clearance
23.	Crankpin diam. (standard)	1·698 to 1·699in.
24.	Min. diam. for regrind—Crankpin	·030in. undersize
25.	Con. Rod—Length between centres	6·123 to 6·127in.
26.	Type of bearing—Con. Rod	... Integral babbitt
27.	Side clearance—Con. Rod	·004 to ·010in.
28.	Diam. clearance—Con. Rod	·0005 to ·002in.
29.	No. of crankshaft brgs.	... 3
30.	Type of bearings—Main	Detachable shells
31.	Standard main journal diam.	1·698 to 1·699in.
32.	Min. diam. for regrind—Main journals	·030in. undersize
33.	Main Bearings—Length	Front, 1·615 to 1·635in.; Centre, 1·310 to 1·314in.; Rear 2·003 to 2·006in.
34.	End clearance—Main Bearing	·002 to ·011in.
35.	Diam. clearance—Main Bearing	·000 to ·0015in.
36.	End thrust taken on—Crankshaft	Split thrust washers each side of rear main bearing
37.	No. of Camshaft bearings	3
38.	Type of bearing—Camshaft	Steel-backed, babbitt liners
39.	Bearing clearance—Camshaft	·001 to ·002in.
40.	End thrust taken on—Camshaft	Thrust washer and spring plunger at front end
41.	Camshaft drive (type)	... Duplex roller chain
42.	Valve timing markings	... Lines on timing gears
43.	Exhaust Valve diam.—Head and Stem	1·05 to 1·06in. and ·3086 to ·3096in.
44.	Inlet Valve diam.—Head and Stem	1·15 to 1·16in. and ·3095 to ·3105in.
45.	Valve seat angle ...	... 45¼°
46.	Tappet type ...	... Mushroom
47.	Inlet Valve clearance for timing	·015in.
48.	Inlet Valve opens—degrees	3° 30′ B.T.D.C.
49.	Inlet Valve opens—No. of flywheel teeth	—
50.	Inlet Valve opens—piston traverse	—
51.	Inlet Valve working clearance	·0115 to ·0135in. (cold)
52.	Exhaust Valve working clearance	·0115 to ·0135in. (cold)
53.	Are guides removable?	... Yes
54.	Carburettor make & type	Downdraught
55.	Choke tube ...	... 21mm.
56.	Main jet	... 110
57.	Main air correction	... 160
58.	Idling jet ...	... 50; air bleed, 1·2mm.
59.	Starter jet ...	... 130; starter air jets, 5mm.

CLUTCH AND GEARBOX

60.	Clutch type ...	... Single dry plate
61.	Type of facing ...	... Asbestos composition
62.	Gear ratio, top ...	... 4·4 : 1 overall
63.	Gear ratio, third ...	... Not fitted
64.	Gear ratio, second	... 8·25 : 1
65.	Gear ratio, first ...	... 15·07 : 1
66.	Gear ratio, reverse	... 19·71 : 1

FRONT AXLE AND STEERING, Etc.

67.	Camber	... 2°
68.	Caster angle ...	... 4°
69.	Toe-in	... ⅛in.
70.	King pin inclination	... 2°
71.	Track	... 3ft. 11in.
72.	Turning circle ...	... 33ft. 0in.
73.	Wheelbase ...	... 8ft.
74.	Tyre size and pressures	... 5·00 × 16; 16 to 18 lb., front and rear

REAR AXLE

75.	Type of axle ...	... Hypoid. Make: Salisbury HA; follow maker's service instructions
76.	Type of drive ...	... Bevel
77.	Ratio or No. of teeth	... 4·4 to 1
78.	Adjustment ...	... Shim
79.	End play—ins. ...	... ·006 to ·008in.
80.	Lash—ins. ...	... Min., ·004in. (etched on drive gear)

BRAKES

81.	Type	... Girling hydraulic
82.	Type of linings ...	... Moulded
83.	Lining size—Front	... } Width, 1¼in., Drum diam. 9in.
84.	Lining size—Rear	... } Width, 1¼in., Drum diam. 9in.

SPRINGS (Front)

85.	Type	... Independent coil.
86.	Length	... 14⅜in.

SPRINGS (Rear)

87.	Type	... Semi-elliptic
88.	Length	... 42¾in.
89.	Width	... 1⅜in.
90.	No. of leaves ...	... 5
91.	Thickness of leaves	... ¼in.
92.	Camber	... 3½in.

ELECTRICAL

93.	Distributor rotation	... Anti-clockwise
94.	Manual advance ...	... None
95.	Automatic advance	... Centrifugal, 14°
96.	Breaker gap ...	... ·014 to ·016in.
97.	Plug make and type	... Champion L10, 14mm.
98.	Plug gap	... ·025in.
99.	Firing order ...	... 1—2—4—3
100.	Ignition timing—degrees	5° B.T.D.C.
101.	Ignition timing—No. of flywheel teeth	—
102.	Charging system ...	... Voltage control
103.	Battery—Capacity	... 12-volt, 40 amp. hr.
104.	Battery earth ...	... Positive

CAPACITIES (Imperial)

105.	Sump—pints ...	... 4½ plus ¾ for filter
106.	Gearbox—pints ...	... 1¾
107.	Rear Axle—pints ...	... 2½
108.	Cooling system—pints	... 12
109.	Radiator hose (top) length. Int. diam.	11·1, 1·25 and 1·84in. (moulded)
110.	Radiator hose (bottom) length. Int. diam.	5·5 and 1·25in. (moulded)
111.	Petrol—galls. ...	... 8

GENERAL DIMENSIONS

112.	Overall Length ...	... 12ft.
113.	Overall Height ...	... 4ft. 4in.
114.	Overall Width ...	... 4ft. 8in.
115.	Ground clearance	... 7in. (approx.
116.	Total weight ...	... 12¾ cwt.

4/4 Series II 1959-1960

NOTE—Current models have the TR3 engine (described below) and details of the modified Standard Vanguard engine, which was an alternative up to 1959, are given on page 96 of the 1958 edition.

ENGINE (Triumph TR3)

1.	No. of Cylinders	4
2.	R.A.C. Rating	17·14 h.p.
3.	Cubic capacity	1,991cc.
4.	B.H.P.	100 at 5,000 r.p.m.
5.	Bore	83mm. (3·268in.)
6.	Stroke	92mm. (3·622in.)
7.	Compression ratio	8·5 to 1
8.	System of Cooling	Pump and fan, thermostat controlled, pressurized
9.	1st O.S. Bore	+ ·020in.
10.	Max. O.S. for boring	+ ·040in.
11.	Firing order	1—3—4—2
12.	Piston clearance top	·023 to ·0207in.
13.	Piston clearance bottom	·0039 to ·0032in.
14.	Ring Gap	·003 to ·001in.
15.	No. of compression rings	2
16.	Width of compression rings	·0625in.
17.	No. of oil rings	1
18.	Width of oil rings	·156 to ·155in.
19.	Oil pressure, lb. sq. in.	70 at 70°C. and 2,000 r.p.m.
20.	Gudgeon pin type	Fully floating, located by circlips
21.	Fit in piston	Floating
22.	Fit in con. rod	Floating
23.	Crankpin dia. (standard)	2·0860 to 2·0866in.
24.	Min. diam. for regrind—Crankpin	— ·040in.
25.	Con. Rod—Length between centres	6·252 to 6·248in.
26.	Type of bearing—Con. Rod	Shell, lead-indium, bronze
27.	Side clearance—Con. Rod	·007 to ·014in.
28.	Diam. clearance—Con. Rod	·0016 to ·0035in.
29.	No. of crankshaft bearings	3
30.	Type of bearings—Main	Steel shell, white metal lined
31.	Standard main journal diam.	2·4790 to 2·4795in.
32.	Min. diam. for regrind—Main journals	— ·040in.
33.	Main Bearings—Length	1·5050 to 1·4950in.
34.	End clearance—Main Bearing	·004 to ·006in.
35.	Diam. clearance—Main Bearing	·001 to ·0025in.
36.	End thrust taken on—Crankshaft	Between flanks of centre main bearing by w. metal-covered steel washers
37.	No. of Camshaft bearings	4
38.	Type of bearing—Camshaft	Front, cast iron; remainder, steel shell, white metal lined, replaceable
39.	Bearing clearance—Camshaft	·0029 to ·0044in.
40.	End thrust taken on—Camshaft	Between the flanged front bearing and a shoulder on the shaft
41.	Camshaft drive (type)	Duplex silent roller chain
42.	Valve timing markings	Centre-punched and scribed lines on timing gears
43.	Exhaust Valve diam.—Head and Stem	1·303/1·299in. and ·3715/·3705in.
44.	Inlet Valve diam.—Head and Stem	1·562/1·558in. and ·310/·311in.
45.	Valve seat angle	45°
46.	Tappet type	Hollow, flat base, chilled face
47.	Inlet Valve clearance for timing	·015in., cold
48.	Inlet Valve opens—degrees	15° B.T.D.C.
49.	Inlet Valve opens—Flywheel	3·8 teeth B.T.D.C.
50.	Inlet Valve opens—piston traverse	·081in. B.T.D.C.
51.	Inlet Valve working clearance	·010in., cold
52.	Exhaust Valve working clearance	·010in., cold
53.	Are guides removable?	Yes
54.	Carburettor make and type	Twin S.U. type H.6
55.	Choke setting	Single jet; SM needle (Needle CIW used with oil-bath air cleaner)
56.	Main setting	
57.	Air correction	
58.	Idling setting	
59.	Starter setting	

CLUTCH AND GEARBOX

60.	Clutch type	Borg & Beck, single dry plate	
61.	Type of facing	B. & B. composition	
62.	Gear ratio, top	Direct	Overall 3·73 : 1
63.	Gear ratio, third	1·40 : 1	,, 5·24 : 1
64.	Gear ratio, second	1·98 : 1	,, 7·38 : 1
65.	Gear ratio, first	3·44 : 1	,, 12·85 : 1
66.	Gear ratio, reverse	3·44 : 1	,, 12·85 : 1

FRONT AXLE AND STEERING, Etc.

67.	Camber	2°
68.	Caster angle	4°
69.	Toe-in	$\frac{1}{8}$in.
70.	King pin inclination	2°
71.	Track	3ft. 11in.
72.	Turning circle	32ft. 0in.
73.	Wheelbase	8ft. 0in.
74.	Tyre size and pressures	5·60 × 15in. 16/18lb. F. and R.

REAR AXLE

75.	Type of axle	—	Make—Salisbury—Follow maker's Service Instructions
76.	Type of drive	Spiral Bevel	
77.	Ratio or No. of teeth	3.73 : 1	
78.	Adjustment	Shim	
79.	End play—ins.	—	
80.	Lash—ins.	·003 to ·006in.	

BRAKES

81.	Type	Girling, hydraulic, two-leading shoe. Front disc brakes available as an extra
82.	Lining size—Front	9in. dia. × $1\frac{3}{4}$in. wide
83.	Lining size—Rear	9in. dia. × $1\frac{1}{4}$in. wide

SPRINGS (Front)

84.	Type	Independent, coil
85.	Length	13in.

SPRINGS (Rear)

86.	Type	Semi-elliptic
87.	Length	$42\frac{3}{4}$in.
88.	Width	$1\frac{3}{8}$in.
89.	No. of leaves	6
90.	Thickness of leaves	$\frac{1}{4}$in.
91.	Camber	$3\frac{1}{2}$in.

ELECTRICAL

92.	Distributor rotation	Anti-clockwise
93.	Manual advance	Micro adjuster
94.	Automatic advance	Centrifugal and vacuum
95.	Breaker gap	·015in.
96.	Plug make and type	Lodge CNY
97.	Plug gap	·025in.
98.	Firing order	1—3—4—2
99.	Ignition timing—degrees	4° before T.D.C., fully retarded (measured at crankshaft)
100.	Ignition timing—flywheel	—
101.	Charging system	C.V.C.
102.	Battery—Capacity and make, type	12v. 54 a.h. Lucas
103.	Battery earth	Positive

CAPACITIES (Imperial)

104.	Sump—pints	11 (Refill, 10 pts.)	
105.	Gearbox—pints	$2\frac{1}{2}$	
106.	Rear Axle—pints	$2\frac{1}{2}$	
107.	Cooling system—pints	16	
108.	Radiator hose (top) length	$6\frac{1}{2}$in. × $\frac{13}{16}$in.	Moulded hose
109.	Radiator hose (bottom) length. Int. diam.	13in. × $1\frac{1}{2}$in.	Moulded hose
110.	Petrol—galls.	11	

GENERAL DIMENSIONS

		2 Str.	Coupé	4 Str.
111.	Overall Length	11ft. 8in.	11ft. 10in.	11ft. 8in.
112.	Overall Height (Hood up)	4ft. $4\frac{1}{2}$in.	4ft. 2in.	4ft. $6\frac{1}{2}$in.
113.	Overall Width	4ft. 8in.	4ft. 8in.	4ft. 8in.
114.	Ground clearance	7in.		
115.	Total Weight (approx) with Tools and full Tank	16 cwt. 3 qr.	17 cwt. 2 qr.	17 cwt. 0 qr.

Plus 4 1959–1960

ENGINE (Ford 109E)

1. Type ... 4-stroke; overhead valves; side camshaft
2. Cooling system ... Water pump, radiator, fan and thermostat
3. Number of cylinders ... 4 (bored direct in block)
4. Firing order ... 1—2—4—3 (No. 1 at front)
5. Bore ... 80·96mm.
6. Stroke ... 65·07mm.
7. Cubic capacity ... 1340c.c.
8. Compression ratio ... 8·5:1
9. Brake horse-power ... 56·5 at 5,000 r.p.m.
10. Torque ... 76lb.ft. at 2,500 r.p.m.
11. Rebore oversizes ... 0·0025in.; 0·005in.; 0·015in.; 0·030in.
12. Piston clearance ... 0·0005 to 0·0011in. at bottom of skirt on thrust side
13. Piston rings—number ... 2 compression, 1 oil
14. Piston rings—width ... Comp., 0·077 to 0·078in.; oil, 0·155 to 0·156in.
15. Piston rings—groove cl'nce Comp., 0·0016 to 0·0036in.; oil, 0·0018 to 0·0038in.
16. Piston rings—gap ... Comp., 0·009 to 0·014in.; oil, 0·009 to 0·014in.
17. Oil pressure ... 35 to 40lb./sq.in.
18. Gudgeon pin diameter ... 0·8120 to 0·8123in.
19. Gudgeon pin fit in piston... 0·0000 to 0·0002in. (selective)
20. Gudgeon pin fit in con. rod 0·0001 to 0·0003in. (selective)
21. Crankpin diameter ... 1·9370 to 1·9375in.
22. Crankpin undersizes ... 0·01; 0·02; 0·03; 0·04in.
23. Con rod length (centres) 4·283 to 4·285in.
24. Con. rod bearings ... Vandervell/Glacier, steel-backed, copper-lead or lead-bronze
25. Con. rod bearing cl'nce ... 0·0005 to 0·0022in.
26. Con. rod side cl'nce ... 0·002 to 0·008in.
27. Main journal diameter ... 2·1255 to 2·1260in.
28. Main journal undersizes... 0·01; 0·02; 0·03in.
29. Main bearings ... 3, Vandervell and Glacier, steel-backed lead-bronze, or copper-lead with lead overlay
30. Main bearings—length 0·724 to 0·800in. (effective)
31. Main bearing cl'nce ... 0·0005 to 0·0020in.
32. Crankshaft end-play ... 0·003 to 0·011in.
33. Crankshaft end thrust on Centre bearing thrust washer
34. Camshaft journal diam ... 1·5615 to 1·5635in.
35. Camshaft bearings ... 3, Vandervell and Glacier, steel-backed, babbit liners
36. Camshaft bearing cl'nce... 0·0010 to 0·0035in.
37. Camshaft end-float ... 0·002 to 0·007in.
38. Camshaft drive type ... Simplex chain; 46 × 0·375in. pitches
39. Valve head diameter ... Inlet, 1·262 to 1·272in.; exhaust, 1·183 to 1·193in.
40. Valve stem diameter ... Inlet, 0·3095 to 0·3105in.; exhaust, 0·3086 to 0·3096in.
41. Valve seat angle ... 45° inlet and exhaust
42. Valve lift ... Inlet, 0·3147in.; exhaust, 0·3181in.
43. Valve guide length ... 1·52 to 1·53in.
44. Valve stem/guide cl'nce... Inlet, 0·0008 to 0·0030in.; exhaust, 0·0017 to 0·0039in.
45. Valve spring free length... 1·48in.
46. Valve spring rate ... Valve open, 0·953in. at 113/123lb.; valve closed, 1·263in. at 44/49lb.
47. Valve working cl'nce ... Inlet (H) 0·010in., (C) 0·008in.; exhaust (H) 0·017in., (C) 0·018in.
48. Valve timing clearance ... Inlet, 0·015in.; exhaust, 0·026in.
49. Valve timing ... Inlet opens 17° B.T.D.C.; closes 51° A.B.D.C. Exhaust opens 51° B.B.D.C.; closes 17° A.T.D.C.
50. Petrol pump ... A.C.-Delco mechanical
51. Carburettor ... Zenith, type 32 VN-2 downdraught
52. Choke tube ... 26
53. Main jet ... 65
54. Compensating jet ... 115
56. Slow-running jet ... 55
57. Pump jet ... 60

TRANSMISSION

101. Type ... Rear wheel drive
102. Clutch ... Ford/Borg & Beck, $7\frac{1}{4}$in. diam. single dry plate
103. Gearbox ... 4 forward speeds; synchromesh on 2nd, 3rd and top gears
105. Gearbox ratios ... Top, 1:1; 3rd, 1·412:1; 2nd, 2·396:1; 1st, 4·118:1; reverse, 5·404:1
107. Driving axles—type ... Semi-floating
108. Final drive gear—type ... Hypoid (Salisbury)
109. Final drive ratio ... 4·56:1
110. Crown wheel/pinion teeth 41/9

BRAKES

201. Front ... Girling hydraulic, 11in. diam. discs
202. Rear ... Girling hydraulic, 9in. diam. drums
203. Handbrake type ... Mechanical, on rear wheels
204. Linings—front ... Ferodo, type 14 caliper
205. Linings—rear ... Ferodo, $1\frac{1}{2}$in.
206. Power assistance ... None

STEERING (see also "Dimensions")

301. Type ... Cam gear
302. Power assistance ... None
303. Camber ... 2° non-adjustable
304. Castor ... 4°; non-adjustable
305. King-pin inclination ... 2°
306. Toe-in ... $\frac{1}{8}$ to $\frac{3}{16}$in. at wheel rim

FRONT SUSPENSION

401. Type ... Independent, vertical coil springs
408. Number of coils ... 21
409. Length—free ... $14\frac{1}{4}$in.
410. Length—fitted ... $8\frac{1}{2}$in.
411. Coil diameter ... $2\frac{1}{4}$in.
412. Wire diameter ... 0·31in.

REAR SUSPENSION

501. Type ... Semi-elliptic leaf springs
504. Number of leaves ... 5
505. Length (eye centres) free $42\frac{3}{4}$in.
506. Width ... $1\frac{3}{8}$in.
507. Thickness ... $\frac{1}{4}$in.
508. Camber—free ... $3\frac{5}{8}$in.

ELECTRICAL

601. Ignition timing (static) 6° B.T.D.C.
602. Sparking plugs ... Champion, N5, 14mm.
603. Sparking plug gap ... 0·023 to 0·028in.
604. Distributor ... Lucas, type DM2
605. Contact breaker gap ... 0·014 to 0·016in.
606. Centrifugal adv. begins at 1,260 r.p.m. (crank)
607. Centrifugal adv.—max.... 32° (crank) at 5,000 r.p.m.
608. Vacuum adv. begins at ... 5in. Hg.
609. Vacuum adv.—max. ... 26° (crank) at 16in. Hg.
610. Ignition coil ... Lucas, type LA12
611. Battery ... 12v. 36a.h.
612. Battery earth ... Positive
613. Dynamo ... Lucas, type C40L
614. Control box ... Lucas, type RB310
615. Starter motor ... Lucas, type M3

CAPACITIES

701. Engine sump ... 4 pints
702. Gearbox ... $1\frac{3}{4}$ pints
703. Differential ... $1\frac{3}{4}$ pints
704. Cooling system ... 12 pints
705. Fuel tank ... $8\frac{1}{2}$ gallons
706. No. of grease gun points ... 9

DIMENSIONS

801. Length ... 12ft.
802. Width ... 4ft. 8in.
803. Height ... 4ft. $3\frac{3}{4}$in.
804. Weight (approx.) ... 1,456lb.
805. Ground clearance ... 7in.
806. Track ... 3ft. 11in. front and rear
807. Wheelbase ... 8ft.
808. Turning circle ... 31ft.
809. Tyre size ... 5·20 × 15in.
810. Tyre pressures ... 16lb./sq.in. front and rear (normal driving)

TORQUE SPANNER SETTINGS

901. Cylinder head nuts ... 65/70lb.ft.
902. Con. rod bearing nuts 20/25lb.ft.
903. Main bearing nuts ... 55/60lb.ft.
905. Flywheel bolts ... 45/50lb.ft.
906. Clutch to flywheel bolts... 12/15lb.ft.

BODY

1001. Type of construction ... Coachbuilt
1002. Material of body panels... Steel

ENGINE (Triumph TR3)

1.	Type	4-stroke; overhead valves; side camshaft
2.	Cooling system	Water pump, radiator, fan and thermostat
3.	Number of cylinders	4 (wet liners)
4.	Firing order	1—3—4—2 (No. 1 at front)
5.	Bore	83mm. (3·268in.)
6.	Stroke	92mm. (3·622in.)
7.	Cubic capacity	1991c.c.
8.	Compression ratio	8·5:1
9.	Brake horse-power	100 at 5,000 r.p.m.
10.	Torque	117·5lb.ft. at 3,000 r.p.m.
11.	Rebore oversizes	0·020; 0·040in.
12.	Piston clearance	Top, 0·023 to 0·0207in.; bottom, 0·0039 to 0·0032in.
13.	Piston rings—number	2 compression, 1 oil
14.	Piston rings—width	Comp., 0·0625in.; oil, 0·156 to 0·155in.
16.	Piston rings—gap	0·003 to 0·001in.
17.	Oil pressure	70lb./sq.in. at 70° C. and 2,000 r.p.m.
19.	Gudgeon pin fit in piston	Floating
20.	Gudgeon pin fit in con. rod	Floating
21.	Crankpin diameter	2·0860 to 2·0866in.
22.	Crankpin undersizes	Max. 0·04in.
23.	Con. rod length (centres)	6·252 to 6·248in.
24.	Con. rod bearings	Shell, lead-indium bronze
25.	Con. rod bearing cl'nce	0·0016 to 0·0035in.
26.	Con. rod side cl'nce	0·007 to 0·014in.
27.	Main journal diameter	2·4790 to 2·4795in.
28.	Main journal undersizes	Max. 0·04in.
29.	Main bearings	3, steel shell, white-metal lined
30.	Main bearings—length	1·5050 to 1·4950in.
31.	Main bearing cl'nce	0·001 to 0·0025in.
32.	Crankshaft end-play	0·004 to 0·006in.
33.	Crankshaft end thrust on	Centre main bearing thrust washers
35.	Camshaft bearings	4; front, cast iron; remainder white-metal, steel shell
36.	Camshaft bearing cl'nce	0·0029 to 0·0044in.
38.	Camshaft drive type	Double roller chain
39.	Valve head diameter	Inlet, 1·562 to 1·558in.; exhaust, 1·303 to 1·299in.
40.	Valve stem diameter	Inlet, 0·310 to 0·311in.; exhaust, 0·3715 to 0·3705in.
41.	Valve seat angle	45° inlet and exhaust
47.	Valve working cl'nce	0·010in. inlet and exhaust (cold)
48.	Valve timing cl'nce	0·015in.
49.	Valve timing	Inlet valve opens 15° B.T.D.C.
50.	Petrol pump	A.C. mechanical
51.	Carburettors	Two S.U. type H6
52.	Diameter	1¾in.
53.	Needle (standard)	SM
57.	Richer needle	RH
58.	Weaker needle	SL

TRANSMISSION

101.	Type	Rear wheel drive
102.	Clutch	Borg and Beck, 9in. diam., single dry plate
103.	Gearbox	4 forward speeds; synchromesh on 2nd, 3rd and top gears
105.	Gearbox ratios	Top, 1:1; 3rd, 1·37:1; 2nd, 1·98:1; 1st and reverse, 3·44:1
107.	Driving axles—type	Semi-floating
108.	Final drive gear—type	Hypoid (Salisbury)
109.	Final drive ratio	3·73:1
110.	Crown wheel/pinion teeth	41/11
111.	Crown wheel/pinion b'lash	0·003 to 0·006in.; shim adjustment
112.	Pinion bearing pre-load	8 to 12lb.in.

BRAKES

201.	Front	Girling hydraulic, 11in. diam. discs
202.	Rear	Girling hydraulic, 9in. diam. drums
203.	Handbrake type	Mechanical, on rear wheels
204.	Linings—front	Ferodo type 14, caliper
205.	Linings—rear	Ferodo, 1½in.
206.	Power assistance	None

STEERING (see also "Dimensions")

301.	Type	Cam gear
302.	Power assistance	None
303.	Camber	2°; non-adjustable
304.	Castor	4°; non-adjustable
305.	Kin-pin inclination	2°
306.	Toe-in	⅛ to 3/16in. at wheel rim

FRONT SUSPENSION

401.	Type	Independent, vertical coil springs
408.	Number of coils	18
409.	Length—free	13in.
410.	Length—fitted	9in.
411.	Coil diameter	1⅜in.
412.	Wire diameter	0·375in.

REAR SUSPENSION

501.	Type	Semi-elliptic leaf-springs
504.	Number of leaves	6
505.	Length (eye centres) free	42¾in.
506.	Width	1⅜in.
507.	Thickness	¼in.
508.	Camber—free	3½in.

ELECTRICAL

601.	Ignition timing (static)	4° B.T.D.C.
602.	Sparking plugs	Lodge, type CNY, 14mm.
603.	Sparking plug gap	0·025in.
604.	Distributor	Lucas
605.	Contact breaker gap	0·05in.
610.	Ignition coil	Lucas, 45054J/M
611.	Battery	Lucas, 12v., 54 a.h.
612.	Battery earth	Positive
613.	Dynamo	Lucas, type C40-1
615.	Starter motor	Lucas, type M35H

CAPACITIES

701.	Engine sump	11 pints
702.	Gearbox	2½ pints
703.	Differential	1¾ pints
704.	Cooling system	16 pints
705.	Fuel tank	11 gallons
706.	No. of grease gun points	9

DIMENSIONS

801.	Length	12ft.
802.	Width	4ft. 8in.
803.	Height	4ft. 4½in.
804.	Weight (approx.)	1,848lb.
805.	Ground clearance	7in.
806.	Track	3ft. 11in. front and rear
807.	Wheelbase	8ft.
808.	Turning circle	32ft.
809.	Tyre size	5·60 × 15in.
810.	Tyre pressures	18lb./sq.in. front and rear (normal driving)

TORQUE SPANNER SETTINGS

901.	Cylinder head nuts	100/105lb.ft.
902.	Con. rod bearing nuts	55/60lb.ft.
903.	Main bearing nuts	85/90lb.ft.

BODY

1001.	Type of construction	Coachbuilt
1002.	Material of body panels	Steel and aluminium

Plus 4 1961-1962

ENGINE (Ford 109E)

1.	Type	4-stroke; overhead valves; side camshaft
2.	Cooling system	Water pump, radiator, fan and thermostat
3.	Number of cylinders ...	4 (bored direct in block)
4.	Firing order and idling speed	1—2—4—3 (No. 1 at front); 500/700 r.p.m.
5.	Bore	80·96mm.
6.	Stroke	65·07mm.
7.	Cubic capacity ...	1340c.c.
8.	Compression ratio ...	8·5:1
9.	Brake horse-power ...	56·5 at 5,000 r.p.m.
10.	Torque	76lb.ft. at 2,500 r.p.m.
11.	Piston oversizes	0·0025in.; 0·005in.; 0·015in.; 0·030in.
12.	Piston clearance	0·0005 to 0·0011in. at bottom of skirt on thrust side
13.	Piston rings—number ...	2 compression, 1 oil
14.	Piston rings—width ...	Comp., 0·077 to 0·078in.; oil, 0·155 to 0·156in.
15.	Piston rings—groove cl'nce	Comp., 0·0016 to 0·0036in.; oil, 0·0018 to 0·0038in.
16.	Piston rings—gap (in cylinder) ...	Comp., 0·009 to 0·014in.; oil, 0·009 to 0·014in.
17.	Oil pressure	35 to 40lb./sq.in.
18.	Gudgeon pin diameter ...	0·8120 to 0·8123in.
19.	Gudgeon pin fit in piston	0·0000 to 0·0002in. clearance (selective)
20.	Gudgeon pin fit in con. rod	0·0001 to 0·0003in. clearance (selective)
21.	Crankpin diameter ...	1·9370 to 1·9375in.
22.	Crankpin undersizes ...	0·010, 0·020, 0·030, 0·040in.
23.	Connecting rod length ...	4·283 to 4·285in. (centres)
24.	Big-end bearings ...	Vandervell/Glacier, steel-backed, copper-lead or lead-bronze
25.	Big-end bearing clearance	0·0005 to 0·0022in.
26.	Big-end side clearance ...	0·002 to 0·008in.
27.	Main journal diameter ...	2·1255 to 2·1260in.
28.	Main journal undersizes	0·010, 0·020, 0·030, 0·040in.
29	Main bearings	3; Vandervell/Glacier, steel-backed lead-bronze or copper-lead with lead overlay
30	Main bearings—length	0·724 to 0·800in. (effective)
31.	Main bearing clearance	0·0005 to 0·0020in.
32.	Crankshaft end-float ...	0·003 to 0·011in.
33.	Crankshaft end-thrust on	Centre bearing thrust washer
34.	Camshaft journal dia. ...	1·5615 to 1·5635in.
35.	Camshaft bearings ...	3; Vandervell/Glacier, steel-backed babbit liners
36.	Camshaft bearing cl'nce	0·0010 to 0·0035in.
37.	Camshaft end-float ...	0·002 to 0·007in.
38.	Camshaft drive type ...	Simplex chain; 46 × 0·375in. pitches
39.	Valve head diameter ...	Inlet, 1·262 to 1·272in.; exhaust, 1·183 to 1·193in.
40.	Valve stem diameter ...	Inlet, 0·3095 to 0·3105in.; exhaust, 0·3086 to 0·3096in.
41.	Valve seat angle	45° inlet and exhaust
42.	Valve lift	Inlet, 0·3147in.; exhaust, 0·3181in.
43.	Valve guide length ...	1·52 to 1·53in.
44.	Valve stem/guide cl'nce ...	Inlet, 0·0008 to 0·0030in.; exhaust, 0·0017 to 0·0039in.
45.	Valve spring free length ...	1·48in.
46.	Valve spring rate ...	Valve open, 0·953in. at 113/123lb.; valve closed, 1·263in. at 44/49lb.
47.	Valve working cl'nce ...	Inlet (H) 0·010in., (C) 0·008in.; exhaust (H) 0·017in., (C) 0·018in.
48.	Valve timing clearance ...	Inlet, 0·015in.; exhaust, 0·026in.
49.	Valve timing	Inlet opens 17° B.T.D.C.; closes 51° A.B.D.C. Exhaust opens 51° B.B.D.C.; closes 17° A.T.D.C.
50.	Petrol pump	A.C.-Delco mechanical
51.	Carburettor	Zenith, type 32 VN-2 downdraught
52.	Choke tube	26
53.	Main jet	65
54.	Air correction jet ...	115 (Compensating jet)
55.	Pilot jet	55 (Slow-running jet)
58.	Pump jet	60

TRANSMISSION

101.	Type	Rear wheel drive
102.	Clutch	Ford/Borg & Beck, 7¼in. diam. single dry plate
103.	Gearbox	4 forward speeds; synchromesh on 2nd, 3rd and top gears
105.	Gearbox ratios	Top, 1:1; 3rd, 1·412:1; 2nd, 2·396:1; 1st, 4·118:1; reverse, 5·404:1
107.	Driving axles	Semi-floating
108.	Final drive gear	Hypoid (Salisbury)
109.	Final drive ratio	4·56:1
110.	Number of gear teeth ...	Crown wheel, 41; pinion, 9
111.	Crown wheel/pinion b'lash	·003 to ·006in.; shim adjustment
112.	Pinion bearing pre-load ...	8 to 12lb.in.; shim adjustment
113.	Differential brg. pre-load	·003 to ·005in.; shim adjustment

BRAKES

201.	Front	Girling hydraulic, 11in. diam. discs
202.	Rear	Girling hydraulic, 9in. diam. drums
203.	Handbrake	Mechanical, on rear wheels
204.	Linings—front	Ferodo, type 14 caliper pads
205.	Linings—rear	Ferodo, 1¾in. wide

STEERING (see also "Dimensions")

301.	Type	Cam gear
302.	Camber angle	2°; non-adjustable
303.	Castor angle	4°; non-adjustable
304.	King-pin inclination ...	2°
305.	Toe-in	⅛ to 3/16in. at wheel rim
306.	Conditions for checking ...	Unladen

FRONT SUSPENSION

401.	Type	Independent, vertical coil springs
402.	Number of coils	21
403.	Coil diameter	2¼in. (external)
404.	Wire diameter	0·31in.
405.	Spring length—free ...	14¼in.
406.	Spring length—fitted ...	8½in.

REAR SUSPENSION

501.	Type	Semi-elliptic leaf springs
504.	Spring length (centres) ...	42¾in. (free)
505.	Number of leaves ...	5
506.	Width of leaves	1⅜in.
507.	Thickness of leaves ...	¼in.
508.	Spring camber—free ...	3⅝in.

ELECTRICAL SYSTEM

601.	Ignition timing (static) ...	6° B.T.D.C. (mark on crankshaft pulley)
602.	Sparking plugs	Champion, N5, 14mm.
603.	Sparking plug gap ...	0·023 to 0·028in.
604.	Distributor	Lucas, type DM2
605.	Contact breaker gap ...	0·014 to 0·016in.
606.	Centrif. advance starts at	1,260 r.p.m. (crank)
607.	Centrif. advance—max. ...	32° (crank) at 5,000 r.p.m.
608.	Vacuum adv. starts at ...	5in. Hg.
609.	Vacuum adv.—max. ...	26° (crank) at 16in. Hg.
610.	Ignition coil	Lucas, type LA12
611.	Battery	Lucas, BT7A; 12v. 36a.h.
612.	Battery earth	Positive
613.	Dynamo	Lucas, type C40L
614.	Control box	Lucas, RB 105/2
615.	Starter motor	Lucas, M35H

CAPACITIES

701.	Engine sump	4 pints
702.	Gearbox	1¾ pints
703.	Differential	1¾ pints
704.	Cooling system	12 pints
705.	Fuel tank	8½ gallons
706.	Tyre size	5·60 × 15in.
707.	Tyre pressures	16 lb./sq.in., front and rear, normal (add 6lb./sq in. for sustained high-speed driving on motorways)
708.	No. of grease gun points ...	13 (Front sliding axles have 'one-shot' lubrication; press foot plunger every 200 miles approx.)
709.	Normal servicing intervals	Greasing, 2000/2500 miles; engine oil change, 5000 miles
710.	Hydraulic fluid	Girling

DIMENSIONS

801.	Length overall	12ft.
802.	Width overall	4ft. 8in.
803.	Height overall	4ft. 3¾in.
804.	Weight (approx.) ...	1,456lb. (dry)
805.	Ground clearance (min.)	7in.
806.	Track	3ft. 11in. front and rear
807.	Wheelbase	8ft.
808.	Turning circle diameter	32ft.

TORQUE SPANNER DATA

901.	Cylinder head nuts ...	65/70lb.ft.
902.	Big-end bearing nuts ...	20/25lb.ft.
903.	Main bearing nuts. ...	55/60lb.ft.
905.	Flywheel bolts	45/50lb.ft.
906.	Clutch to flywheel bolts ...	12/15lb.ft.
910.	Road wheel nuts ...	70/80lb.ft.

BODY AND CHASSIS

1001.	Type of construction ...	Coachbuilt
1002.	Material of body panels ...	Steel and aluminium

4/4 Series IV 1962-1963

Models: 1963

ENGINE (Triumph TR4). See the 1962 edition of this publication for details of previous model fitted with TR3 engine

1.	Type	4-stroke; overhead valves; side camshaft
2.	Cooling system	Water pump, fan and thermostat
3.	Number of cylinders	4 (wet liners)
4.	Firing order & idl'g speed	1—3—4—2 (No. 1 at front); 550 r.p.m.
5.	Bore	86mm. (3·386in.)
6.	Stroke	92mm. (3·622in.)
7.	Cubic capacity	2,138c.c. (130·5cu.in.)
8.	Comp. ratio and pressure	9:1; max. variation 5lb./sq.in.
9.	Brake horse-power	105 at 4,750 r.p.m.
10.	Torque	128·3lb.ft. at 3,350 r.p.m.
11.	Piston oversizes	0·010, 0·020, 0·030, 0·040in.
12.	Piston clearance	0·0047 to 0·0054in. measured at top; and 0·0032 to 0·0039in. at bottom
13.	Piston rings—number	2 compression, 1 oil
14.	Piston rings—width	Comp., 0·0615 to 0·0625in.; oil, 0·1552 to 0·1562in.
15.	Piston rings—groove cl'nce	Comp., 0·0010 to 0·0030in.; oil, 0·0010 to 0·0030in.
16.	Piston rings—gap (in cylinder)	Comp., 0·010 to 0·015in.; oil, 0·010 to 0·015in.
17.	Oil pressure	65 to 75lb./sq.in. (above 30 m.p.h.)
18.	Gudgeon pin diameter	0·87485 to 0·87510in.
19.	Gudgeon pin fit in piston	0·0002in. clearance at 68°F.
20.	Gudgeon pin fit in con. rod	Selective
21.	Crankpin diameter	2·0861 to 2·0866in.
22.	Crankpin undersizes	0·010, 0·020, 0·030in.
23.	Connecting rod length	6·248 to 6·252in. (centres)
24.	Big-end bearings	Vandervell, 3-layer
25.	Big-end bearing clearance	0·0028 to 0·0040in.
26.	Big-end side clearance	0·007 to 0·014in.
27.	Main journal diameter	2·4790 to 2·4795in.
28.	Main journal undersizes	0·010, 0·020, 0·030, 0·040in.
29.	Main bearings	3, Vandervell, bimetal (micro) and steel
30.	Main bearings—length	1·495 to 1·505in. [back
31.	Main bearing clearance	0·0015 to 0·0025in.
32.	Crankshaft end-float	0·0048 to 0·0117in.
33.	Crankshaft end-thrust on	Centre main bearing thrust washers
34.	Camshaft journal dia.	Front, 1·871 to 1·872in.; others 1·7153 to 1·7158in.
35.	Camshaft bearings	4; Vandervell (2, 3 & 4); front, cast
36.	Camshaft bearing cl'nce	Front, 0·0028 to 0·0047in.; others 0·0015 to 0·0029in.
37.	Camshaft end-float	0·0040 to 0·0075in.
38.	Camshaft drive type	Duplex chain, $21\frac{3}{4}$in. long
39.	Valve head diameter	Inlet, 1·558 to 1·562in.; exhaust, 1·299 to 1·303in.
40.	Valve stem diameter	Inlet, 0·310 to 0·311in.; exhaust, 0·3705
41.	Valve seat angle	Inlet, 45°; exhaust, 45° [to 0·3715in.
42.	Valve lift	0·250in. (cold)
43.	Valve guide length	Inlet, 2·0525 to 2·0725in.; exhaust, 2·05 to 2·07in.
44.	Valve stem/guide cl'nce	Inlet, 0·001 to 0·003in.; exhaust, 0·003 to 0·005in.
45.	Valve spring free length	Outer, 1·94 to 1·96in.; inner, 1·88 to 1·90in. (auxiliary inner, exhaust only, 1·55 to 1·57in.)
46.	Valve spring rate	Outer, 178lb./in.; inner, 93lb./in. (auxiliary inner, exhaust only, 28·5lb./in.)
47.	Valve working cl'nce	Inlet, 0·010in.; exhaust, 0·010in. (cold)
48.	Valve timing clearance	0·0165in.
49.	Valve timing	Inlet opens 17° B.T.D.C.; closes 57° A.B.D.C. Exhaust opens 57° B.B.D.C.; closes 17° A.T.D.C.
50.	Petrol pump	AC-Delco, mechanical
51.	Carburettors	2, S.U., type H6
52.	Diameter	$1\frac{3}{4}$in.
53.	Needle (standard)	SM
54.	Jet diameter	100

TRANSMISSION

101.	Type	Rear wheel drive
102.	Clutch	Borg and Beck, 9in. diam., single dry plate
103.	Gearbox	4 forward speeds; synchromesh on 2nd, 3rd and top gears
105.	Gearbox ratios	Top, 1:1; 3rd, 1·37:1; 2nd, 1·98:1; 1st and reverse, 3·44:1
106.	Overdrive	Not fitted
107.	Driving axles	Semi-floating
108.	Final drive gear	Hypoid (Salisbury)
109.	Final drive ratio	3·73:1
110.	Number of gear teeth	41/11
111.	Crown wheel/pinion b'lash	0·003 to 0·006in.; shim adjustment
112.	Pinion bearing pre-load	8 to 12lb.in.
113.	Differential brg. pre-load	·003 to ·005in.; shim adjustment

BRAKES

201.	Front	Girling hydraulic, 11in. diam. discs
202.	Rear	Girling hydraulic, 9in. diam. drums
203.	Handbrake	Mechanical, on rear wheels
204.	Linings—front	Ferodo, type 14, caliper pads
205.	Linings—rear	Ferodo, $1\frac{3}{4}$in. wide

STEERING (see also "Dimensions")

301.	Type	Cam gear
302.	Camber angle	2°, non-adjustable
303.	Castor angle	4°; non-adjustable
304.	King-pin inclination	2°
305.	Toe-in	$\frac{1}{8}$ to $\frac{3}{16}$ at wheel rims
306.	Conditions for checking	Unladen

FRONT SUSPENSION

401.	Type	Independent, vertical coil springs
402.	Number of coils	18
403.	Coil diameter	$2\frac{3}{8}$in. (external)
404.	Wire diameter	0·375in.
405.	Spring length—free	13in.
406.	Spring length—fitted	9in.

REAR SUSPENSION

501.	Type	Semi-elliptic leaf-springs
504.	Spring length (centres)	$42\frac{3}{4}$in. (free)
505.	Number of leaves	6
506.	Width of leaves	$1\frac{3}{8}$in.
507.	Thickness of leaves	$\frac{1}{4}$in.
508.	Spring camber—free	$3\frac{1}{2}$°

ELECTRICAL SYSTEM

601.	Ignition timing (static)	4° B.T.D.C. (Hole in pulley and pointer on timing case, T.D.C.)
602.	Sparking plugs	Lodge, type CNY; $\frac{1}{2}$in. reach, 14mm.
603.	Sparking plug gap	0·025in.
604.	Distributor	Lucas, type 25D4
605.	Contact breaker gap	0·015in.; dwell angle, 30°±3°
606.	Centrif. advance starts at	275 r.p.m.
607.	Centrif. advance—max.	$9\frac{3}{4}$° at 800 r.p.m.
608.	Vacuum adv. starts at	2in. Hg.
609.	Vacuum adv.—max.	2·6° at 6in. Hg.
610.	Ignition coil	Lucas, type HA12
611.	Battery	Lucas, two SG9E in series; 12v. 54a.h.
612.	Battery earth	Positive
613.	Dynamo	Lucas, C40/1
614.	Control box	Lucas, RB105/2
615.	Starter motor	Lucas, M418G

CAPACITIES

701.	Engine sump	11 pints (dry); 10 pints (drain and refill)
702.	Gearbox	$2\frac{1}{2}$ pints
703.	Differential	$1\frac{3}{4}$ pints
704.	Cooling system	16 pints
705.	Fuel tank	11 gallons
706.	Tyre size	5·60 × 15in.
707.	Tyre pressures	18lb./sq.in. normal (add 6lb./sq.in. for sustained high-speed driving on motorways)
708.	No. of grease gun points	15 (front sliding axles have 'one-shot' lubrication; press foot plunger every 200 miles approx.)
709.	Normal servicing intervals	Greasing 2000/2500 miles; engine oil change, 2,500 miles
710.	Hydraulic fluid	Girling

DIMENSIONS

801.	Length overall	12ft.
802.	Width overall	4ft. 8in.
803.	Height overall	4ft. $4\frac{1}{2}$in.
804.	Weight	1848lb. (approx.)
805.	Ground clearance (min.)	7in.
806.	Track	3ft. 11in. front and rear
807.	Wheelbase	8ft.
808.	Turning circle diameter	32ft.

TORQUE SPANNER DATA

901.	Cylinder head nuts	100/105lb.ft.
902.	Big-end bearing nuts	55/60lb.ft.
903.	Main bearing nuts.	85/90lb.ft.
905.	Flywheel bolts	42/46lb.ft.
906.	Clutch to flywheel bolts	20lb.ft.
910.	Road wheel nuts	70/80lb.ft.

BODY AND CHASSIS

1001.	Type of construction	Coachbuilt
1002.	Material of body panels	Steel and aluminium

Models: Standard and Competition versions

ENGINE (Ford 116E and G.T. version)

1.	Type	4-stroke; overhead valves; side camshaft
2.	Cooling system	Water pump, radiator, fan and thermostat
3.	Number of cylinders ...	4 (bored direct in block)
4.	Firing order & idling speed	1—2—4—3 (No. 1 at front); 600 r.p.m.
5.	Bore	80·96mm.
6.	Stroke	72·75mm.
7.	Cubic capacity	1498c.c.
8.	Comp'n ratio and press....	8·3:1; 175lb./sq.in. (Comp. 9:1; 179)
9.	Brake hore-power (gross)	64 at 4,600 r.p.m. (Comp. 83·5 at 5,200)
10.	Torque (gross)	85·5lb.ft. at 2,300 r.p.m.; (Comp. 97 at 3,600)
11.	Piston oversizes	0·0025in.; 0·005in.; 0·015in.; 0·030in.
12.	Piston clearance (bottom)	0·0005 to 0·0011in. on thrust side
13.	Piston rings—number ...	2 compression, 1 oil
14.	Piston rings—width ...	Comp., 0·077 to 0·078in.; oil, 0·155 to 0·156in.
15.	Piston rings—groove clearance	Comp., 0·0016 to 0·0036in.; oil, 0·0018 to 0·0038in.
16.	Piston rings—gap (in cyl.)	0·009 to 0·014in.
17.	Oil pressure	35 to 40lb./sq.in.
18.	Gudgeon pin diameter ...	0·8120 to 0·8123in.
19.	Gudgeon pin fit in piston	0·0000 to 0·0002in. clearance (selective)
20.	Gudgeon pin fit in con. rod	0·0001 to 0·0003in. clearance (selective)
21.	Crankpin diameter ...	1·9370 to 1·9375in.
22.	Crankpin undersizes ...	0·010, 0·020, 0·030, 0·040in.
23.	Connecting rod length ...	4·799 to 4·801in. (centres)
24.	Big-end bearings ...	Vandervell or Glacier, steel-backed, copper-lead or lead-bronze with lead overlay
25.	Big-end bearing clearance	0·0005 to 0·0022in.
26.	Big-end side clearance ...	·004 to ·010in.
27.	Main journal diameter ...	2·1255 to 2·1260in.
28.	Main journal undersizes...	0·010, 0·020, 0·030in.
29.	Main bearings	5; Vandervell or Glacier, steel-backed lead-bronze or copper-lead with lead overlay
30.	Main bearings—length ...	0·724 to 0·800in. (effective)
31.	Main bearing clearance ...	0·0005 to 0·0020in.
32.	Crankshaft end-float ...	0·003 to 0·011in.
33.	Crankshaft end-thrust on	Centre bearing thrust washer
34.	Camshaft journal dia. ...	1·560 to 1·5605
35.	Camshaft bearings ...	3; Vandervell or Glacier, steel-backed babbit liners
36.	Camshaft bearing cl'nce...	0·0010 to 0·0020in.
37.	Camshaft end-float ...	0·002 to 0·007in.
38.	Camshaft drive type ...	Simplex chain; 46 × 0·375in. pitches
39.	Valve head diameter ...	Inlet, 1·432/1·442in. (Comp. 1·405/1·415in.); exhaust, 1·183/1·193in. (Comp., 1·240/1·250in.)
40.	Valve stem diameter ...	Inlet, 0·3095 to 0·3105in.; exhaust, 0·3086 to 0·3096in.
41.	Valve seat angle	45° to 45° 15′ inlet and exhaust
42.	Valve lift	Inlet, 0·315in. (Comp., 0·3436in.); exhaust, 0·319in. (Comp., 0·3354in.)
43.	Valve guide length ...	1·52 to 1·53in.; fitted height, 0·28in. above head
44.	Valve stem/guide cl'nce ...	Inlet, 0·0008 to 0·0030in.; exhaust, 0·0017 to 0·0039in.
45.	Valve spring free length ...	1·48in.
46.	Valve spring rate ...	Valve open, 0·953in. at 113/123lb.; valve closed, 1·263in. at 44/49lb.
47.	Valve work'g cl'nce (cold)	Inlet, ·008in. (Comp., ·010in.); exhaust, ·018in. (Comp., ·023in.)
48.	Valve timing clearance ...	Inlet, ·015in. (Comp., ·016in.); exh., ·027in. (Comp., ·026in.)
49.	Valve timing ... ... (*Competition model)	Inlet opens 17° (*27°) B.T.D.C.; closes 51° (*65°) A.B.D.C. Exhaust opens 51° (*65°) B.B.D.C.; closes 17° (*27°) A.T.D.C.
50.	Petrol pump	A.C.-Delco mechanical
51.	Carburettor	Zenith 33VN (Comp. Weber DCD1)

52.	Choke tube	29	Primary	26	Secondary	27
53.	Main jet	92	,,	140	,,	155
54.	Air correction jet ...	112	,,	230	,,	180
55.	Pilot jet	55	,,	50	,,	70
58.	Pump jet	50		60		
60.	Starter air jet	—		100		
61.	Starter petrol jet ...	—		190		
62.	Needle valve	—		1·75mm.		

TRANSMISSION

101.	Type	Rear wheel drive
102.	Clutch	Ford/Borg & Beck, 7¼in. diam. single dry [plate
103.	Gearbox	4 forward speeds, all synchromesh
105.	Gearbox ratios	Top, 1:1; 3rd, 1·412:1; 2nd, 2·396:1; 1st, 3·543:1; reverse, 3·963:1
107.	Driving axles	Semi-floating
108.	Final drive gear	Hypoid (Salisbury)
109.	Final drive ratio	4·56:1 (41/9)
111.	Crown wheel/pinion b'lash	·003 to ·006in.; shim adjustment
112.	Pinion bearing pre-load...	8 to 12lb.in.; shim adjustment
113.	Differential brg. pre-load	·003 to ·005in.; shim adjustment

BRAKES

201.	Front	Girling hydraulic, 11in. diam. discs
202.	Rear	Girling hydraulic, 9in. diam. drums
203.	Handbrake	Mechanical, on rear wheels
204.	Linings—front	Ferodo, type 14 caliper pads
205.	Linings—rear	Ferodo, 1¾in. wide

STEERING (see also "Dimensions")

301.	Type	Cam Gear
302.	Camber angle	2°; non-adjustable
303.	Castor angle	4°; non-adjustable
304.	King-pin inclination ...	2°
305.	Toe-in	⅛ to 3/16in. at wheel rim
306.	Conditions for checking ...	Unladen

FRONT SUSPENSION

401.	Type	Independent, vertical coil springs
402.	Number of coils	21
403.	Coil diameter	2¼in. (external)
404.	Wire diameter	0·31in.
405.	Spring length—free ...	14¼in.
406.	Spring length—fitted ...	8½in.

REAR SUSPENSION

501.	Type	Semi-elliptic leaf springs
504.	Spring length (centres) ...	42¾in. (free)
505.	Number of leaves ...	5
506.	Width of leaves	1⅜in.
507.	Thickness of leaves ...	¼in.
508.	Spring camber—free ...	3⅝in.

ELECTRICAL SYSTEM

601.	Ignition timing (static) ...	8° (Comp., 10°) B.T.D.C. (mark on pulley and cover)
602.	Sparking plugs	Champion, N5 (Comp., N4) 14mm.
603.	Sparking plug gap ...	0·023 to 0·028in.
604.	Distributor	Lucas, type 25D4
605.	Contact breaker gap ...	0·014 to 0·016in.; dwell angle 57°/63°
606.	Centrif. advance starts at	1,200 r.p.m. (crank)
607.	Centrif. advance—max. ...	37° (crank) at 5,000 r.p.m.
608.	Vacuum advance starts at	5·2in. Hg.
609.	Vacuum advance—max.	28° (crank) at 15in. Hg.
610.	Ignition coil	Lucas or A.C.-Delco type LA12
611.	Battery	Lucas, BT7A; 12v. 36a.h.
612.	Battery earth	Positive
613.	Dynamo	Lucas, type C40L; max. output 300w.
614.	Control box	Lucas, RB 106/2
615.	Starter motor	Lucas, M35G

CAPACITIES

701.	Engine sump	6¼ pints
702.	Gearbox	1¾ pints
703.	Differential	1¾ pints
704.	Cooling system	12 pints
705.	Fuel tank	8½ gallons
706.	Tyre size	5·60 × 15in.
707.	Tyre pressures	16/17lb./sq.in., front and rear, normal (add 6lb./sq.in. for sustained high-speed driving on motorways)
708.	No. of greasegun points ...	13 (Front sliding axles have 'one-shot' lubrication; press foot plunger every 200 miles approx.)
709.	Servicing intervals ...	Greasing, 2000/2500 miles; engine oil change, 5000 miles
710.	Hydraulic fluid	Girling/Castrol

DIMENSIONS

801.	Length overall	12ft.
802.	Width overall	4ft. 8in.
803.	Height overall	4ft. 3¾in.
804.	Weight (approx.) ...	1,456lb. (dry)
805.	Ground clearance (min.)	7in.
806.	Track	3ft. 11in. front and rear
807.	Wheelbase	8ft.
808.	Turning circle diameter ...	32ft.

TORQUE SPANNER DATA

		lb.ft.			lb.ft.
901.	Cylinder head nuts ...	65/70	907.	Final drive pinion bearing nut ...	95/100
902.	Big-end bearing nuts ...	20/25	908.	Crown wheel bolts	50/60
903.	Main bearing nuts ...	55/60	909.	Diff. bearing nuts	35/40
905.	Flywheel bolts	45/50	910.	Road wheel nuts	65/70
906.	Clutch to flywheel bolts ...	12/15			

BODY AND CHASSIS

1001.	Type of construction ...	Coachbuilt; steel 'Z'-section chassis; ash frame
1002.	Material of body panels ...	Steel and aluminium
1003.	Type of glass.	Toughened (laminated optional extra)

Plus 4 & Plus 4 Plus 1963-1964

ENGINE (Triumph TR4). See the 1962 edition of this publication for details of previous model fitted with TR3 engine

1. Type ... ... 4-stroke; overhead valves; side camshaft
2. Cooling system ... ... Water pump, fan and thermostat
3. Number of cylinders ... 4 (wet liners)
4. Firing order & idl'g speed 1—3—4—2 (No. 1 at front); 550 r.p.m.
5. Bore ... ... 86mm. (3·386in.)
6. Stroke ... ... 92mm. (3·622in.)
7. Cubic capacity ... ... 2,138c.c. (130·5cu.in.)
8. Compression ratio and pressure 9:1 (7·0:1 optional); max. variation 5lb./sq.in.
9. Brake horse-power ... 105 at 4,750 r.p.m.
10. Torque ... ... 128·3lb.ft. at 3,350 r.p.m.
11. Piston oversizes ... ... 0·010, 0·020, 0·030, 0·040in.
12. Piston clearance ... ... 0·0047 to 0·0054in. measured at top; and 0·0032 to 0·0039in. at bottom
13. Piston rings—number ... 2 compression, 1 oil
14. Piston rings—width ... Comp., 0·0615 to 0·0625in.; oil, 0·1552 to 0·1562in.
15. Piston rings—groove cl'nce Comp., 0·0010 to 0·0030in.; oil, 0·0010 to 0·0030in.
16. Piston rings—gap (in cylinder) ... Comp., 0·010 to 0·015in.; oil, 0·010 to 0·015in.
17. Oil pressure ... ... 65 to 75lb./sq.in. (above 30 m.p.h.)
18. Gudgeon pin diameter ... 0·87485 to 0·87510in.
19. Gudgeon pin fit in piston 0·0002in. clearance at 68°F.
20. Gudgeon pin fit in con. rod Selective
21. Crankpin diameter ... 2·0861 to 2·0866in.
22. Crankpin undersizes ... 0·010, 0·020, 0·030in.
23. Connecting rod length ... 6·248 to 6·252in. (centres)
24. Big-end bearings ... Vandervell, 3-layer
25. Big-end bearing clearance 0·0028 to 0·0040in.
26. Big-end side clearance ... 0·007 to 0·014in.
27. Main journal diameter ... 2·4790 to 2·4795in.
28. Main journal undersizes 0·010, 0·020, 0·030, 0·040in.
29. Main bearings ... ... 3, Vandervell, bimetal (micro) and steel [back
30. Main bearings—length ... 1·495 to 1·505in.
31. Main bearing clearance 0·0015 to 0·0025in.
32. Crankshaft end-float ... 0·0048 to 0·0117in.
33. Crankshaft end-thrust on Centre main bearing thrust washers
34. Camshaft journal dia. ... Front, 1·871 to 1·872in.; others 1·7153 to 1·7158in.
35. Camshaft bearings ... 4; Vandervell (2, 3 & 4); front, cast
36. Camshaft bearing cl'nce Front, 0·0028 to 0·0047in.; others 0·0015 to 0·0029in.
37. Camshaft end-float ... 0·0040 to 0·0075in.
38. Camshaft drive type ... Duplex chain, 21¾in. long
39. Valve head diameter ... Inlet, 1·558 to 1·562in.; exhaust, 1·299 to 1·303in.
40. Valve stem diameter ... Inlet, 0·310 to 0·311in.; exhaust, 0·3705 [to 0·3715in.
41. Valve seat angle ... ... Inlet, 45°; exhaust, 45°
42. Valve lift ... ... 0·250in. (cold)
43. Valve guide length ... Inlet, 2·0525 to 2·0725in.; exhaust, 2·05 to 2·07in.
44. Valve stem/guide cl'nce ... Inlet, 0·001 to 0·003in.; exhaust, 0·003 to 0·005in.
45. Valve spring free length ... Outer, 1·94 to 1·96in.; inner, 1·88 to 1·90in. (auxiliary inner, exhaust only, 1·55 to 1·57in.)
46. Valve spring rate... ... Outer, 178lb./in.; inner, 93lb./in. (auxiliary inner, exhaust only, 28·5lb./in.)
47. Valve working cl'nce ... Inlet, 0·010in.; exhaust, 0·010in. (cold)
48. Valve timing clearance ... 0·0165in.
49. Valve timing ... ... Inlet opens 17° B.T.D.C.; closes 57° A.B.D.C. Exhaust opens 57° B.B.D.C.; closes 17° A.T.D.C.
50. Petrol pump ... ... AC-Delco, mechanical
51. Carburettors ... ... 2 Stromberg, type 175 CD
53. Needle ... ... 2 O

TRANSMISSION

101. Type ... ... Rear wheel drive
102. Clutch ... ... Borg and Beck, 9in. diam., single dry plate
103. Gearbox ... ... 4 forward speeds; synchromesh on 2nd, 3rd and top gears
105. Gearbox ratios ... ... Top, 1:1; 3rd 1·205:1; 2nd, 1·745:1; 1st and reverse, 2·97:1
106. Overdrive ... ... Not fitted
107. Driving axles ... ... Semi-floating
108. Final drive gear ... ... Hypoid (Salisbury)
109. Final drive ratio ... ... 3·73:1
110. Number of gear teeth ... 41/11
111. Crown wheel/pinion b'lash 0·003 to 0·006in.; shim adjustment
112. Pinion bearing pre-load ... 8 to 12lb.in.
113. Differential brg. pre-load ·003 to ·005in.; shim adjustment

BRAKES

201. Front ... ... Girling hydraulic, 11in. diam. discs
202. Rear ... ... Girling hydraulic, 9in. diam. drums
203. Handbrake ... ... Mechanical, on rear wheels
204. Linings—front ... ... Ferodo, type 14, caliper pads
205. Linings—rear ... ... Ferodo, 1¾in. wide

STEERING (see also "Dimensions")

301. Type ... ... Cam Gear
302. Camber angle ... ... 2°, non-adjustable
303. Castor angle ... ... 4°; non-adjustable
304. King-pin inclination ... 2°
305. Toe-in ... ... ⅛ to 3/16 at wheel rims
306. Conditions for checking ... Unladen

FRONT SUSPENSION

401. Type ... ... Independent, vertical coil springs
402. Number of coils 18
403. Coil diameter ... ... 2⅜in. (external)
404. Wire diameter ... ... 0·375in.
405. Spring length—free ... 13in.
406. Spring length—fitted ... 9in.

REAR SUSPENSION

501. Type ... ... Semi-elliptic leaf-springs
504. Spring length (centres) ... 42¾in. (free)
505. Number of leaves ... 6 (2-str. and 'Plus'); 7 (4-str.)
506. Width of leaves ... ... 1⅜in.
507. Thickness of leaves ... ¼in.
508. Spring camber—free ... 3½°

ELECTRICAL SYSTEM

601. Ignition timing (static) ... 4° B.T.D.C. (Hole in pulley and pointer on timing case, T.D.C.)
602. Sparking plugs ... ... Lodge, type CNY; ½in. reach, 14mm.
603. Sparking plug gap ... 0·025in.
604. Distributor ... ... Lucas, type 25D4
605. Contact breaker gap ... 0·015in.; dwell angle, 60°±3°
606. Centrif. advance starts at 275 r.p.m.
607. Centrif. advance—max.... 9¾° at 800 r.p.m.
608. Vacuum adv. starts at ... 2in. Hg.
609. Vacuum adv.—max. ... 2·6° at 6in. Hg.
610. Ignition coil ... ... Lucas, type HA12
611. Battery ... ... Lucas, two SG9E in series; 12 v. 54a.h.
612. Battery earth ... ... Positive
613. Dynamo ... ... Lucas, C40/1; max. output, 22 amp.
614. Control box ... ... Lucas, RB106/2
615. Starter motor ... ... Lucas, M418G

CAPACITIES

701. Engine sump ... ... 11 pints (dry); 10 pints (drain and refill)
702. Gearbox ... ... 2½ pints
703. Differential ... ... 1¾ pints
704. Cooling system ... ... 16 pints
705. Fuel tank ... ... 11 gallons ('Plus', 10)
706. Tyre size ... ... 5·60 × 15in.
707. Tyre pressures ... ... 18lb./sq.in. normal (add 6lb./sq.in. for sustained high-speed driving on motorways)
708. No. of grease gun points ... 15 (front sliding axles have 'one-shot' lubrication; press foot plunger every 200 miles approx.)
709. Normal servicing intervals Greasing 2000/2500 miles; engine oil change, 2,500 miles
710. Hydraulic fluid ... ... Girling

DIMENSIONS

801. Length overall ... ... 12ft. ('Plus', 12ft. 8in.)
802. Width overall ... ... 4ft. 8in. ('Plus', 5ft. 1in.)
803. Height overall ... ... 4ft. 4½in. ('Plus', 4ft. 2in.)
804. Weight (approx.) ... 1848lb. ('Plus', 1820lb.)
805. Ground clearance (min.) 7in.
806. Track ... ... 47in. front and rear ('Plus': front, 47¾in.; rear, 49¾in.)
807. Wheelbase ... ... 8ft.
808. Turning circle diameter 32ft.

TORQUE SPANNER DATA

	lb.ft.		lb.ft.
901. Cylinder head nuts ...	100/105	907. Final drive pinion bearing nut ...	95/100
902. Big-end bearing nuts ...	55/60	908. Crown wheel bolts	50/60
903. Main bearing nuts. ...	85/90	909. Diff. bearing nuts	35/40
905. Flywheel bolts	42/46	910. Road wheel nuts ...	65/70
906. Clutch to flywheel bolts ...	20		

BODY AND CHASSIS

1001. Type of construction ... Steel 'Z'-section chassis
1002. Material of body panels ... Steel and aluminium ('Plus', glass fibre)
1003. Type of glass ... ... Windscreen toughened (laminated optional on 'Plus 4', standard on 'Plus 4 Plus')

4/4 Series V 1963-1964

Models: Standard and Competition versions

ENGINE (Ford 116E and G.T. version)

1. Type ... ... 4-stroke; overhead valves; side camshaft
2. Cooling system ... ... Water pump, radiator, fan and thermostat
3. Number of cylinders ... 4 (bored direct in block)
4. Firing order & idling speed 1—2—4—3 (No. 1 at front); 600 r.p.m.
5. Bore ... ... 80·96mm.
6. Stroke ... ... 72·75mm.
7. Cubic capacity ... ... 1498c.c.
8. Comp'n ratio and press.... 8·3:1; 175lb./sq.in. (Comp. 9:1; 179)
9. Brake hore-power (gross) 64 at 4,600 r.p.m. (Comp. 83·5 at 5,200)
10. Torque (gross) ... ... 85·5lb.ft. at 2,300 r.p.m.; (Comp. 97 at 3,600)
11. Piston oversizes ... ... 0·0025in.; 0·005in.; 0·015in.; 0·030in.
12. Piston clearance (bottom) 0·0005 to 0·0011in. on thrust side
13. Piston rings—number ... 2 compression, 1 oil
14. Piston rings—width ... Comp., 0·077 to 0·078in.; oil, 0·155 to 0·156in.
15. Piston rings—groove clearance Comp., 0·0016 to 0·0036in.; oil, 0·0018 to 0·0038in.
16. Piston rings—gap (in cyl.) 0·009 to 0·014in.
17. Oil pressure ... ... 35 to 40lb./sq.in.
18. Gudgeon pin diameter ... 0·8120 to 0·8123in.
19. Gudgeon pin fit in piston 0·0000 to 0·0002in. clearance (selective)
20. Gudgeon pin fit in con. rod 0·0001 to 0·0003in. clearance (selective)
21. Crankpin diameter ... 1·9370 to 1·9375in.
22. Crankpin undersizes ... 0·010, 0·020, 0·030, 0·040in.
23. Connecting rod length ... 4·799 to 4·801in. (centres)
24. Big-end bearings ... Vandervell or Glacier, steel-backed, copper-lead or lead-bronze with lead overlay
25. Big-end bearing clearance 0·0005 to 0·0022in.
26. Big-end side clearance ... ·004 to ·010in.
27. Main journal diameter ... 2·1255 to 2·1260in.
28. Main journal undersizes... 0·010, 0·020, 0·030in.
29. Main bearings ... ... 5; Vandervell or Glacier, steel-backed lead-bronze or copper-lead with lead overlay
30. Main bearings—length ... 0·724 to 0·800in. (effective)
31. Main bearing clearance ... 0·0005 to 0·0020in.
32. Crankshaft end-float ... 0·003 to 0·011in.
33. Crankshaft end-thrust on Centre bearing thrust washer
34. Camshaft journal dia. ... 1·560 to 1·5605
35. Camshaft bearings ... 3; Vandervell or Glacier, steel-backed babbit liners
36. Camshaft bearing cl'nce... 0·0010 to 0·0020in.
37. Camshaft end-float ... 0·002 to 0·007in.
38. Camshaft drive type ... Simplex chain; 46 × 0·375in. pitches
39. Valve head diameter ... Inlet, 1·432/1·442in. (Comp. 1·405/1·415in.); exhaust, 1·183/1·193in. (Comp., 1·240/1·250in.)
40. Valve stem diameter ... Inlet, 0·3095 to 0·3105in.; exhaust, 0·3086 to 0·3096in.
41. Valve seat angle ... ... 45° to 45° 15′ inlet and exhaust
42. Valve lift ... ... Inlet, 0·315in. (Comp., 0·3436in.); exhaust, 0·319in. (Comp., 0·3354in.)
43. Valve guide length ... 1·52 to 1·53in.; fitted height, 0·28in. above head
44. Valve stem/guide cl'nce ... Inlet, 0·0008 to 0·0030in.; exhaust, 0·0017 to 0·0039in.
45. Valve spring free length ... 1·48in.
46. Valve spring rate ... Valve open, 0·953in. at 113/123lb.; valve closed, 1·263in. at 44/49lb.
47. Valve work'g cl'nce (cold) Inlet, ·008in. (Comp., ·010in.); exhaust, ·018in. (Comp., ·023in.)
48. Valve timing clearance ... Inlet, ·015in. (Comp., ·016in.); exh., ·027in. (Comp., ·026in.)
49. Valve timing ... ... (*Competition model) Inlet opens 17° (*27°) B.T.D.C.; closes 51° (*65°) A.B.D.C. Exhaust opens 51° (*65°) B.B.D.C.; closes 17° (*27°) A.T.D.C.
50. Petrol pump ... ... A.C.-Delco mechanical
51. Carburettor ... ... Zenith 33VN (Comp. Weber DCD1)

52. Choke tube	29	Primary 26	Secondary 27
53. Main jet	92	,, 140	,, 155
54. Air correction jet ...	112	,, 230	,, 180
55. Pilot jet	55	,, 50	,, 70
56. Pump jet	50	60	
60. Starter air jet ...	—	100	
61. Starter petrol jet ...	—	190	
62. Needle valve	—	1·75mm.	

TRANSMISSION

101. Type ... ... Rear wheel drive [plate
102. Clutch ... ... Ford/Borg & Beck, $7\frac{1}{4}$in. diam. single dry
103. Gearbox ... ... 4 forward speeds, all synchromesh
105. Gearbox ratios ... ... Top, 1:1; 3rd, 1·412:1; 2nd, 2·396:1; 1st, 3·543:1; reverse, 3·963:1
107. Driving axles ... ... Semi-floating
108. Final drive gear ... ... Hypoid (Salisbury)
109. Final drive ratio ... ... 4·56:1 (41/9)
111. Crown wheel/pinion b'lash ·003 to ·006in.; shim adjustment
112. Pinion bearing pre-load... 8 to 12lb.in.; shim adjustment
113. Differential brg. pre-load ·003 to ·005in.; shim adjustment

BRAKES

201. Front ... ... Girling hydraulic, 11in. diam. discs
202. Rear ... ... Girling hydraulic, 9in. diam. drums
203. Handbrake ... ... Mechanical, on rear wheels
204. Linings—front ... ... Ferodo, type 14 caliper pads
205. Linings—rear ... ... Ferodo, $1\frac{3}{4}$in. wide

STEERING (see also "Dimensions")

301. Type ... ... Cam Gear
302. Camber angle ... ... 2°; non-adjustable
303. Castor angle ... ... 4°; non-adjustable
304. King-pin inclination ... 2°
305. Toe-in ... ... $\frac{1}{8}$ to $\frac{3}{16}$in. at wheel rim
306. Conditions for checking ... Unladen

FRONT SUSPENSION

401. Type ... ... Independent, vertical coil springs
402. Number of coils 21
403. Coil diameter ... ... $2\frac{1}{4}$in. (external)
404. Wire diameter ... ... 0·31in.
405. Spring length—free ... $14\frac{1}{4}$in.
406. Spring length—fitted ... $8\frac{1}{2}$in.

REAR SUSPENSION

501. Type ... ... Semi-elliptic leaf springs
504. Spring length (centres) ... $42\frac{3}{4}$in. (free)
505. Number of leaves ... 5
506. Width of leaves ... ... $1\frac{3}{8}$in.
507. Thickness of leaves ... $\frac{1}{4}$in.
508. Spring camber—free ... $3\frac{5}{8}$in.

ELECTRICAL SYSTEM

601. Ignition timing (static) ... 8° (Comp., 10°) B.T.D.C. (mark on pulley and cover)
602. Sparking plugs ... ... Champion, N5 (Comp., N4) 14mm.
603. Sparking plug gap ... 0·023 to 0·028in.
604. Distributor ... ... Lucas, type 25D4
605. Contact breaker gap ... 0·014 to 0·016in.; dwell angle 57°/63°
606. Centrif. advance starts at 1,200 r.p.m. (crank)
607. Centrif. advance—max.... 37° (crank) at 5,000 r.p.m.
608. Vacuum advance starts at 5·2in. Hg.
609. Vacuum advance—max. 28° (crank) at 15in. Hg.
610. Ignition coil ... ... Lucas or A.C.-Delco type LA12
611. Battery ... ... Lucas, BT7A; 12v. 36a.h.
612. Battery earth ... ... Positive
613. Dynamo ... ... Lucas, type C40L; max. output 300w.
614. Control box ... ... Lucas, RB 106/2
615. Starter motor ... ... Lucas, M35G

CAPACITIES

701. Engine sump ... ... $6\frac{1}{4}$ pints
702. Gearbox ... ... $1\frac{3}{4}$ pints
703. Differential ... ... $1\frac{3}{4}$ pints
704. Cooling system ... ... 12 pints
705. Fuel tank ... ... $8\frac{1}{2}$ gallons
706. Tyre size ... ... 5·60 × 15in.
707. Tyre pressures ... ... 16/17lb./sq.in., front and rear, normal (add 6lb./sq.in. for sustained high-speed driving on motorways)
708. No. of greasegun points ... 13 (Front sliding axles have 'one-shot' lubrication; press foot plunger every 200 miles approx.)
709. Servicing intervals ... Greasing, 2000/2500 miles; engine oil change, 5000 miles
710. Hydraulic fluid ... ... Girling/Castrol

DIMENSIONS

801. Length overall ... ... 12ft.
802. Width overall ... ... 4ft. 8in.
803. Height overall ... ... 4ft. $3\frac{3}{4}$in.
804. Weight (approx.) ... 1,456lb. (dry)
805. Ground clearance (min.) 7in.
806. Track ... ... 3ft. 11in. front and rear
807. Wheelbase ... ... 8ft.
808. Turning circle diameter ... 32ft.

TORQUE SPANNER DATA

	lb.ft.		lb.ft.
901. Cylinder head nuts ...	65/70	907. Final drive pinion bearing nut ...	95/100
902. Big-end bearing nuts ...	20/25	908. Crown wheel bolts	50/60
903. Main bearing nuts ...	55/60	909. Diff. bearing nuts	35/40
905. Flywheel bolts	45/50	910. Road wheel nuts	65/70
906. Clutch to flywheel bolts ...	12/15		

BODY AND CHASSIS

1001. Type of construction ... Coachbuilt; steel 'Z'-section chassis; ash frame
1002. Material of body panels ... Steel and aluminium
1003. Type of glass. ... ... Toughened (laminated optional extra)

4/4 Series V 1964-1965

ENGINE (Triumph TR4). See the 1962 edition of this publication for details of previous model fitted with TR3 engine

1. Type ... ... 4-stroke; overhead valves; side camshaft
2. Cooling system ... ... Water pump, fan and thermostat
3. Number of cylinders ... 4 (wet liners)
4. Firing order & idl'g speed 1—3—4—2 (No. 1 at front); 550 r.p.m.
5. Bore ... ... 86mm. (3·386in.)
6. Stroke ... ... 92mm. (3·622in.)
7. Cubic capacity ... ... 2,138c.c. (130·5cu.in.)
8. Compression ratio and pressure 9:1 (7·0:1 optional); max. variation 5lb./sq.in.
9. Brake horse-power ... 105 at 4,750 r.p.m.
10. Torque ... ... 128·3lb.ft. at 3,350 r.p.m.
11. Piston oversizes ... ... 0·010, 0·020, 0·030, 0·040in.
12. Piston clearance ... ... 0·0047 to 0·0054in. measured at top; and 0·0032 to 0·0039in. at bottom
13. Piston rings—number ... 2 compression, 1 oil
14. Piston rings—width ... Comp., 0·0615 to 0·0625in.; oil, 0·1552 to 0·1562in.
15. Piston rings—groove cl'nce Comp., 0·0010 to 0·0030in.; oil, 0·0010 to 0·0030in.
16. Piston rings—gap (in cylinder) ... Comp., 0·010 to 0·015in.; oil, 0·010 to 0·015in.
17. Oil pressure ... ... 65 to 75lb./sq.in. (above 30 m.p.h.)
18. Gudgeon pin diameter ... 0·87485 to 0·87510in.
19. Gudgeon pin fit in piston 0·0002in. clearance at 68°F.
20. Gudgeon pin fit in con. rod Selective
21. Crankpin diameter ... 2·0861 to 2·0866in.
22. Crankpin undersizes ... 0·010, 0·020, 0·030in.
23. Connecting rod length ... 6·248 to 6·252in. (centres)
24. Big-end bearings ... Vandervell, 3-layer
25. Big-end bearing clearance 0·0028 to 0·0040in.
26. Big-end side clearance ... 0·007 to 0·014in.
27. Main journal diameter ... 2·4790 to 2·4795in.
28. Main journal undersizes 0·010, 0·020, 0·030, 0·040in.
29. Main bearings ... ... 3, Vandervell, bimetal (micro) and steel
30. Main bearings—length ... 1·495 to 1·505in. [back
31. Main bearing clearance 0·0015 to 0·0025in.
32. Crankshaft end-float ... 0·0048 to 0·0117in.
33. Crankshaft end-thrust on Centre main bearing thrust washers
34. Camshaft journal dia. ... Front, 1·871 to 1·872in.; others 1·7153 to 1·7158in.
35. Camshaft bearings ... 4; Vandervell (2, 3 & 4); front, cast
36. Camshaft bearing cl'nce Front, 0·0028 to 0·0047in.; others 0·0015 to 0·0029in.
37. Camshaft end-float ... 0·0040 to 0·0075in.
38. Camshaft drive type ... Duplex chain, 21¾in. long
39. Valve head diameter ... Inlet, 1·558 to 1·562in.; exhaust, 1·299 to 1·303in.
40. Valve stem diameter ... Inlet, 0·310 to 0·311in.; exhaust, 0·3705
41. Valve seat angle ... ... Inlet, 45°; exhaust, 45° [to 0·3715in.
42. Valve lift ... ... 0·250in. (cold)
43. Valve guide length ... Inlet, 2·0525 to 2·0725in.; exhaust, 2·05 to 2·07in.
44. Valve stem/guide cl'nce ... Inlet, 0·001 to 0·003in.; exhaust, 0·003 to 0·005in.
45. Valve spring free length ... Outer, 1·94 to 1·96in.; inner, 1·88 to 1·90in. (auxiliary inner, exhaust only, 1·55 to 1·57in.)
46. Valve spring rate... ... Outer, 178lb./in.; inner, 93lb./in. (auxiliary inner, exhaust only, 28·5lb./in.)
47. Valve working cl'nce ... Inlet, 0·010in.; exhaust, 0·010in. (cold)
48. Valve timing clearance ... 0·0165in.
49. Valve timing ... ... Inlet opens 17° B.T.D.C.; closes 57° A.B.D.C. Exhaust opens 57° B.B.D.C.; closes 17° A.T.D.C.
50. Petrol pump ... ... AC-Delco, mechanical
51. Carburettors ... ... 2 Stromberg, type 175 CD
53. Needle ... ... 2 O

TRANSMISSION

101. Type ... ... Rear wheel drive
102. Clutch ... ... Borg and Beck, 9in. diam., single dry plate
103. Gearbox ... ... 4 forward speeds; synchromesh on 2nd, 3rd and top gears
105. Gearbox ratios ... ... Top, 1:1; 3rd 1·205:1; 2nd, 1·745:1; 1st and reverse, 2·97:1
106. Overdrive ... ... Not fitted
107. Driving axles ... ... Semi-floating
108. Final drive gear ... ... Hypoid (Salisbury)
109. Final drive ratio ... ... 3·73:1
110. Number of gear teeth ... 41/11
111. Crown wheel/pinion b'lash 0·003 to 0·006in.; shim adjustment
112. Pinion bearing pre-load ... 8 to 12lb.in.
113. Differential brg. pre-load ·003 to ·005in.; shim adjustment

BRAKES

201. Front ... ... Girling hydraulic, 11in. diam. discs
202. Rear ... ... Girling hydraulic, 9in. diam. drums
203. Handbrake ... ... Mechanical, on rear wheels
204. Linings—front ... ... Ferodo, type 14, caliper pads
205. Linings—rear ... ... Ferodo, 1¾in. wide

STEERING (see also "Dimensions")

301. Type ... ... Cam Gear
302. Camber angle ... ... 2°, non-adjustable
303. Castor angle ... ... 4°; non-adjustable
304. King-pin inclination ... 2°
305. Toe-in ... ... ⅛ to $\frac{3}{16}$ at wheel rims
306. Conditions for checking ... Unladen

FRONT SUSPENSION

401. Type ... ... Independent, vertical coil springs
402. Number of coils 18
403. Coil diameter ... ... 2⅜in. (external)
404. Wire diameter ... ... 0·375in.
405. Spring length—free ... 13in.
406. Spring length—fitted ... 9in.

REAR SUSPENSION

501. Type ... ... Semi-elliptic leaf-springs
504. Spring length (centres) ... 42¾in. (free)
505. Number of leaves ... 6 (2-str. and 'Plus'); 7 (4-str.)
506. Width of leaves ... ... 1⅜in.
507. Thickness of leaves ... ¼in.
508. Spring camber—free ... 3½°

ELECTRICAL SYSTEM

601. Ignition timing (static) ... 4° B.T.D.C. (Hole in pulley and pointer on timing case, T.D.C.)
602. Sparking plugs ... ... Lodge, type CNY; ½in. reach, 14mm.
603. Sparking plug gap ... 0·025in.
604. Distributor ... ... Lucas, type 25D4
605. Contact breaker gap ... 0·015in.; dwell angle, 60°±3°
606. Centrif. advance starts at 275 r.p.m.
607. Centrif. advance—max.... 9¾° at 800 r.p.m.
608. Vacuum adv. starts at ... 2in. Hg.
609. Vacuum adv.—max. ... 2·6° at 6in. Hg.
610. Ignition coil ... ... Lucas, type HA12
611. Battery ... ... Lucas, two SG9E in series; 12 v. 54a.h.
612. Battery earth ... ... Positive
613. Dynamo ... ... Lucas, C40/1; max. output, 22 amp.
614. Control box ... ... Lucas, RB106/2
615. Starter motor ... ... Lucas, M418G

CAPACITIES

701. Engine sump ... ... 11 pints (dry); 10 pints (drain and refill)
702. Gearbox ... ... 2½ pints
703. Differential ... ... 1¾ pints
704. Cooling system ... ... 16 pints
705. Fuel tank ... ... 11 gallons ('Plus', 10)
706. Tyre size ... ... 5·60 × 15in.
707. Tyre pressures ... ... 18lb./sq.in. normal (add 6lb./sq.in. for sustained high-speed driving on motorways)
708. No. of grease gun points ... 15 (front sliding axles have 'one-shot' lubrication; press foot plunger every 200 miles approx.)
709. Normal servicing intervals Greasing 2,000/2,500 miles; engine oil change, 3,000 miles
710. Hydraulic fluid ... ... Girling

DIMENSIONS

801. Length overall ... ... 12ft. ('Plus', 12ft. 8in.)
802. Width overall ... ... 4ft. 8in. ('Plus', 5ft. 1in.)
803. Height overall ... ... 4ft. 4½in. ('Plus', 4ft. 2in.)
804. Weight (approx.) ... 1848lb. ('Plus', 1820lb.)
805. Ground clearance (min.) 7in.
806. Track ... ... 47in. front and rear ('Plus': front, 47¾in.; rear, 49¾in.)
807. Wheelbase ... ... 8ft.
808. Turning circle diameter 32ft.

TORQUE SPANNER DATA

	lb.ft.		lb.ft.
901. Cylinder head nuts ...	100/105	907. Final drive pinion bearing nut ...	95/100
902. Big-end bearing nuts ...	55/60	908. Crown wheel bolts	50/60
903. Main bearing nuts. ...	85/90	909. Diff. bearing nuts	35/40
905. Flywheel bolts	42/46	910. Road wheel nuts ...	65/70
906. Clutch to flywheel bolts ...	20		

BODY AND CHASSIS

1001. Type of construction ... Steel 'Z'-section chassis
1002. Material of body panels ... Steel and aluminium ('Plus', glass fibre)
1003. Type of glass ... ... Windscreen toughened (laminated optional on 'Plus 4', standard on 'Plus 4 Plus')

Plus 4 & Plus 4 Plus 1964-1965

Models: Standard and Competition versions

ENGINE (Ford 116E and G.T. version)

1. Type ... ... 4-stroke; overhead valves; side camshaft
2. Cooling system ... ... Water pump, radiator, fan and thermostat
3. Number of cylinders ... 4 (bored direct in block)
4. Firing order & idling speed 1—2—4—3 (No. 1 at front); 600 r.p.m.
5. Bore ... ... 80·96mm.
6. Stroke ... ... 72·75mm.
7. Cubic capacity ... ... 1498c.c.
8. Comp'n ratio and press.... 8·3:1; 175lb./sq.in. (Comp. 9:1; 179)
9. Brake hore-power (gross) 64 at 4,600 r.p.m. (Comp. 83·5 at 5,200)
10. Torque (gross) ... ... 85·5lb.ft. at 2,300 r.p.m.; (Comp. 97 at 3,600)
11. Piston oversizes ... ... 0·0025in.; 0·005in.; 0·015in.; 0·030in.
12. Piston clearance (bottom) 0·0005 to 0·0011in. on thrust side
13. Piston rings—number ... 2 compression, 1 oil
14. Piston rings—width ... Comp., 0·077 to 0·078in.; oil, 0·155 to 0·156in.
15. Piston rings—groove clearance Comp., 0·0016 to 0·0036in.; oil, 0·0018 to 0·0038in.
16. Piston rings—gap (in cyl.) 0·009 to 0·014in.
17. Oil pressure ... ... 35 to 40lb./sq.in.
18. Gudgeon pin diameter ... 0·8120 to 0·8123in.
19. Gudgeon pin fit in piston 0·0000 to 0·0002in. clearance (selective)
20. Gudgeon pin fit in con. rod 0·0001 to 0·0003in. clearance (selective)
21. Crankpin diameter ... 1·9370 to 1·9375in.
22. Crankpin undersizes ... 0·010, 0·020, 0·030, 0·040in.
23. Connecting rod length ... 4·799 to 4·801in. (centres)
24. Big-end bearings ... Vandervell or Glacier, steel-backed, copper-lead or lead-bronze with lead overlay
25. Big-end bearing clearance 0·0005 to 0·0022in.
26. Big-end side clearance ... ·004 to ·010in.
27. Main journal diameter ... Red, 2·1261/2·1257, blue, 2·1257/2·1253
28. Main journal undersizes... 0·010, 0·020, 0·030in.
29. Main bearings ... ... 5; Vandervell or Glacier, steel-backed lead-bronze or copper-lead with lead overlay
30. Main bearings—length ... 0·724 to 0·800in. (effective)
31. Main bearing clearance ... 0·0005 to 0·0020in.
32. Crankshaft end-float ... 0·003 to 0·011in.
33. Crankshaft end-thrust on Centre bearing thrust washer
34. Camshaft journal dia. ... 1·560 to 1·5605
35. Camshaft bearings ... 3; Vandervell or Glacier, steel-backed babbit liners
36. Camshaft bearing cl'nce... 0·0010 to 0·0020in.
37. Camshaft end-float ... 0·002 to 0·007in.
38. Camshaft drive type ... Simplex chain; 46 × 0·375in. pitches
39. Valve head diameter ... Inlet, 1·432/1·442in. (Comp. 1·405/1·415in.); exhaust, 1·183/1·193in. (Comp., 1·240/1·250in.)
40. Valve stem diameter ... Inlet, 0·3095 to 0·3105in.; exhaust, 0·3086 to 0·3096in.
41. Valve seat angle ... ... 45° to 45° 15′ inlet and exhaust
42. Valve lift ... ... Inlet, 0·315in. (Comp., 0·3436in.); exhaust, 0·319in. (Comp., 0·3354in.)
43. Valve guide length ... 1·52 to 1·53in.; fitted height, 0·28in. above head
44. Valve stem/guide cl'nce ... Inlet, 0·0008 to 0·0030in.; exhaust, 0·0017 to 0·0039in.
45. Valve spring free length ... 1·48in.
46. Valve spring rate ... Valve open, 0·953in. at 113/123lb.; valve closed, 1·263in. at 44/49lb.
47. Valve work'g cl'nce (cold) Inlet, ·008in. (Comp., ·010in.); exhaust, ·018in. (Comp., ·023in.)
48. Valve timing clearance ... Inlet, ·015in. (Comp., ·016in.); exh., ·027in. (Comp., ·026in.)
49. Valve timing ... ... Inlet opens 17° (*27°) B.T.D.C.; closes 51° (*65°) A.B.D.C. Exhaust opens 51° (*65°) B.B.D.C.; closes 17° (*27°) A.T.D.C.
(*Competition model)
50. Petrol pump ... ... A.C.-Delco mechanical
51. Carburettor ... ... Zenith 33VN (Comp. Weber DCD22)

52. Choke tube	29	Primary	26	Secondary	27
53. Main jet	92	,,	140	,,	155
54. Air correction jet ...	112	,,	230	,,	180
55. Pilot jet	55	,,	50	,,	70
58. Pump jet	50		60		
60. Starter air jet ...	—		100		
61. Starter petrol jet ...	—		190		
62. Needle valve	—		1·75mm.		

TRANSMISSION

101. Type ... ... Rear wheel drive
102. Clutch ... ... Ford/Borg & Beck, 7¼in. diam. single dry [plate
103. Gearbox ... ... 4 forward speeds, all synchromesh
105. Gearbox ratios ... ... Top, 1:1; 3rd, 1·412:1; 2nd, 2·396:1; 1st, 3·543:1; reverse, 3·963:1
107. Driving axles ... ... Semi-floating
108. Final drive gear ... ... Hypoid (Salisbury)
109. Final drive ratio ... ... 4·56:1 (41/9)
111. Crown wheel/pinion b'lash ·003 to ·006in.; shim adjustment
112. Pinion bearing pre-load... 8 to 12lb.in.; shim adjustment
113. Differential brg. pre-load ·003 to ·005in.; shim adjustment

BRAKES

201. Front ... ... Girling hydraulic, 11in. diam. discs
202. Rear ... ... Girling hydraulic, 9in. diam. drums
203. Handbrake ... ... Mechanical, on rear wheels
204. Linings—front ... ... Ferodo, type 14 caliper pads
205. Linings—rear ... ... Ferodo, 1¾in. wide

STEERING (see also "Dimensions")

301. Type ... ... Cam Gear
302. Camber angle ... ... 2°; non-adjustable
303. Castor angle ... ... 4°; non-adjustable
304. King-pin inclination ... 2°
305. Toe-in ... ... ⅛ to 3/16in. at wheel rim
306. Conditions for checking ... Unladen

FRONT SUSPENSION

401. Type ... ... Independent, vertical coil springs
402. Number of coils 21
403. Coil diameter ... ... 2¼in. (external)
404. Wire diameter ... ... 0·31in.
405. Spring length—free ... 14½in.
406. Spring length—fitted ... 8½in.

REAR SUSPENSION

501. Type ... ... Semi-elliptic leaf springs
504. Spring length (centres) ... 42¾in. (free)
505. Number of leaves ... 5
506. Width of leaves ... ... 1⅜in.
507. Thickness of leaves ... ¼in.
508. Spring camber—free ... 3⅝in.

ELECTRICAL SYSTEM

601. Ignition timing (static) ... 6° (Comp., 10°) B.T.D.C. (mark on pulley and cover)
602. Sparking plugs ... ... Autolite AG32 (Comp. AG22)
603. Sparking plug gap ... 0·025in.
604. Distributor ... ... Lucas, type 25D4
605. Contact breaker gap ... 0·014 to 0·016in.; dwell angle 57°/63°
606. Centrif. advance starts at 1,200 r.p.m. (crank)
607. Centrif. advance—max.... 37° (crank) at 5,000 r.p.m.
608. Vacuum advance starts at 5·2in. Hg.
609. Vacuum advance—max. 28° (crank) at 15in. Hg.
610. Ignition coil ... ... Lucas or A.C.-Delco type LA12
611. Battery ... ... Lucas, BT7A; 12v. 36a.h.
612. Battery earth ... ... Positive
613. Dynamo ... ... Lucas, type C40L; max. output 300w.
614. Control box ... ... Lucas, RB 106/2
615. Starter motor ... ... Lucas, M35G

CAPACITIES

701. Engine sump ... ... 7 pints (including filter)
702. Gearbox ... ... 1¾ pints
703. Differential ... ... 1¾ pints
704. Cooling system ... ... 12 pints
705. Fuel tank ... ... 8½ gallons
706. Tyre size ... ... 5·60 × 15in.
707. Tyre pressures ... ... 16/17lb./sq.in., front and rear, normal (add 6lb./sq.in. for sustained high-speed driving on motorways)
708. No. of greasegun points ... 13 (Front sliding axles have 'one-shot' lubrication; press foot plunger every 200 miles approx.)
709. Servicing intervals ... Greasing, 2000/2500 miles; engine oil change, 5000 miles
710. Hydraulic fluid ... ... Girling/Castrol

DIMENSIONS

801. Length overall ... ... 12ft.
802. Width overall ... ... 4ft. 8in.
803. Height overall ... ... 4ft. 3¾in.
804. Weight (approx.) ... 1,456lb. (dry)
805. Ground clearance (min.) 7in.
806. Track ... ... 3ft. 11in. front and rear
807. Wheelbase ... ... 8ft.
808. Turning circle diameter ... 32ft.

TORQUE SPANNER DATA

	lb.ft.		lb.ft.
901. Cylinder head nuts ...	65/70	907. Final drive pinion bearing nut ...	95/100
902. Big-end bearing nuts ...	20/25	908. Crown wheel bolts	50/60
903. Main bearing nuts ...	55/60	909. Diff. bearing nuts	35/40
905. Flywheel bolts	45/50	910. Road wheel nuts	65/70
906. Clutch to flywheel bolts ...	12/15		

BODY AND CHASSIS

1001. Type of construction ... Coachbuilt; steel 'Z'-section chassis; ash frame
1002. Material of body panels ... Steel and aluminium
1003. Type of glass. ... ... Toughened (laminated optional extra)

Plus 4 & Plus 4 Plus 1965-1966

ENGINE (Triumph TR4)

1.	Type	4-stroke; overhead valves; side camshaft
2.	Cooling system	Water pump, fan and thermostat
3.	Number of cylinders ...	4 (wet liners)
4.	Firing order & idl'g speed	1—3—4—2 (No. 1 at front); 550 r.p.m.
5.	Bore	86mm. (3·386in.)
6.	Stroke	92mm. (3·622in.)
7.	Cubic capacity	2,138c.c. (130·5cu.in.)
8.	Compression ratio and pressure	9:1 (7·0:1 optional); max. variation 5lb./sq.in.
9.	Brake horse-power ...	105 at 4,750 r.p.m.
10.	Torque	128·3lb.ft. at 3,350 r.p.m.
11.	Piston oversizes	0·010, 0·020, 0·030, 0·040in.
12.	Piston clearance	0·0047 to 0·0054in. measured at top; and 0·0032 to 0·0039in. at bottom
13.	Piston rings—number ...	2 compression, 1 oil
14.	Piston rings—width ...	Comp., 0·0615 to 0·0625in.; oil, 0·1552 to 0·1562in.
15.	Piston rings—groove cl'nce	Comp., 0·0010 to 0·0030in.; oil, 0·0010 to 0·0030in.
16.	Piston rings—gap (in cylinder) ...	Comp., 0·010 to 0·015in.; oil, 0·010 to 0·015in.
17.	Oil pressure	65 to 75lb./sq.in. (above 30 m.p.h.)
18.	Gudgeon pin diameter ...	0·87485 to 0·87510in.
19.	Gudgeon pin fit in piston	0·0002in. clearance at 68°F.
20.	Gudgeon pin fit in con. rod	Selective
21.	Crankpin diameter ...	2·0861 to 2·0866in.
22.	Crankpin undersizes ...	0·010, 0·020, 0·030in.
23.	Connecting rod length ...	6·248 to 6·252in. (centres)
24.	Big-end bearings ...	Vandervell, 3-layer
25.	Big-end bearing clearance	0·0028 to 0·0040in.
26.	Big-end side clearance ...	0·007 to 0·014in.
27.	Main journal diameter ...	2·4790 to 2·4795in.
28.	Main journal undersizes	0·010, 0·020, 0·030, 0·040in.
29.	Main bearings	3, Vandervell, bimetal (micro) and steel
30.	Main bearings—length ...	1·495 to 1·505in. [back
31.	Main bearing clearance	0·0015 to 0·0025in.
32.	Crankshaft end-float ...	0·0048 to 0·0117in.
33.	Crankshaft end-thrust on	Centre main bearing thrust washers
34.	Camshaft journal dia. ...	Front, 1·871 to 1·872in.; others 1·7153 to 1·7158in.
35.	Camshaft bearings ...	4; Vandervell (2, 3 & 4); front, cast
36.	Camshaft bearing cl'nce	Front, 0·0028 to 0·0047in.; others 0·0015 to 0·0029in.
37.	Camshaft end-float ...	0·0040 to 0·0075in.
38.	Camshaft drive type ...	Duplex chain, 21¾in. long
39.	Valve head diameter ...	Inlet, 1·558 to 1·562in.; exhaust, 1·299 to 1·303in.
40.	Valve stem diameter ...	Inlet, 0·310 to 0·311in.; exhaust, 0·3705
41.	Valve seat angle	Inlet, 45°; exhaust, 45° [to 0·3715in.
42.	Valve lift	0·250in. (cold)
43.	Valve guide length ...	Inlet, 2·0525 to 2·0725in.; exhaust, 2·05 to 2·07in.
44.	Valve stem/guide cl'nce ...	Inlet, 0·001 to 0·003in.; exhaust, 0·003 to 0·005in.
45.	Valve spring free length ...	Outer, 1·94 to 1·96in.; inner, 1·88 to 1·90in. (auxiliary inner, exhaust only, 1·55 to 1·57in.)
46.	Valve spring rate... ...	Outer, 178lb./in.; inner, 93lb./in. (auxiliary inner, exhaust only, 28·5lb./in.)
47.	Valve working cl'nce ...	Inlet, 0·010in.; exhaust, 0·010in. (cold)
48.	Valve timing clearance ...	0·0165in.
49.	Valve timing	Inlet opens 17° B.T.D.C.; closes 57° A.B.D.C. Exhaust opens 57° B.B.D.C.; closes 17° A.T.D.C.
50.	Petrol pump	AC-Delco, mechanical
51.	Carburettors	2 Stromberg, type 175 CD (later S.U.,
53.	Needle	2 H (S.U., SM) [HS6)

TRANSMISSION

101.	Type	Rear wheel drive
102.	Clutch	Borg and Beck, 9in. diam., single dry plate
103.	Gearbox	4 forward speeds; synchromesh on 2nd, 3rd and top gears
105.	Gearbox ratios	Top, 1:1; 3rd 1·205:1; 2nd, 1·745:1; 1st and reverse, 2·97:1
106.	Overdrive	Not fitted
107.	Driving axles	Semi-floating
108.	Final drive gear	Hypoid (Salisbury)
109.	Final drive ratio	3·73:1
110.	Number of gear teeth ...	41/11
111.	Crown wheel/pinion b'lash	0·003 to 0·006in.; shim adjustment
112.	Pinion bearing pre-load ...	8 to 12lb.in.
113.	Differential brg. pre-load	·003 to ·005in.; shim adjustment

BRAKES

201.	Front	Girling hydraulic, 11in. diam. discs
202.	Rear	Girling hydraulic, 9in. diam. drums
203.	Handbrake	Mechanical, on rear wheels
204.	Linings—front	Ferodo, type 14, caliper pads
205.	Linings—rear	Ferodo, 1¾in. wide

STEERING (see also "Dimensions")

301.	Type	Cam Gear
302.	Camber angle	2°, non-adjustable
303.	Castor angle	4°; non-adjustable
304.	King-pin inclination ...	2°
305.	Toe-in	⅛ to $\frac{3}{16}$ at wheel rims
306.	Conditions for checking ...	Unladen

FRONT SUSPENSION

401.	Type	Independent, vertical coil springs
402.	Number of coils	18
403.	Coil diameter	2⅜in. (external)
404.	Wire diameter	0·375in.
405.	Spring length—free ...	13in.
406.	Spring length—fitted ...	9in.

REAR SUSPENSION

501.	Type	Semi-elliptic leaf-springs
504.	Spring length (centres) ...	42¾in. (free)
505.	Number of leaves ...	6 (2-str. and 'Plus'); 7 (4-str.)
506.	Width of leaves	1⅜in.
507.	Thickness of leaves ...	¼in.
508.	Spring camber—free ...	3½°

ELECTRICAL SYSTEM

601.	Ignition timing (static) ...	4° B.T.D.C. (Hole in pulley and pointer on timing case, T.D.C.)
602.	Sparking plugs	Lodge, type CNY; ½in. reach, 14mm.
603.	Sparking plug gap ...	0·025in.
604.	Distributor	Lucas, type 25D4
605.	Contact breaker gap ...	0·015in.; dwell angle, 60°±3°
606.	Centrif. advance starts at	275 r.p.m.
607.	Centrif. advance—max. ...	9¾° at 800 r.p.m.
608.	Vacuum adv. starts at ...	2in. Hg.
609.	Vacuum adv.—max. ...	2·6° at 6in. Hg.
610.	Ignition coil	Lucas, type HA12
611.	Battery	Lucas, two SG9E in series; 12v. 54a.h.
612.	Battery earth	Positive
613.	Dynamo	Lucas, C40/1; max. output, 22 amp.
614.	Control box	Lucas, RB106/2
615.	Starter motor	Lucas, M418G

CAPACITIES

701.	Engine sump	11 pints (dry); 10 pints (drain and refill)
702.	Gearbox	2½ pints
703.	Differential	1¾ pints
704.	Cooling system	16 pints
705.	Fuel tank	11 gallons ('Plus', 10)
706.	Tyre size	5·60 × 15in.
707.	Tyre pressures	18lb./sq.in. normal (add 6lb./sq.in. for sustained high-speed driving on motorways)
708.	No. of grease gun points ...	15 (front sliding axles have 'one-shot' lubrication; press foot plunger every 200 miles approx.)
709.	Normal servicing intervals	Greasing 2,000/2,500 miles; engine oil change, 3,000 miles
710.	Hydraulic fluid	Girling

DIMENSIONS

801.	Length overall	12ft.	('Plus', 12ft. 8in.)
802.	Width overall	4ft. 8in.	('Plus', 5ft. 1in.)
803.	Height overall	4ft. 4½in.	('Plus', 4ft. 2in.)
804.	Weight (approx.) ...	1848lb.	('Plus', 1820lb.)
805.	Ground clearance (min.)	7in.	
806.	Track	47in. front and rear ('Plus': front, 47¾in.; rear, 49¾in.)	
807.	Wheelbase	8ft.	
808.	Turning circle diameter	32ft.	

TORQUE SPANNER DATA

		lb.ft.			lb.ft.
901.	Cylinder head nuts ...	100/105	907.	Final drive pinion bearing nut ...	95/100
902.	Big-end bearing nuts ...	55/60	908.	Crown wheel bolts	50/60
903.	Main bearing nuts. ...	85/90	909.	Diff. bearing nuts	35/40
905.	Flywheel bolts	42/46	910.	Road wheel nuts ...	65/70
906.	Clutch to flywheel bolts ...	20			

BODY AND CHASSIS

1001.	Type of construction ...	Steel 'Z'-section chassis
1002.	Material of body panels ...	Steel and aluminium ('Plus', glass fibre)
1003.	Type of glass	Windscreen toughened (laminated optional on 'Plus 4', standard on 'Plus 4 Plus')

4/4 Series V 1965-1966

Models: Standard and Competition versions

ENGINE (Ford 116E and G.T. version)

1. Type ... ... 4-stroke; overhead valves; side camshaft
2. Cooling system ... ... Water pump, radiator, fan and thermostat
3. Number of cylinders ... 4 (bored direct in block)
4. Firing order & idling speed 1—2—4—3 (No. 1 at front); 600r.p.m.
5. Bore ... ... 3·1878in. (80·97mm.)
6. Stroke ... ... 2·867in. (72·82mm.)
7. Cubic capacity ... ... 91·53cu.in. (1499·9c.c.)
8. Comp'n ratio and press.... 9·0:1; 179lb./sq.in.
9. Brake hore-power (gross) 65 at 4,800 r.p.m. (Comp., 83·5 at 5,200)
10. Torque (gross) b.ft. ... 88·5 at 2,500 r.p.m. (Comp. 97·0 at 3,600)
11. Piston oversizes ... ... 0·0025in.; 0·005in.; 0·015in.; 0·030in.
12. Piston clearance ... ... ·0008/·0014in. (Comp., ·0011/·0017in.) at grade plane
13. Piston rings—number ... 2 compression, 1 oil
14. Piston rings—width ... Comp., 0·077 to 0·078in.; oil, 0·155 to 0·156in.
15. Piston rings—groove clearance Comp., 0·0016 to 0·0036in.; oil, 0·0018 to 0·0038in.
16. Piston rings—gap (in cyl.) 0·009 to 0·014in.
17. Oil pressure ... ... 35 to 40lb./sq.in.
18. Gudgeon pin diameter ... 0·8120 to 0·8123in.
19. Gudgeon pin fit in piston 0·0000 to 0·0002in. clearance (selective)
20. Gudgeon pin fit in con. rod 0·0001 to 0·0003in. clearance (selective)
21. Crankpin diameter ... 1·9370 to 1·9375in.
22. Crankpin undersizes ... 0·010, 0·020, 0·030, 0·040in.
23. Connecting rod length ... 4·799 to 4·801in. (centres)
24. Big-end bearings ... Vandervell or Glacier, steel-backed, copper-lead or lead-bronze with lead [overlay
25. Big-end bearing clearance 0·0005 to 0·0022in.
26. Big-end side clearance ... ·004 to ·010in.
27. Main journal diameter ... Red, 2·1261/2·1257, blue, 2·1257/2·1253
28. Main journal undersizes... 0·010, 0·020, 0·030in.
29. Main bearings ... ... 5; Vandervell or Glacier, steel-backed lead-bronze or copper-lead with lead overlay
30. Main bearings—length ... 0·724 to 0·800in. (effective)
31. Main bearing clearance ... 0·0005 to 0·0020in.
32. Crankshaft end-float ... 0·003 to 0·011in.
33. Crankshaft end-thrust on Centre bearing thrust washer
34. Camshaft journal dia. ... 1·560 to 1·5605
35. Camshaft bearings ... 3; Vandervell or Glacier, steel-backed babbit liners
36. Camshaft bearing cl'nce... 0·0010 to 0·0020in.
37. Camshaft end-float ... 0·002 to 0·007in.
38. Camshaft drive type ... Simplex chain; 46 × 0·375in. pitches
39. Valve head diameter ... Inlet, 1·432/1·442in. (Comp. 1·405/1·415in.); exhaust, 1·183/1·193in. (Comp., 1·240/1·250in.)
40. Valve stem diameter ... Inlet, 0·3095 to 0·3105in.; exhaust, 0·3086 to 0·3096in.
41. Valve seat angle ... ... 45° to 45° 15′ inlet and exhaust
42. Valve lift ... ... Inlet, 0·315in. (Comp., 0·3436in.); exhaust, 0·319in. (Comp., 0·3354in.)
43. Valve guide length ... 1·52 to 1·53in.; fitted height, 0·28in. above head
44. Valve stem/guide cl'nce ... Inlet, 0·0008 to 0·0030in.; exhaust, 0·0017 to 0·0039in.
45. Valve spring free length ... 1·48in.
46. Valve spring rate ... Valve open, 0·953in. at 113/123lb.; valve closed, 1·263in. at 44/49lb.
47. Valve work'g cl'nce (cold) Inlet, ·008in. (Comp., ·010in.); exhaust, ·018in. (Comp., ·023in.)
48. Valve timing clearance ... Inlet, ·015in. (Comp., ·016in.); exh., ·027in. (Comp., ·026in.)
49. Valve timing ... ... Inlet opens 17° (*27°) B.T.D.C.; closes (*Competition model) 51° (*65°) A.B.D.C. Exhaust opens 51° (*65°) B.B.D.C.; closes 17° (*27°) A.T.D.C.
50. Petrol pump ... ... A.C.-Delco mechanical
51. Carburettor ... ... Zenith 33VN (Comp. Weber DCD22)

52. Choke tube	29	Primary 26	Secondary 27
53. Main jet	92	„ 140	„ 155
54. Air correction jet ...	112	„ 230	„ 180
55. Pilot jet	55	„ 50	„ 70
58. Pump jet	50	60	
60. Starter air jet	—	100	
61. Starter petrol jet ...	—	190	
62. Needle valve	—	1·75mm.	

TRANSMISSION

101. Type ... ... Rear wheel drive [plate
102. Clutch ... ... Ford/Borg & Beck, $7\frac{1}{4}$in. diam. single dry
103. Gearbox ... ... 4 forward speeds, all synchromesh
105. Gearbox ratios ... ... Top, 1:1; 3rd, 1·412:1; 2nd, 2·396:1; 1st, 3·543:1; reverse, 3·963:1
107. Driving axles ... ... Semi-floating
108. Final drive gear ... ... Hypoid (Salisbury)
109. Final drive ratio ... ... 4·56:1 (41/9)
111. Crown wheel/pinion b'lash ·003 to ·006in.; shim adjustment
112. Pinion bearing pre-load... 8 to 12lb.in.; shim adjustment
113. Differential brg. pre-load ·003 to ·005in.; shim adjustment

BRAKES

201. Front ... ... Girling hydraulic, 11in. diam. discs
202. Rear ... ... Girling hydraulic, 9in. diam. drums
203. Handbrake ... ... Mechanical, on rear wheels
204. Linings—front ... ... Ferodo, type 14 caliper pads
205. Linings—rear ... ... Ferodo, $1\frac{3}{4}$in. wide

STEERING (see also "Dimensions")

301. Type ... ... Cam Gear
302. Camber angle ... ... 2°; non-adjustable
303. Castor angle ... ... 4°; non-adjustable
304. King-pin inclination ... 2°
305. Toe-in ... ... $\frac{1}{8}$ to $\frac{3}{16}$in. at wheel rim
306. Conditions for checking ... Unladen

FRONT SUSPENSION

401. Type ... ... Independent, vertical coil springs
402. Number of coils 21
403. Coil diameter ... ... $2\frac{1}{4}$in. (external)
404. Wire diameter ... ... 0·31in.
405. Spring length—free ... $14\frac{1}{4}$in.
406. Spring length—fitted ... $8\frac{1}{2}$in.

REAR SUSPENSION

501. Type ... ... Semi-elliptic leaf springs
504. Spring length (centres) ... $42\frac{3}{4}$in. (free)
505. Number of leaves ... 5
506. Width of leaves ... ... $1\frac{3}{8}$in.
507. Thickness of leaves ... $\frac{1}{4}$in.
508. Spring camber—free ... $3\frac{5}{8}$in.

ELECTRICAL SYSTEM

601. Ignition timing (static) ... 8° (Comp., 10°) B.T.D.C. (mark on pulley and cover)
602. Sparking plugs ... ... Autolite AG32 (Comp. AG22)
603. Sparking plug gap ... 0·025in.
604. Distributor ... ... Lucas, type 25D4
605. Contact breaker gap ... 0·014 to 0·016in.; dwell angle 57°/63°
606. Centrif. advance starts at 1,200 r.p.m. (crank)
607. Centrif. advance—max. ... 37° (crank) at 5,000 r.p.m.
608. Vacuum advance starts at 5·2in. Hg.
609. Vacuum advance—max. 28° (crank) at 15in. Hg.
610. Ignition coil ... ... Lucas or A.C.-Delco type LA12
611. Battery ... ... Lucas, BT7A; 12v. 36a.h.
612. Battery earth ... ... Positive
613. Dynamo ... ... Lucas, type C40L; max. output 300w.
614. Control box ... ... Lucas, RB 106/2
615. Starter motor ... ... Lucas, M35G

CAPACITIES

701. Engine sump ... ... 7 pints (including filter)
702. Gearbox ... ... $1\frac{3}{4}$ pints
703. Differential ... ... $1\frac{3}{4}$ pints
704. Cooling system ... ... 12 pints
705. Fuel tank ... ... $8\frac{1}{2}$ gallons
706. Tyre size ... ... 5·60 × 15in.
707. Tyre pressures ... ... 16/17lb./sq.in., front and rear, normal (add 6lb./sq.in. for sustained high-speed driving on motorways)
708. No. of greasegun points ... 13 (Front sliding axles have 'one-shot' lubrication; press foot plunger every 200 miles approx.)
709. Servicing intervals ... Greasing, 2000/2500 miles; engine oil change, 5000 miles
710. Hydraulic fluid ... ... Girling/Castrol

DIMENSIONS

801. Length overall ... ... 12ft.
802. Width overall ... ... 4ft. 8in.
803. Height overall ... ... 4ft. $3\frac{3}{4}$in.
804. Weight (approx.) ... 1,456lb. (dry)
805. Ground clearance (min.) 7in.
806. Track ... ... 3ft. 11in. front and rear
807. Wheelbase ... ... 8ft.
808. Turning circle diameter ... 32ft.

TORQUE SPANNER DATA

	lb.ft.		lb.ft.
901. Cylinder head nuts ...	65/70	907. Final drive pinion bearing nut ...	95/100
902. Big-end bearing nuts ...	20/25	908. Crown wheel bolts	50/60
903. Main bearing nuts ...	55/60	909. Diff. bearing nuts	35/40
905. Flywheel bolts	45/50	910. Road wheel nuts	65/70
906. Clutch to flywheel bolts ...	12/15		

BODY AND CHASSIS

1001. Type of construction ... Coachbuilt; steel 'Z'-section chassis; ash frame
1002. Material of body panels ... Steel and aluminium
1003. Type of glass. ... ... Toughened (laminated optional extra)

4/4 Series V 1966-1967

ENGINE (Triumph TR4A)

1.	Type	4-stroke; overhead valves; side camshaft
2.	Cooling system	Water pump, fan and thermostat
3.	Number of cylinders	4 (wet liners)
4.	Firing order & idl'g speed	1—3—4—2 (No. 1 at front); 550 r.p.m.
5.	Bore	86mm. (3·386in.)
6.	Stroke	92mm. (3·622in.)
7.	Cubic capacity	2,138c.c. (130·5cu.in.)
8.	Compression ratio and pressure	9:1 (7·0:1 optional); max. variation 5lb./sq.in.
9.	Brake horse-power	105 at 4,750 r.p.m.
10.	Torque	128·3lb.ft. at 3,350 r.p.m.
11.	Piston oversizes	0·010, 0·020, 0·030, 0·040in.
12.	Piston clearance	0·0047 to 0·0054in. measured at top; and 0·0032 to 0·0039in. at bottom
13.	Piston rings—number	2 compression, 1 oil
14.	Piston rings—width	Comp., 0·0615 to 0·0625in.; oil, 0·1552 to 0·1562in.
15.	Piston rings—groove cl'nce	Comp., 0·0010 to 0·0030in.; oil, 0·0010 to 0·0030in.
16.	Piston rings—gap (in cylinder)	Comp., 0·010 to 0·015in.; oil, 0·010 to 0·015in.
17.	Oil pressure	65 to 75lb./sq.in. (above 30 m.p.h.)
18.	Gudgeon pin diameter	0·87485 to 0·87510in.
19.	Gudgeon pin fit in piston	0·0002in. clearance at 68°F.
20.	Gudgeon pin fit in con. rod	Selective
21.	Crankpin diameter	2·0861 to 2·0866in.
22.	Crankpin undersizes	0·010, 0·020, 0·030in.
23.	Connecting rod length	6·248 to 6·252in. (centres)
24.	Big-end bearings	Vandervell, 3-layer
25.	Big-end bearing clearance	0·0028 to 0·0040in.
26.	Big-end side clearance	0·007 to 0·014in.
27.	Main journal diameter	2·4790 to 2·4795in.
28.	Main journal undersizes	0·010, 0·020, 0·030, 0·040in.
29.	Main bearings	3, Vandervell, bimetal (micro) and steel [back
30.	Main bearings—length	1·495 to 1·505in.
31.	Main bearing clearance	0·0015 to 0·0025in.
32.	Crankshaft end-float	0·0048 to ·006in.
33.	Crankshaft end-thrust on	Centre main bearing thrust washers
34.	Camshaft journal dia.	Front, 1·871 to 1·872in.; others 1·7153 to 1·7158in.
35.	Camshaft bearings	4; Vandervell (2, 3 & 4); front, cast
36.	Camshaft bearing cl'nce	Front, 0·0028 to 0·0047in.; others 0·0015 to 0·0029in.
37.	Camshaft end-float	0·0040 to 0·0075in.
38.	Camshaft drive type	Duplex chain, 21¾in. long
39.	Valve head diameter	Inlet, 1·558 to 1·562in.; exhaust, 1·299 to 1·303in.
40.	Valve stem diameter	Inlet, 0·310 to 0·311in.; exhaust, 0·3705 [to 0·3715in.
41.	Valve seat angle	Inlet, 45°; exhaust, 45°
42.	Valve lift	0·334in. (cold)
43.	Valve guide length	Inlet, 2·0525 to 2·0725in.; exhaust, 2·05 to 2·07in.
44.	Valve stem/guide cl'nce	Inlet, 0·001 to 0·003in.; exhaust, 0·003 to 0·005in.
45.	Valve spring free length	Outer, 1·94 to 1·96in.; inner, 1·88 to 1·90in. (auxiliary inner, exhaust only, 1·55 to 1·57in.)
46.	Valve spring rate	Outer, 178lb./in.; inner, 93lb./in. (auxiliary inner, exhaust only, 28·5lb./in.)
47.	Valve working cl'nce	Inlet, 0·010in.; exhaust, 0·010in. (cold)
48.	Valve timing clearance	0·0165in.
49.	Valve timing	Inlet opens 17° B.T.D.C.; closes 57° A.B.D.C. Exhaust opens 57° B.B.D.C.; closes 17° A.T.D.C.
50.	Petrol pump	AC-Delco, mechanical
51.	Carburettors	2 S.U. type HS6
53.	Needle	TW

TRANSMISSION

101.	Type	Rear wheel drive
102.	Clutch	Borg and Beck, 9in. diam., single dry plate
103.	Gearbox	4 forward speeds; synchromesh on 2nd, 3rd and top gears
105.	Gearbox ratios	Top, 1:1; 3rd 1·205:1; 2nd, 1·745:1; 1st and reverse, 2·97:1
106.	Overdrive	Not fitted
107.	Driving axles	Semi-floating
108.	Final drive gear	Hypoid (Salisbury)
109.	Final drive ratio	3·73:1
110.	Number of gear teeth	41/11
111.	Crown wheel/pinion b'lash	0·003 to 0·006in.; shim adjustment
112.	Pinion bearing pre-load	8 to 12lb.in.
113.	Differential brg. pre-load	·003 to ·005in.; shim adjustment

BRAKES

201.	Front	Girling hydraulic, 11in. diam. discs
202.	Rear	Girling hydraulic, 9in. diam. drums
203.	Handbrake	Mechanical, on rear wheels
204.	Linings—front	Ferodo, type 14, caliper pads
205.	Linings—rear	Ferodo, 1¾in. wide

STEERING (see also "Dimensions")

301.	Type	Cam Gear
302.	Camber angle	2°, non-adjustable
303.	Castor angle	4°; non-adjustable
304.	King-pin inclination	2°
305.	Toe-in	⅛ to 3/16 at wheel rims
306.	Conditions for checking	Unladen

FRONT SUSPENSION

401.	Type	Independent, vertical coil springs
402.	Number of coils	18
403.	Coil diameter	2⅜in. (external)
404.	Wire diameter	0·375in.
405.	Spring length—free	13in.
406.	Spring length—fitted	9in.

REAR SUSPENSION

501.	Type	Semi-elliptic leaf-springs
504.	Spring length (centres)	42¾in. (free)
505.	Number of leaves	6 (2-str.); 7 (4-str.)
506.	Width of leaves	1⅜in.
507.	Thickness of leaves	¼in.
508.	Spring camber—free	3½°

ELECTRICAL SYSTEM

601.	Ignition timing (static)	4° B.T.D.C. (Hole in pulley and pointer on timing case, T.D.C.)
602.	Sparking plugs	Champion L87Y, 14mm.
603.	Sparking plug gap	0·025in.
604.	Distributor	Lucas, type 25D4
605.	Contact breaker gap	0·015in.; dwell angle, 60°±3°
606.	Centrif. advance starts at	275 r.p.m.
607.	Centrif. advance—max.	9° to 11° at 800 r.p.m.
608.	Vacuum adv. starts at	2in. Hg.
609.	Vacuum adv.—max.	2·6° at 6in. Hg.
610.	Ignition coil	Lucas, type HA12
611.	Battery	Lucas, two SG9E in series; 12v. 54a.h.
612.	Battery earth	Negative (formerly positive)
613.	Dynamo	Lucas, C40/1; max. output, 22 amp.
614.	Control box	Lucas, RB106/2
615.	Starter motor	Lucas, M418G

CAPACITIES

701.	Engine sump	11 pints (dry); 10 pints (drain and refill)
702.	Gearbox	2½ pints
703.	Differential	1¾ pints
704.	Cooling system	16 pints
705.	Fuel tank	11 gallons
706.	Tyre size	5·60 × 15in.
707.	Tyre pressures	18lb./sq.in. normal (add 6lb./sq.in. for sustained high-speed driving on motorways)
708.	No. of grease gun points	15 (front sliding axles have 'one-shot' lubrication; press foot plunger every 200 miles approx.)
709.	Normal servicing intervals	Greasing 2,000/2,500 miles; engine oil change, 3,000 miles
710.	Hydraulic fluid	Castrol-Girling

DIMENSIONS

801.	Length overall	12ft.
802.	Width overall	4ft. 8in.
803.	Height overall	4ft. 4½in.
804.	Weight (approx.)	1848lb.
805.	Ground clearance (min.)	7in.
806.	Track	47in. front and rear
807.	Wheelbase	8ft.
808.	Turning circle diameter	32ft.

TORQUE SPANNER DATA

		lb.ft.
901.	Cylinder head nuts	100/105
902.	Big-end bearing nuts	55/60
903.	Main bearing nuts.	85/90
905.	Flywheel bolts	42/46
906.	Clutch to flywheel bolts	20
907.	Final drive pinion bearing nut	95/100
908.	Crown wheel bolts	50/60
909.	Diff. bearing nuts	35/40
910.	Road wheel nuts	65/70

BODY AND CHASSIS

1001.	Type of construction	Steel 'Z'-section chassis
1002.	Material of body panels	Steel and aluminium
1003.	Type of glass	Windscreen toughened (laminated optional)

Models: Standard 2-seater Tourer and Competition model

ENGINE (Ford)

1.	Type	4-stroke; overhead valves; side camshaft
2.	Cooling system	Water pump, radiator, fan and thermostat
3.	Number of cylinders ...	4 (bored direct in block)
4.	Firing order and idling speed	1—2—4—3 (No. 1 at front); 580/620 r.p.m. (Comp., 680/720)
5.	Bore	3·1881in. (80·977mm.) mean grade
6.	Stroke	3·056in. (77·62mm.)
7.	Cubic capacity	1598c.c.
8.	Compression ratio and pressure	9:1; 188lb./sq.in. at 300 r.p.m.
9.	Brake horse-power (gross)	75 at 5,000 r.p.m. (Comp., 93 at 5,400)
10.	Torque (lb.ft.)	97 at 2,500 (Comp., 102 at 3,600)
11.	Piston oversizes	·0025, ·015, ·030in.
12.	Piston clearance	3/7lb. pull on ·0015in.×½in. feeler (·0007/·0013in.)
13.	Piston rings—number ...	2 compression, 1 oil
14.	Piston rings—width ...	Comp., ·077 to ·078in.; oil, 0·155 to 0·156in.
15.	Piston rings—groove cl'nce	Comp., ·0016 to ·0036in.; oil, ·0018 to ·0038in.
16.	Piston rings—gap (in bore)	·009 to ·014in.
17.	Oil pressure	35 to 40lb./sq.in.
18.	Gudgeon pin diameter ...	0·8119 to 0·8123in.
19.	Gudgeon pin fit in piston	Interference, selective (heated piston)
20.	Gudgeon pin fit in con. rod	·0001 to ·0003in. clearance (selective)
21.	Crankpin diameter ...	1·9368 to 1·9376in.
22.	Crankpin undersizes ...	·010, ·020, ·030, ·040in.
23.	Connecting rod length ...	4·927 to 4·929in. (centres)
24.	Big-end bearings ...	Steel-backed, copper/lead, lead/bronze or aluminium/tin
25.	Big-end bearing clearance	·0004 to ·0024in.
26.	Big-end side clearance ...	·004 to ·010in.
27.	Main journal diameter ...	Blue, 2·1253/57in.; red, 2·1257/61in.
28.	Main journal undersizes...	·010, ·020, ·030in.
29.	Main bearings	5; steel-backed copper/lead, lead/bronze or aluminium/tin.
31.	Main bearing clearance ...	·0005 to ·0020in.
32.	Crankshaft end-float ...	·003 to ·011in.
33.	Crankshaft end-thrust on	Centre main bearing
34.	Camshaft journal dia. ...	1·5597 to 1·5605in.
35.	Camshaft bearings ...	3; steel-backed white metal
36.	Camshaft bearing cl'nce...	·001 to ·0023in.
37.	Camshaft end-float ...	·0025 to ·0075in.
38.	Camshaft drive type ...	Chain (single); 46×0·375in. pitches
39.	Valve head diameter ...	Inlet, 1·497/1·507in.; exhaust, 1·240/1·250in.
40.	Valve stem diameter ...	Inlet, 0·3095/0·3105; exhaust, 0·3086 to 0·3096in.
41.	Valve seat angle ...	45°
42.	Valve lift	Inlet, 0·315in. (Comp., 0·3420); exhaust, 0·319in. (Comp., 0·3367)
43.	Valve guide length ...	Integral with head (bushes available)
44.	Valve stem/guide cl'nce...	Inlet, ·0008 to ·003in.; exhaust, ·0017 to ·0039in.
45.	Valve spring free length...	1·48in.
46.	Valve spring rate ...	Length (closed), 1·263in. at 44/49lb.
47.	Valve work'g cl'nce (hot)	Inlet, ·010in. (Comp., ·012); exhaust, ·017in. (Comp., ·022)
48.	Valve timing clearance ...	Inlet, ·015in. (Comp., ·016); exhaust, ·027in. (Comp., ·026)
49.	Valve timing (See 48) (*Competition model)	Inlet opens 17° (*27° B.T.D.C.; closes 51° (*65°) A.B.D.C. Exhaust opens 51° (*65°) B.B.D.C.; closes 17° (*27°) A.T.D.C.
50.	Petrol pump	Mechanical
51.	Carburettor	Ford C7BH-B (Comp., Weber dual choke)

		Ford	Weber Primary	Weber Secondary
52.	Choke tube	28mm.	26mm.	27mm.
53.	Main jet	1·50mm.	150	150
54.	Compensating jet ...	1·50mm.	160	140
55.	Slow-running jet ...	0·65mm.	50	45
56.	Slow-running air bleed ...	1st, 1·05mm. 2nd, 0·6mm.	180 (bush)	70
57.	Emulsion tube		F6	F6
58.	Accelerator pump jet ...	0·55mm.	65	
62.	Needle valve		1·75mm.	

TRANSMISSION

101.	Type	Rear wheel drive
102.	Clutch	7½in. diam., diaphragm type
103.	Gearbox	4 forward speeds, all synchromesh
105.	Gearbox ratios	Top, 1:1; 3rd, 1·397:1; 2nd, 2·010:1; 1st, 2·972:1; reverse, 3·324:1
107.	Driving axles	Semi-floating
108.	Final drive gear ...	Hypoid (Salisbury)
109.	Final drive ratio ...	4·56:1 (41/9); Comp., 4·1:1 (41/10)
111.	Crownwheel/pinion b'lash	·003 to ·006in.; shim adjustment
112.	Pinion bearing pre-load ...	8 to 12lb.in.; shim adjustment
113.	Differential brg. pre-load	·003 to ·005in.; shim adjustment

BRAKES

201.	Front	Girling hydraulic, 11in. diam. discs
202.	Rear	Girling hydraulic, 9in. diam. drums
203.	Handbrake	Mechanical, on rear wheels
204.	Linings—front	Ferodo, type 14 calliper pads
205.	Linings—rear	Ferodo, 1¾in. wide

STEERING (see also "Dimensions")

301.	Type	Cam Gear
302.	Camber angle	2°; non-adjustable
303.	Castor angle	4°; non-adjustable
304.	King-pin inclination ..	2°
305.	Toe-in	⅛ to $\frac{3}{16}$in. at wheel rim
306.	Conditions for checking...	Unladen

FRONT SUSPENSION

401.	Type	Independent, vertical coil springs
402.	Number of coils	21
403.	Coil diameter	2¼in. (external)
404.	Wire diameter	0·31in.
405.	Spring length—free ...	14½in.
406.	Spring length—fitted ...	8½in.

REAR SUSPENSION

501.	Type	Semi-elliptic leaf springs
504.	Spring length (centres) ..	42¾in. (free)
505.	Number of leaves ...	5
506.	Width of leaves	1⅜in.
507.	Thickness of leaves ...	¼in.
508.	Spring camber—free ...	3⅜in.

ELECTRICAL SYSTEM

601.	Ignition timing (static) ...	10° (Comp., 8°B.T.D.C.) (97 oct.); mark on pulley
602.	Sparking plugs	Autolite AG22
603.	Sparking plug gap ...	·023in.
604.	Distributor	Ford, C7BH-A (Comp., –C)
605.	Contact breaker gap ...	·025in.; dwell angle 39°/42°

		Standard	Competition
606.	Centrif. adv. starts at ...	½°/2½° at 700 r.p.m.	1°/3° at 700
607.	Centrif. advance—max....	13°/15° at 2,700	8°/10° at 2,600
608.	Vacuum adv. starts at ...	.... ½° to 3½° at 7in. Hg.	
609.	Vacuum advance—max.	.... 5½° to 8½° at 17in. Hg. .... (Distributor degrees and r.p.m.)	

610.	Ignition coil	12v. oil-filled
611.	Battery	Lucas BT7A or C7, 12v. 36a.h.
612.	Battery earth	Negative
613.	Dynamo	Lucas C40
614.	Control box	Cut-in, 12·6/13·4v.; drop-off, 9·25/11·25v.; V.R. open circuit setting, 14·4/15·6v. at 68°F
615.	Starter motor	Inertia type

CAPACITIES

701.	Engine sump	7·2 pints (including filter)
702.	Gearbox	2·13 pints
703.	Differential	1¾ pints
704.	Cooling system	12 pints
705.	Fuel tank	8½ gallons
706.	Tyre size	5·60×15in.
707.	Tyre pressures	16/17lb./sq.in., front and rear, normal (add 6lb./sq.in. for sustained high-speed driving on motorways)
708.	No. of grease gun points ...	13 (Front sliding axles have 'one-shot' lubrication; press foot plunger every 200 miles approx.)
709.	Normal servicing intervals	Greasing, 2000/2500 miles; engine oil change, 5000 miles
710.	Hydraulic fluid	Castrol/Girling

DIMENSIONS

801.	Length overall	12ft.
802.	Width overall	4ft. 8in.
803.	Height overall	4ft. 3¾in.
804.	Weight (approx.) ...	1,456lb. (13cwt.) dry
805.	Ground clearance (min.)	7in.
806.	Track	3ft. 11in. front and rear
807.	Wheelbase	8ft.
808.	Turning circle diameter...	32ft.

TORQUE SPANNER DATA

		lb.ft.			lb.ft.
901.	Cylinder head nuts ...	65/70	907.	Final drive pinion bearing nut	95/100
902.	Big-end bearing nuts ...	30/35	908.	Crown wheel bolts	50/60
903.	Main bearing nuts ...	65/70	909.	Diff. bearing nuts	35/40
905.	Flywheel bolts ...	45/50	910.	Road wheel nuts	65/70
906.	Clutch to flywheel bolts ...	12/15			

BODY AND CHASSIS

1001.	Type of construction ...	Coachbuilt; steel 'Z'-section chassis; ash body frame
1002.	Material of body panels...	Steel and aluminium
1003.	Type of glass	Toughened (laminated optional extra)

4/4 1600 1968

ENGINE (Triumph TR4A)

1. Type ... ... 4-stroke; overhead valves; side camshaft
2. Cooling system ... ... Water pump, fan and thermostat
3. Number of cylinders ... 4 (wet liners)
4. Firing order & idl'g speed 1—3—4—2 (No. 1 at front); 550 r.p.m.
5. Bore ... ... 86mm. (3·386in.)
6. Stroke ... ... 92mm. (3·622in.)
7. Cubic capacity ... ... 2,138c.c. (130·5cu.in.)
8. Compression ratio and pressure 9:1 (7·0:1 optional); max. variation 5lb./sq.in.
9. Brake horse-power ... 105 at 4,750 r.p.m.
10. Torque ... ... 128·3lb.ft. at 3,350 r.p.m.
11. Piston oversizes ... ... 0·010, 0·020, 0·030, 0·040in.
12. Piston clearance ... ... 0·0047 to 0·0054in. measured at top; and 0·0032 to 0·0039in. at bottom
13. Piston rings—number ... 2 compression, 1 oil
14. Piston rings—width ... Comp., 0·0615 to 0·0625in.; oil, 0·1552 to 0·1562in.
15. Piston rings—groove cl'nce Comp., 0·0010 to 0·0030in.; oil, 0·0010 to 0·0030in.
16. Piston rings—gap (in bore) ... Comp., 0·010 to 0·015in.; oil, 0·010 to 0·015in.
17. Oil pressure ... ... 65 to 75lb./sq.in. (above 30 m.p.h.)
18. Gudgeon pin diameter ... 0·87485 to 0·87510in.
19. Gudgeon pin fit in piston 0·0002in. clearance at 68°F.
20. Gudgeon pin fit in con. rod Selective
21. Crankpin diameter ... 2·0861 to 2·0866in.
22. Crankpin undersizes ... 0·010, 0·020, 0·030in.
23. Connecting rod length ... 6·248 to 6·252in. (centres)
24. Big-end bearings ... Vandervell, 3-layer
25. Big-end bearing clearance 0·0028 to 0·0040in.
26. Big-end side clearance ... 0·007 to 0·014in.
27. Main journal diameter ... 2·4790 to 2·4795in.
28. Main journal undersizes 0·010, 0·020, 0·030, 0·040in.
29. Main bearings ... ... 3, Vandervell, bimetal (micro) and steel [back
30. Main bearings—length ... 1·495 to 1·505in.
31. Main bearing clearance 0·0015 to 0·0025in.
32. Crankshaft end-float ... 0·0048 to ·006in.
33. Crankshaft end-thrust on Centre main bearing thrust washers
34. Camshaft journal dia. ... Front, 1·871 to 1·872in.; others 1·7153 to 1·7158in.
35. Camshaft bearings ... 4; Vandervell (2, 3 & 4); front, cast
36. Camshaft bearing cl'nce Front, 0·0028 to 0·0047in.; others 0·0015 to 0·0029in.
37. Camshaft end-float ... 0·0040 to 0·0075in.
38. Camshaft drive type ... Duplex chain, 21¾in. long
39. Valve head diameter ... Inlet, 1·558 to 1·562in.; exhaust, 1·299 to 1·303in.
40. Valve stem diameter ... Inlet, 0·310 to 0·311in.; exhaust, 0·3705 [to 0·3715in.
41. Valve seat angle ... ... Inlet, 45°; exhaust, 45°
42. Valve lift ... ... 0·334in. (cold)
43. Valve guide length ... Inlet, 2·0525 to 2·0725in.; exhaust, 2·05 to 2·07in.
44. Valve stem/guide cl'nce ... Inlet, 0·001 to 0·003in.; exhaust, 0·003 to 0·005in.
45. Valve spring free length ... Outer, 1·94 to 1·96in.; inner, 1·88 to 1·90in. (auxiliary inner, exhaust only, 1·55 to 1·57in.)
46. Valve spring rate... ... Outer, 178lb./in.; inner, 93lb./in. (auxiliary inner, exhaust only, 28·5lb./in.)
47. Valve working cl'nce ... Inlet, 0·010in.; exhaust, 0·010in. (cold)
48. Valve timing clearance ... 0·0165in.
49. Valve timing ... ... Inlet opens 17° B.T.D.C.; closes 57° A.B.D.C. Exhaust opens 57° B.B.D.C.; closes 17° A.T.D.C.
50. Petrol pump ... ... AC-Delco, mechanical
51. Carburettors ... ... 2 S.U. type HS6
53. Needle ... ... TW

TRANSMISSION

101. Type ... ... Rear wheel drive
102. Clutch ... ... Borg and Beck, 9in. diam., single dry plate
103. Gearbox ... ... 4 forward speeds; synchromesh on 2nd, 3rd and top gears
105. Gearbox ratios ... ... Top, 1:1; 3rd 1·205:1; 2nd, 1·745:1; 1st and reverse, 2·97:1
106. Overdrive ... ... Not fitted
107. Driving axles ... ... Semi-floating
108. Final drive gear ... ... Hypoid (Salisbury)
109. Final drive ratio ... ... 3·73:1
110. Number of gear teeth ... 41/11
111. Crown wheel/pinion b'lash 0·003 to 0·006in.; shim adjustment
112. Pinion bearing pre-load ... 8 to 12lb.in.
113. Differential brg. pre-load ·003 to ·005in.; shim adjustment

BRAKES

201. Front ... ... Girling hydraulic, 11in. diam. discs
202. Rear ... ... Girling hydraulic, 9in. diam. drums
203. Handbrake ... ... Mechanical, on rear wheels
204. Linings—front ... ... Ferodo, type 14, caliper pads
205. Linings—rear ... ... Ferodo, 1¾in. wide

STEERING (see also "Dimensions")

301. Type ... ... Cam Gear
302. Camber angle ... ... 2°, non-adjustable
303. Castor angle ... ... 4°; non-adjustable
304. King-pin inclination ... 2°
305. Toe-in ... ... ⅛ to 3/16 at wheel rims
306. Conditions for checking ... Unladen

FRONT SUSPENSION

401. Type ... ... Independent, vertical coil springs
402. Number of coils 18
403. Coil diameter ... ... 2⅜in. (external)
404. Wire diameter ... ... 0·375in.
405. Spring length—free ... 13in.
406. Spring length—fitted ... 9in.

REAR SUSPENSION

501. Type ... ... Semi-elliptic leaf-springs
504. Spring length (centres) ... 42¾in. (free)
505. Number of leaves ... 6 (2-str.); 7 (4-str.)
506. Width of leaves ... ... 1⅜in.
507. Thickness of leaves ... ¼in.
508. Spring camber—free ... 3½°

ELECTRICAL SYSTEM

601. Ignition timing (static) ... 4° B.T.D.C. (Hole in pulley and pointer on timing case, T.D.C.)
602. Sparking plugs ... ... Champion L87Y, 14mm.
603. Sparking plug gap ... 0·025in.
604. Distributor ... ... Lucas, type 25D4
605. Contact breaker gap ... 0·015in.; dwell angle, 60°±3°
606. Centrif. advance starts at 275 r.p.m.
607. Centrif. advance—max.... 9° to 11° at 800 r.p.m.
608. Vacuum adv. starts at ... 2in. Hg.
609. Vacuum adv.—max. ... 2·6° at 6in. Hg.
610. Ignition coil ... ... Lucas, type HA12
611. Battery ... ... Lucas, two SG9E in series; 12v. 54a.h.
612. Battery earth ... ... Negative
613. Dynamo ... ... Lucas, C40/1; max. output, 22 amp.
614. Control box ... ... Lucas, RB106/2
615. Starter motor ... ... Lucas, M418G

CAPACITIES

701. Engine sump ... ... 11 pints (dry); 10 pints (drain and refill)
702. Gearbox ... ... 2½ pints
703. Differential ... ... 1¾ pints
704. Cooling system ... ... 16 pints
705. Fuel tank ... ... 11 gallons
706. Tyre size ... ... 5·60 × 15in.
707. Tyre pressures ... ... 18lb./sq.in. normal (add 6lb./sq.in. for sustained high-speed driving on motorways)
708. No. of grease gun points ... 15 (front sliding axles have 'one-shot' lubrication; press foot plunger every 200 miles approx.)
709. Normal servicing intervals Greasing 2,000/2,500 miles; engine oil change, 3,000 miles
710. Hydraulic fluid ... ... Castrol-Girling

DIMENSIONS

801. Length overall ... ... 12ft.
802. Width overall ... ... 4ft. 8in.
803. Height overall ... ... 4ft. 4½in.
804. Weight (approx.) ... 1848lb.
805. Ground clearance (min.) 7in.
806. Track ... ... 47in. front and rear
807. Wheelbase ... ... 8ft.
808. Turning circle diameter 32ft.

TORQUE SPANNER DATA

	lb.ft.		lb.ft.
901. Cylinder head nuts ...	100/105	907. Final drive pinion bearing nut ...	95/100
902. Big-end bearing nuts ...	55/60	908. Crown wheel bolts	50/60
903. Main bearing nuts. ...	85/90	909. Diff. bearing nuts	35/40
905. Flywheel bolts	42/46	910. Road wheel nuts ...	65/70
906. Clutch to flywheel bolts ...	20		

BODY AND CHASSIS

1001. Type of construction ... Steel 'Z'-section chassis
1002. Material of body panels ... Steel and aluminium
1003. Type of glass ... ... Windscreen toughened (laminated optional)

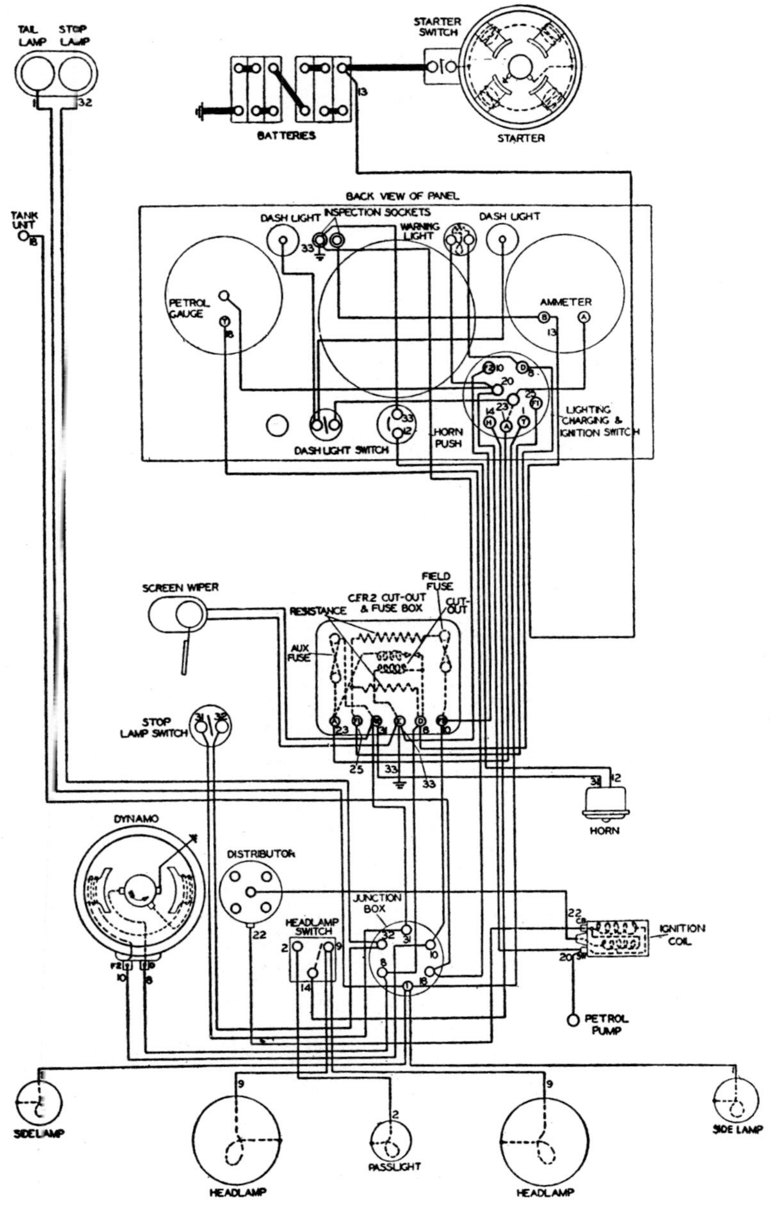

INTERNAL CONNECTIONS OF LIGHTING, CHARGING AND IGNITION SWITCH.

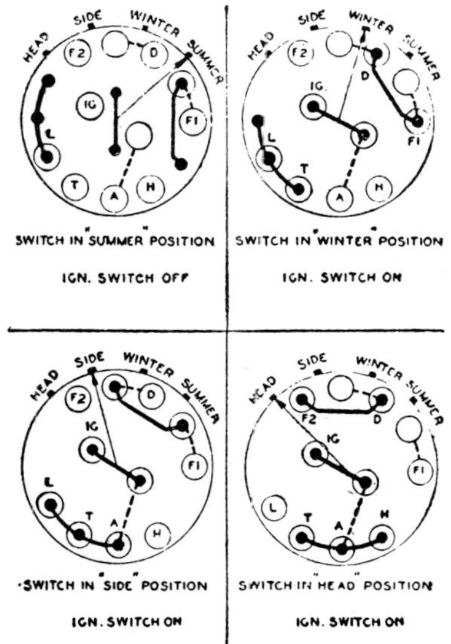

WIRING DIAGRAM MORGAN FOUR/4 SERIES I (1936)

Key to cable colours: 1 Red 2 Red and yellow 3 Red and blue 4 Red and white 5 Red and green 6 Red and brown 7 Red and black 8 Yellow 9 Yellow and blue 10 Yellow and green 11 Yellow and brown 12 Yellow and purple 13 Yellow and black 14 Blue 15 Blue and white 16 Blue and green 17 Blue and brown 18 Blue and purple 19 Blue and black 20 White 21 White and green 22 White and brown 23 White and purple 24 White and black 25 Green 26 Green and brown 27 Green and purple 28 Green and black 29 Brown 30 Brown and purple 31 Purple 32 Purple and black 33 Black

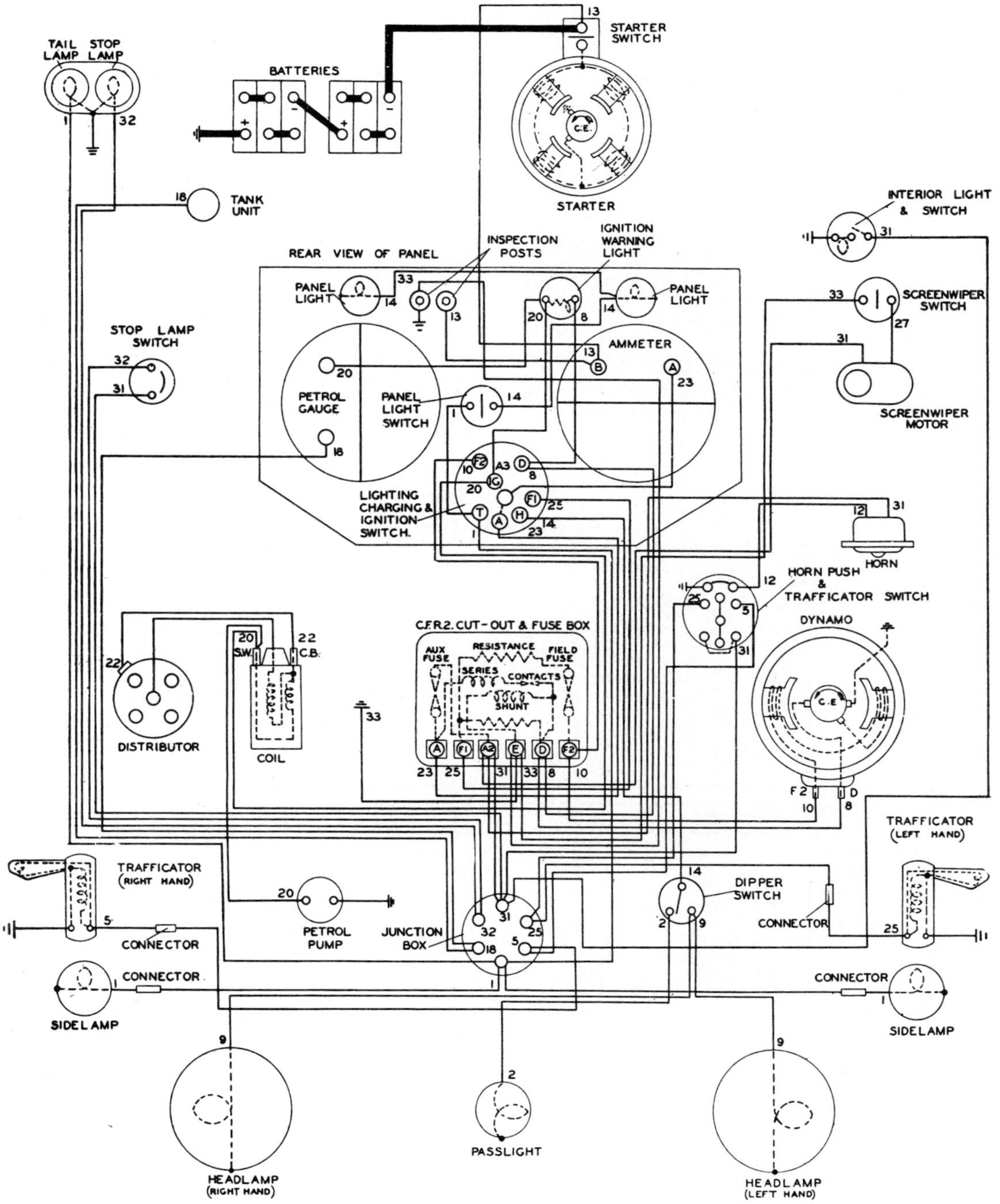

WIRING DIAGRAM MORGAN FOUR/4 SERIES I COUPE (1939)

Key to cable colours: 1 Red 2 Red and yellow 3 Red and blue 4 Red and white 5 Red and green 6 Red and brown 7 Red and black 8 Yellow 9 Yellow and blue 10 Yellow and green 11 Yellow and brown 12 Yellow and purple 13 Yellow and black 14 Blue 15 Blue and white 16 Blue and green 17 Blue and brown 18 Blue and purple 19 Blue and black 20 White 21 White and green 22 White and brown 23 White and purple 24 White and black 25 Green 26 Green and brown 27 Green and purple 28 Green and black 29 Brown 30 Brown and purple 31 Purple 32 Purple and black 33 Black

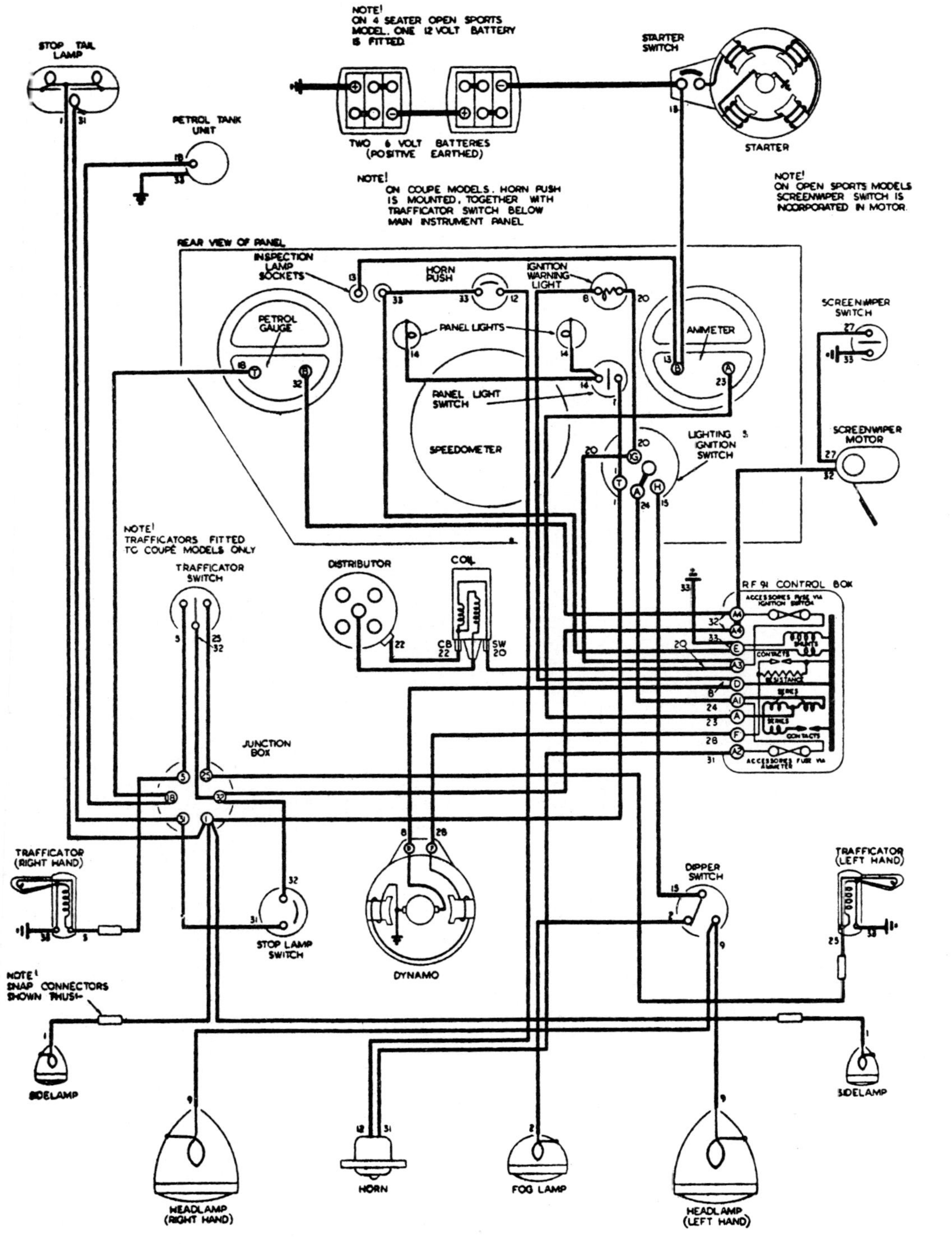

WIRING DIAGRAM MORGAN FOUR/4 SERIES I (1946-47)

Key to cable colours: 1 Red 2 Red and yellow 3 Red and blue 4 Red and white 5 Red and green 6 Red and brown 7 Red and black 8 Yellow 9 Yellow and blue 10 Yellow and green 11 Yellow and brown 12 Yellow and purple 13 Yellow and black 14 Blue 15 Blue and white 16 Blue and green 17 Blue and brown 18 Blue and purple 19 Blue and black 20 White 21 White and green 22 White and brown 23 White and purple 24 White and black 25 Green 26 Green and brown 27 Green and purple 28 Green and black 29 Brown 30 Brown and purple 31 Purple 32 Purple and black 33 Black

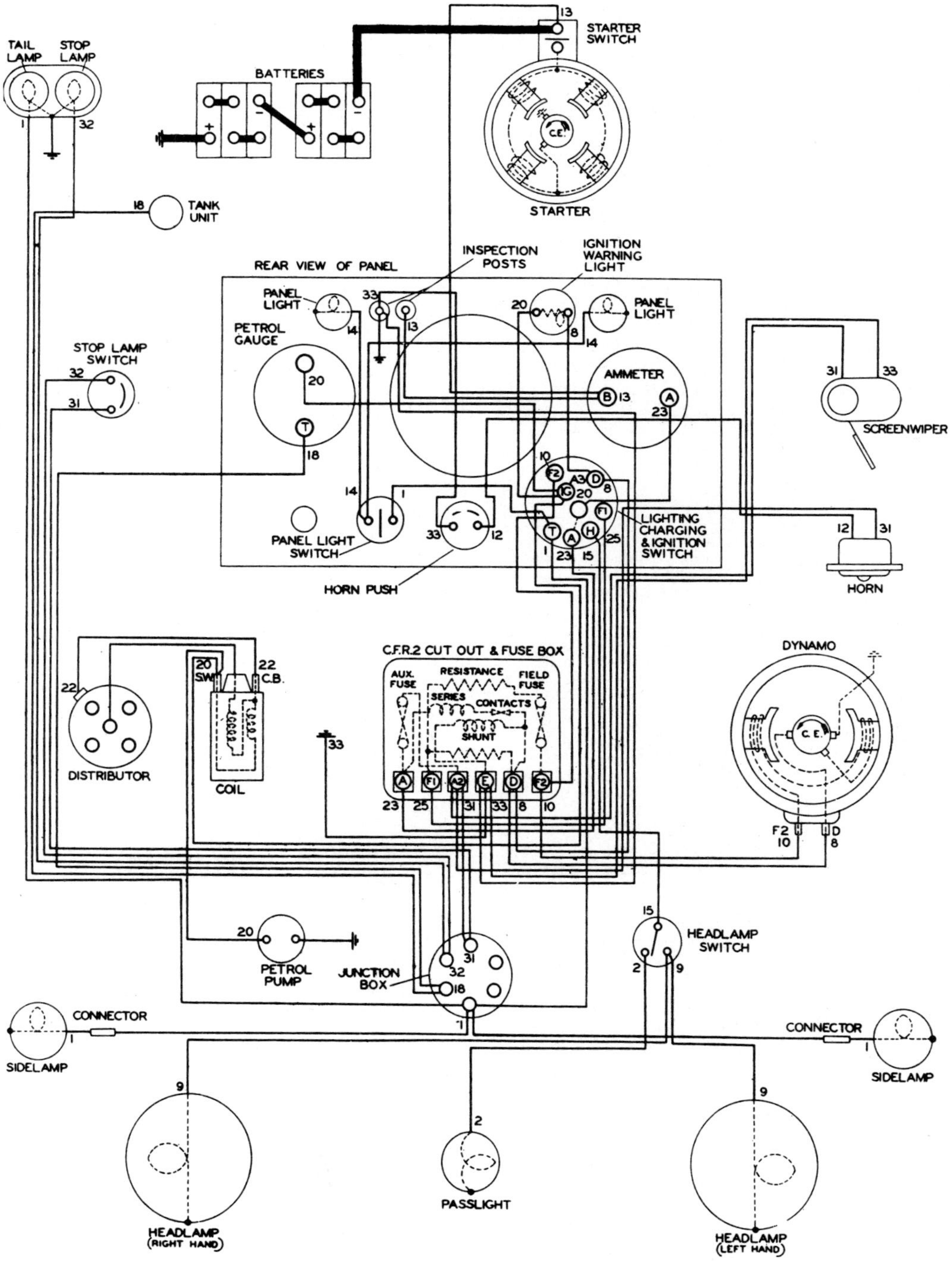

WIRING DIAGRAM MORGAN FOUR/4 SERIES I (1939)

Key to cable colours: 1 Red 2 Red and yellow 3 Red and blue 4 Red and white 5 Red and green 6 Red and brown 7 Red and black 8 Yellow 9 Yellow and blue 10 Yellow and green 11 Yellow and brown 12 Yellow and purple 13 Yellow and black 14 Blue 15 Blue and white 16 Blue and green 17 Blue and brown 18 Blue and purple 19 Blue and black 20 White 21 White and green 22 White and brown 23 White and purple 24 White and black 25 green 26 Green and brown 27 Green and purple 28 Green and black 29 Brown 30 Brown and purple 31 Purple 32 Purple and black 33 Black

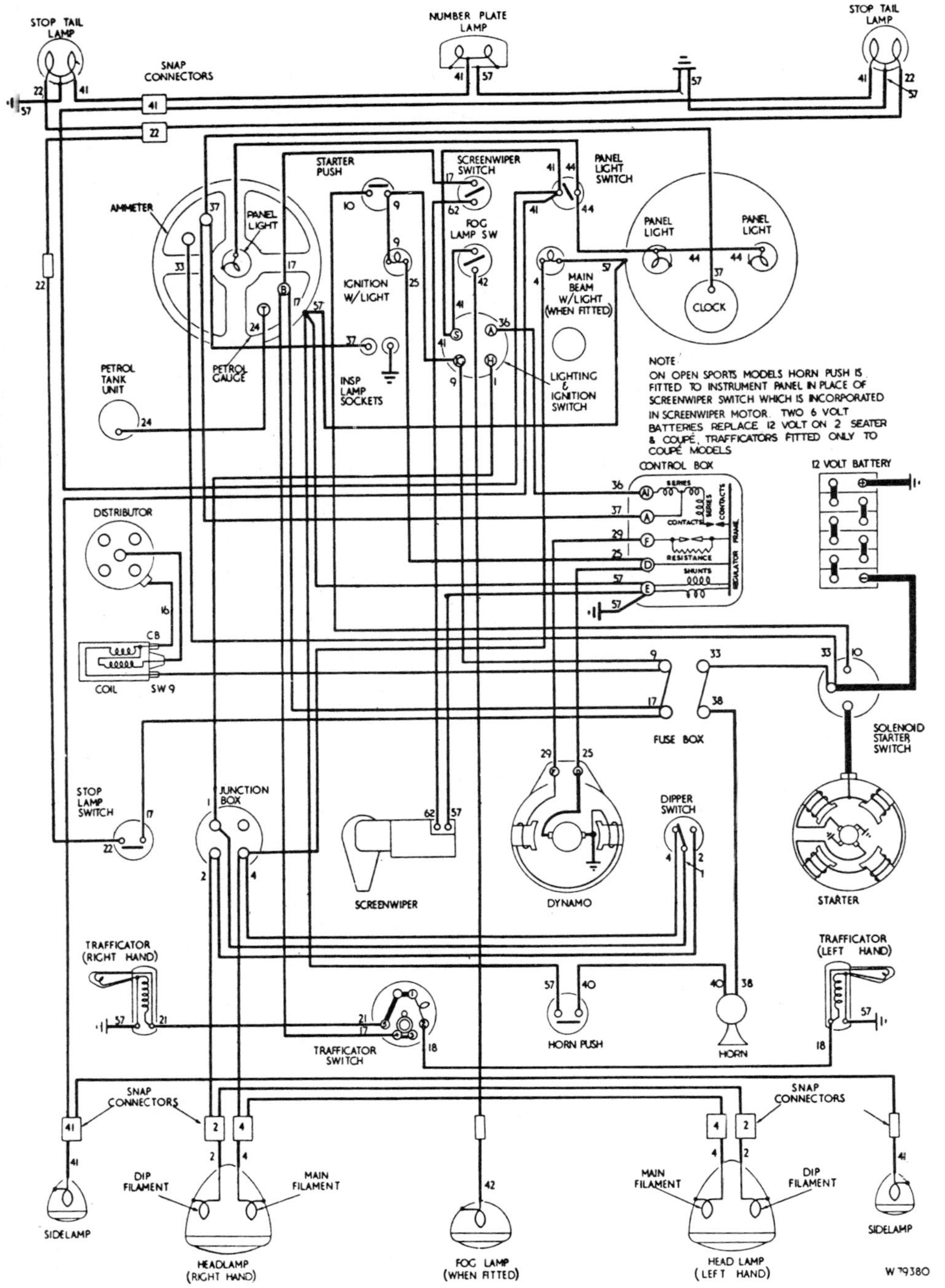

WIRING DIAGRAM MORGAN PLUS 4 MODELS (1950-51)

Key to cable colours: 1 Blue 2 Blue and red 3 Blue and yellow 4 Blue and white 5 Blue and green 6 Blue and purple 7 Blue and brown 8 blue and black 9 White 10 White and red 11 White and yellow 12 White and blue 13 White and green 14 White and purple 15 White and brown 16 White and black 17 Green 18 Green and red 19 Green and yellow 20 Green and blue 21 Green and white 22 Green and purple 23 Green and brown 24 Green and black 25 Yellow 26 Yellow and red 27 Yellow and blue 28 Yellow and white 29 Yellow and green 30 Yellow and purple 31 Yellow and brown 32 Yellow and black 33 Brown 34 Brown and red 35 Brown and yellow 36 Brown and blue 37 Brown and white 38 Brown and green 39 Brown and purple 40 Brown and black 41 Red 42 Red and yellow 43 Red and blue 44 Red and white 45 Red and green 46 Red and purple 47 Red and brown 48 Red and black 49 Purple 50 Purple and red 51 Purple and yellow 52 Purple and blue 53 Purple and white 54 Purple and green 55 Purple and brown 56 Purple and black 57 Black 58 Black and red 59 Black and yellow 60 Black and blue 61 Black and white 62 Black and green 63 Black and purple 64 Black and brown.

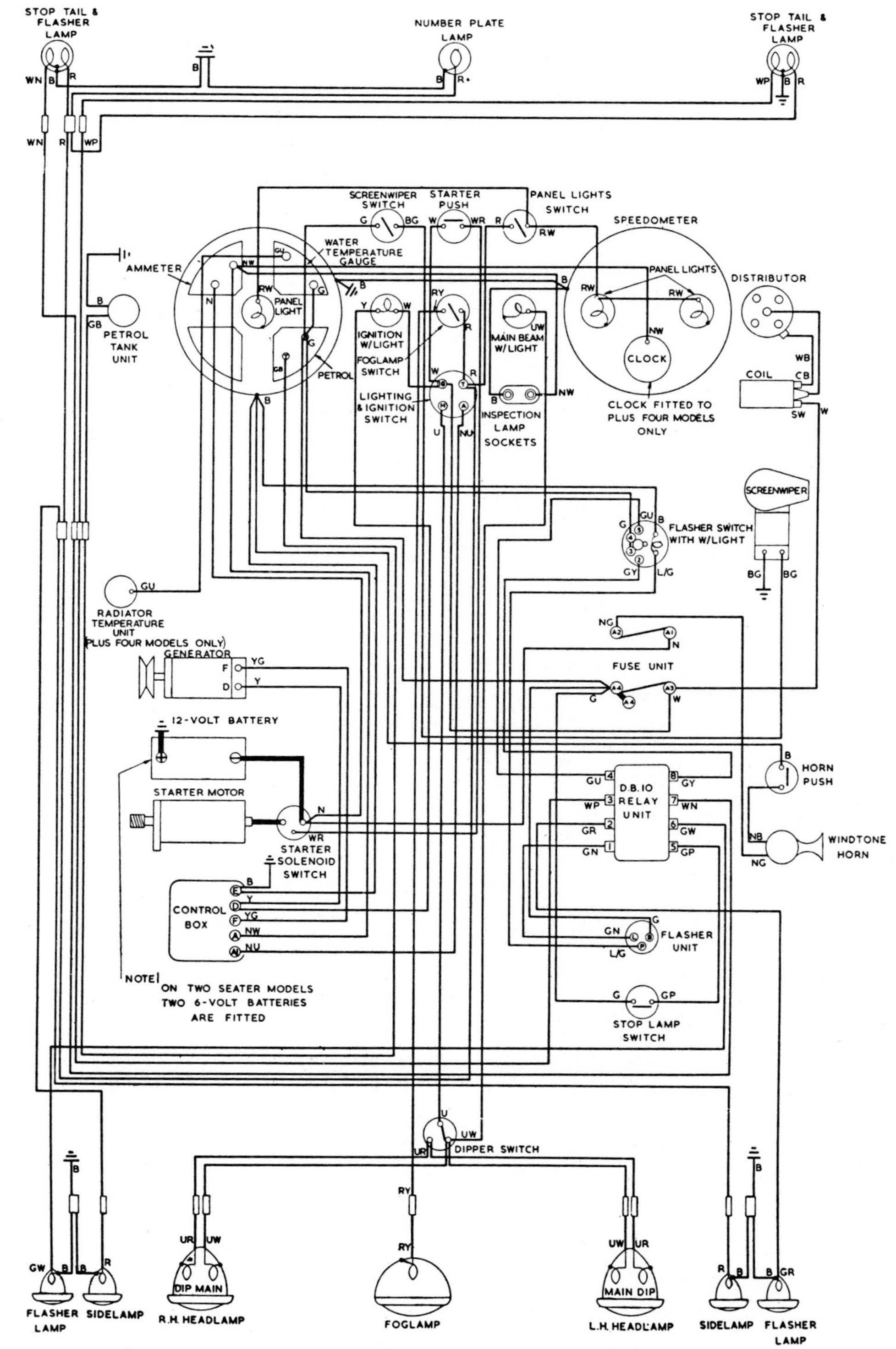

WIRING DIAGRAM MORGAN PLUS 4 AND FOUR/4 MODELS (1956)

Key to cable colours: B Black U Blue N Brown R Red P Purple G Green S Slate W White Y Yellow
D Dark L Light M Medium.

When a cable has two colour code letters, the first denotes the main colour and the second denotes the tracer colour.

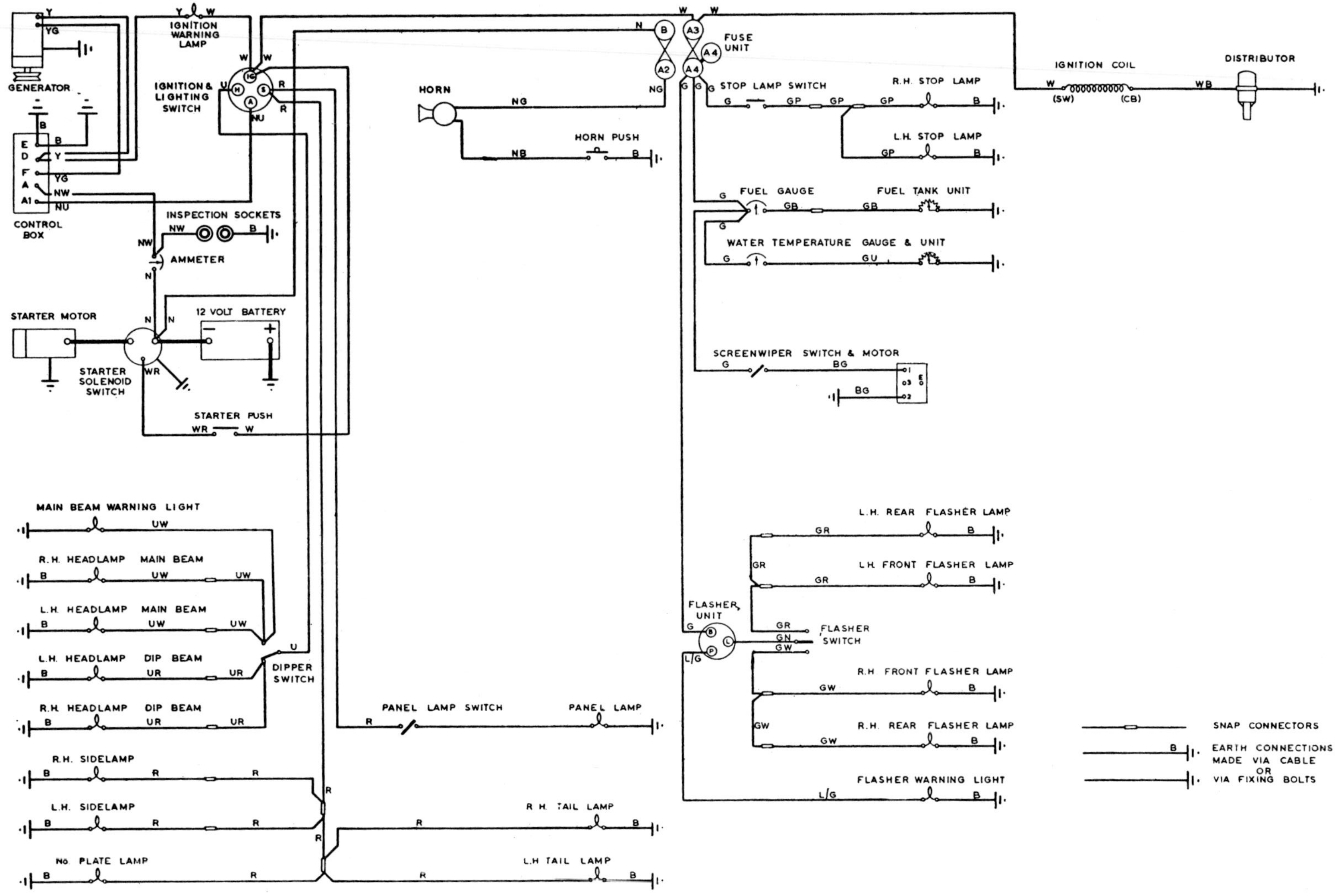

WIRING DIAGRAM MORGAN PLUS 4 AND 4/4 SERIES II CARS (1960)

Key to cable colours: B Black U Blue N Brown R Red P Purple G Green S Slate W White Y Yellow
D Dark L Light M Medium.
When a cable has two colour code letters, the first denotes the main colour and the second denotes the tracer colour.

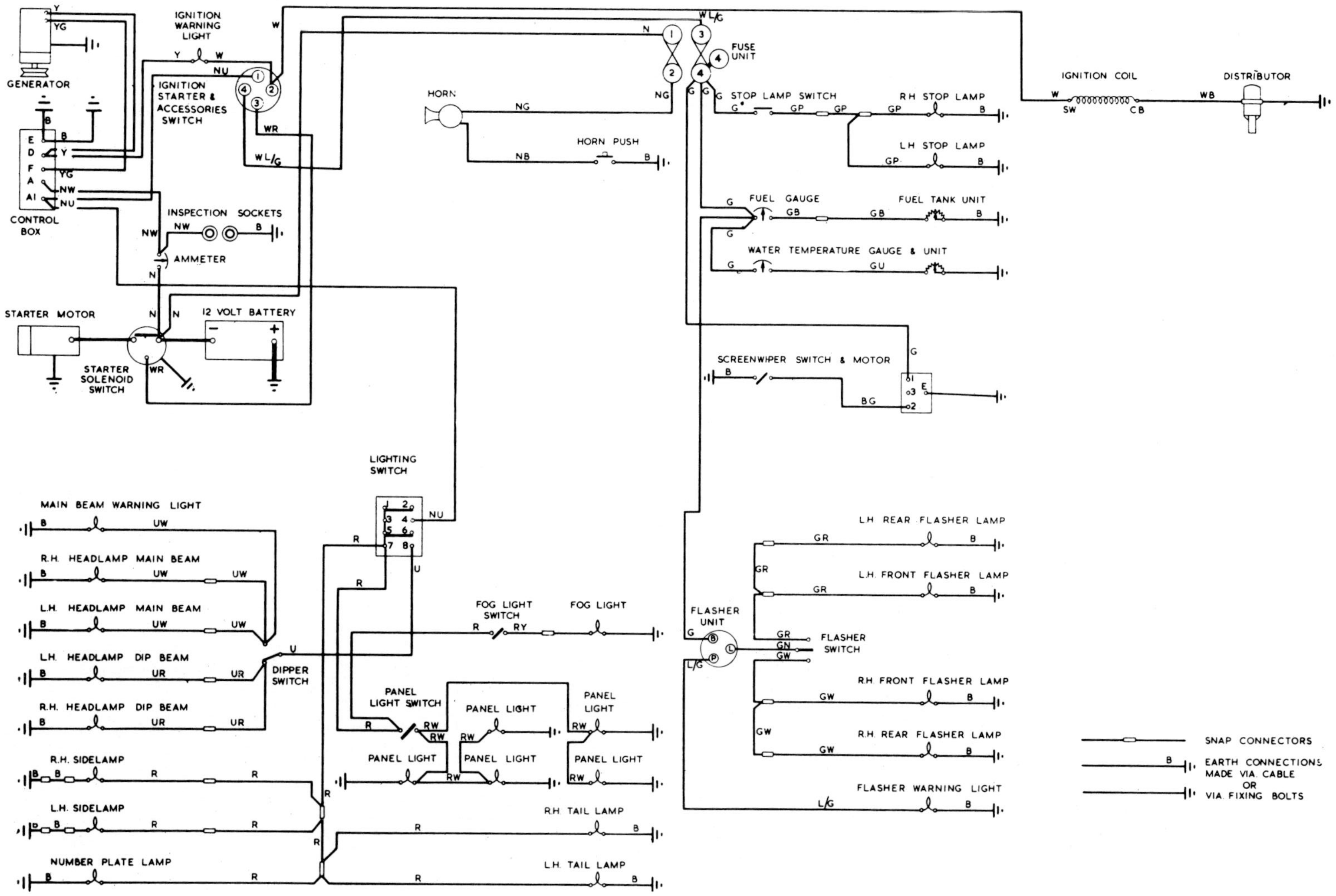

WIRING DIAGRAM MORGAN PLUS 4 AND 4/4 SERIES IV CARS (1961-62)

Key to cable colours: B Black U Blue N Brown R Red P Purple G Green S Slate W White Y Yellow
D Dark L Light M Medium.
When a cable has two colour code letters, the first denotes the main colour and the second denotes the tracer colour:

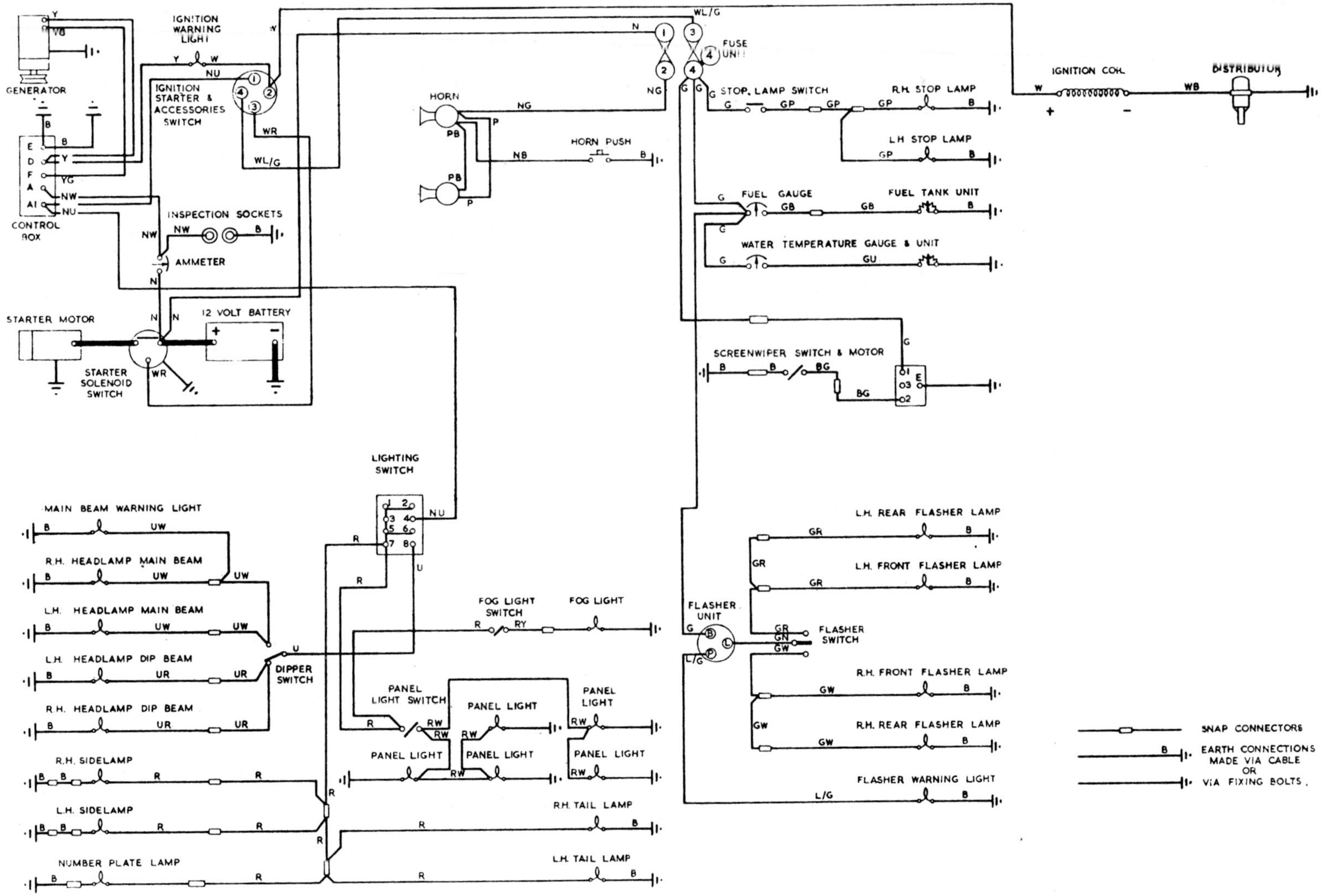

WIRING DIAGRAM MORGAN PLUS-4-PLUS SPORTS CARS (1965-onwards)

Key to cable colours: B Black U Blue N Brown R Red P Purple G Green S Slate W White Y Yellow
D Dark L Light M Medium.
When a cable has two colour code letters, the first denotes the main colour and the second denotes the tracer colour.

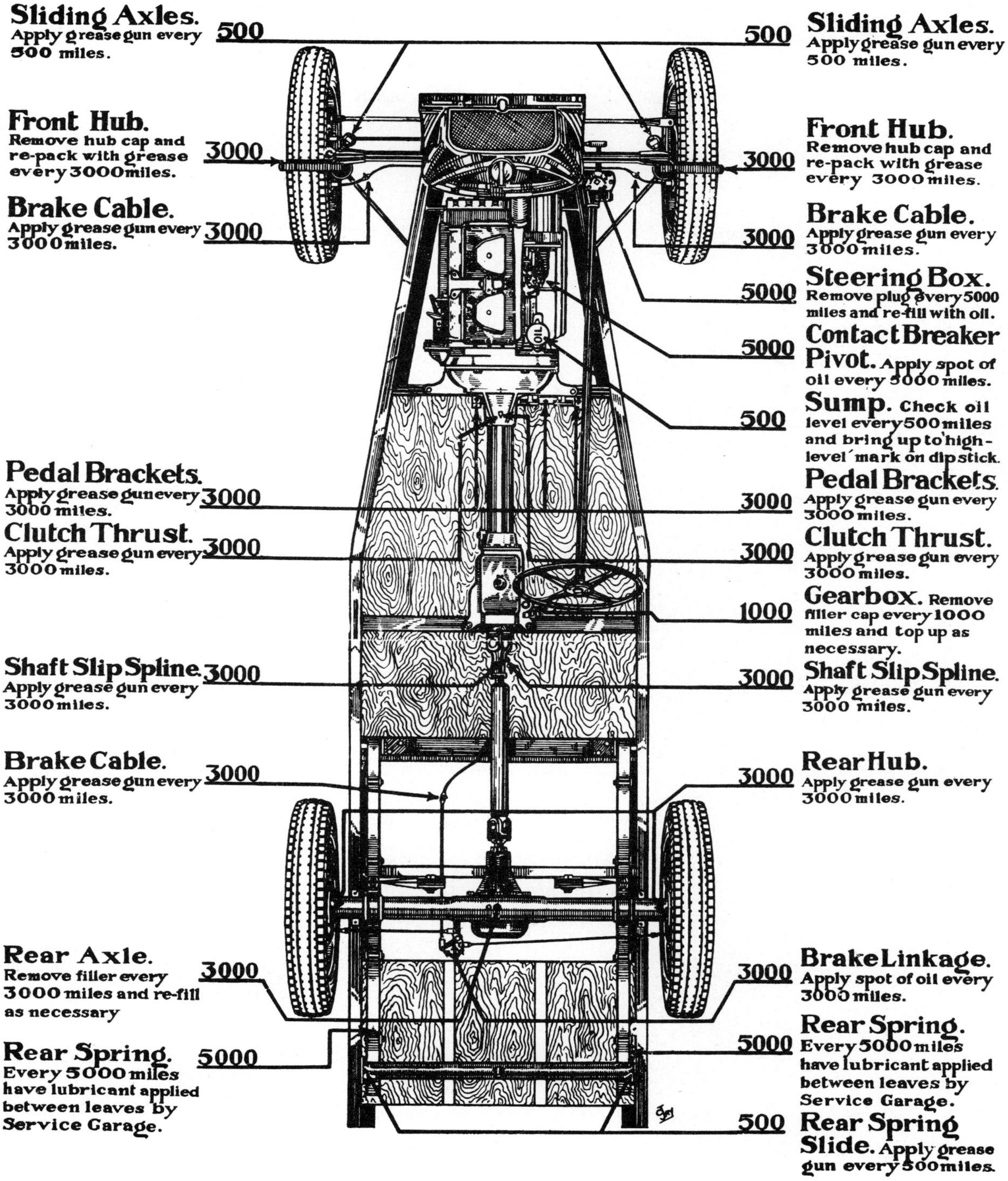

LUBRICATION CHART FOUR/4 SERIES I
LUBRICATE AT MILEAGE SHOWN

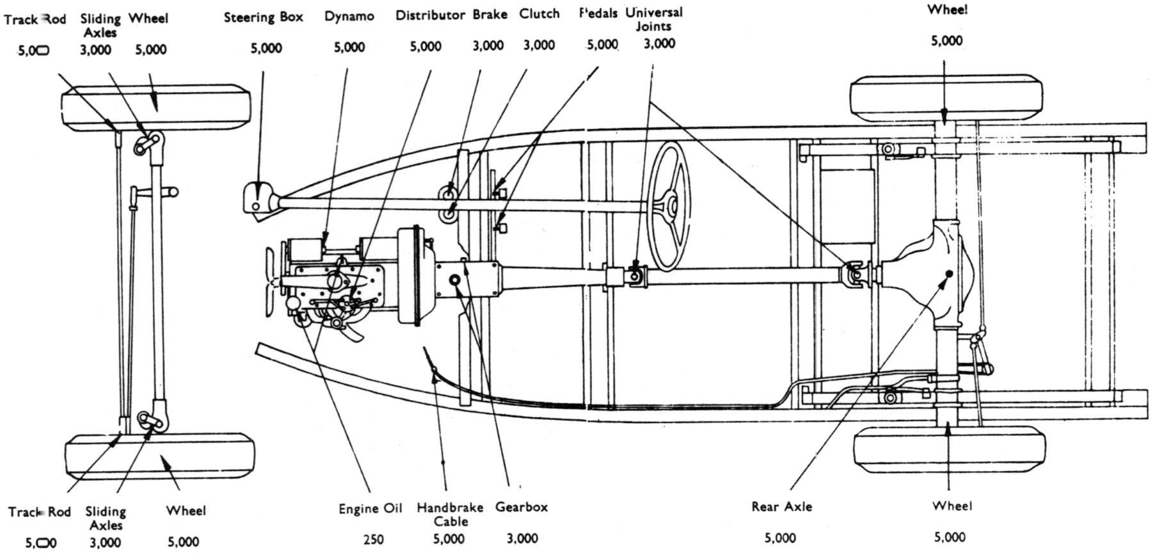

LUBRICATION CHART FOUR/4 SERIES II
LUBRICATE AT MILEAGE SHOWN

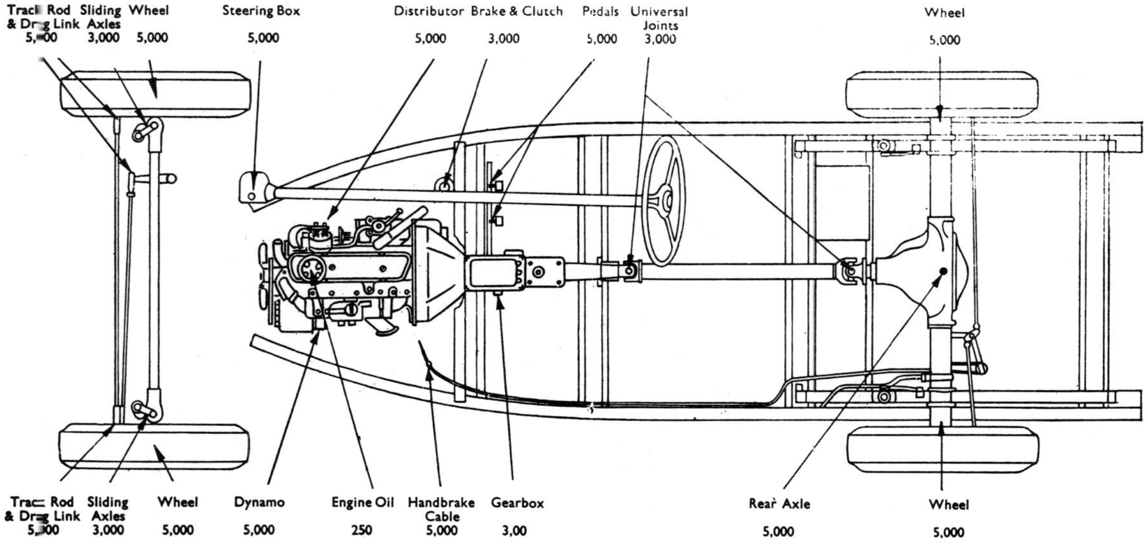

LUBRICATION CHART FOUR/4 SERIES IV and V
LUBRICATE AT MILEAGE SHOWN

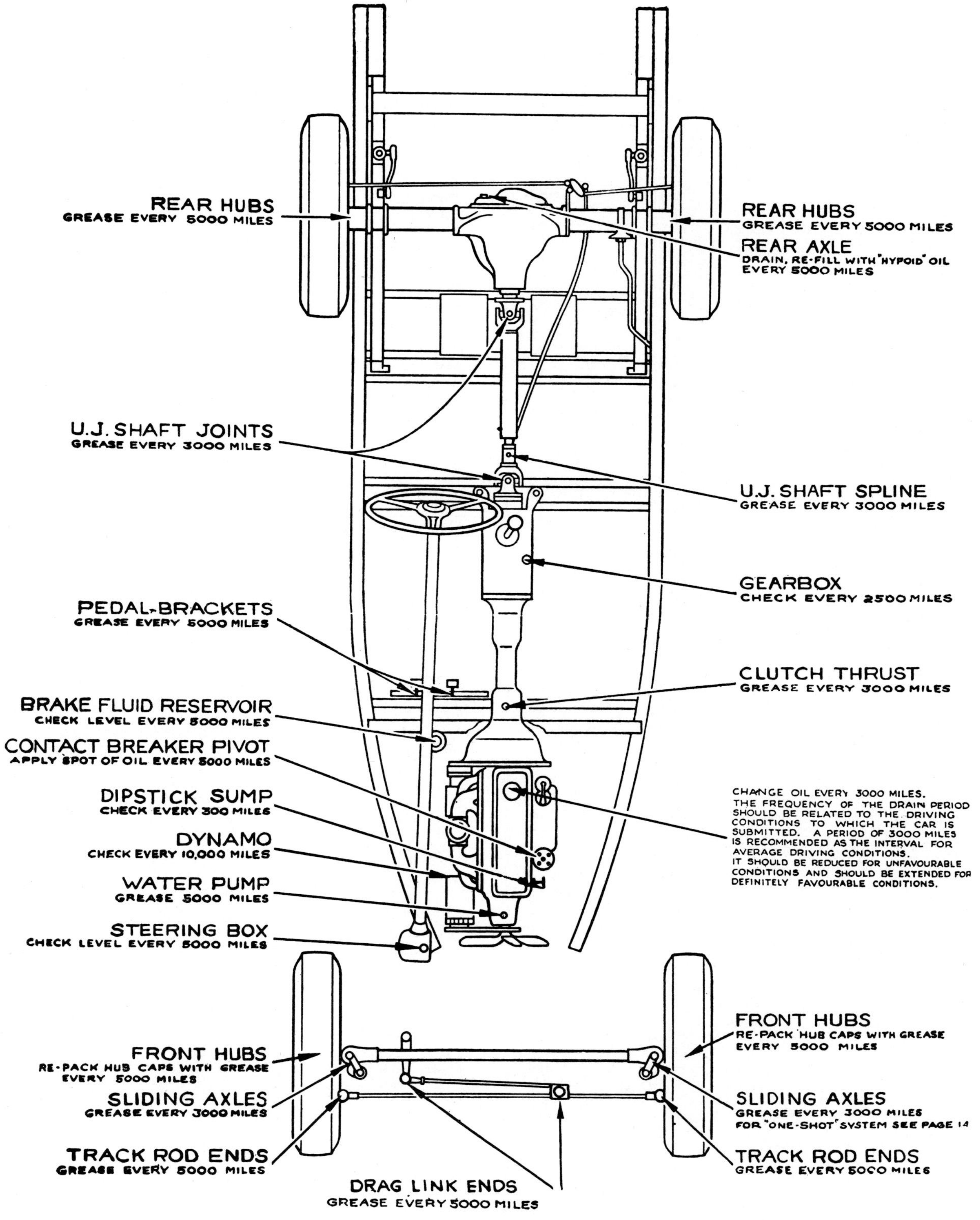

LUBRICATION CHART PLUS/4
LUBRICATE AT MILEAGE SHOWN

Description		Year	Part No
Ammeter		1946–50	CZU20
Ammeter		1951–54	CZU20
Battery (4-seater)		1946–54	SLG11A
Battery (2-seater and Coupé)		1950–51	SLG11E
Battery (2-seater and Coupé)		1952–54	SG9E
Coil, ignition		1939–54	B12
Control box		1939–48	CFR2
Control box		1949–50	RF95/2
Control box		1951–53	RB106/1
Control box	Later	1953–54	RB106/1
Distributor		1939–47	DKH4A
Distributor		1948–50	DKYH4A
Distributor		1951–54	DVX4A
Fuse box		1951–53	SF4
Fuse box	Later	1953–54	SF6
Generator		1939–48	C45YV-4
Generator		1949–50	C45YV-4
Generator		1951–53	C39PV-2
Generator	Later	1953–54	C39PV-2
Horn		1939–52	HF1235
Horn		1953–54	WT614
Horn push (2 and 4-seater)		1946–54	HP19
Horn push (Coupé)		1939–54	HP19
Inspection lamp sockets		1946–54	HP19
Junction box		1946–50	6J
Junction box		1951–54	4J
LAMPS			
Fog		1946–51	SFT462
Fog		1952–54	SFT462
Head			
Home, L.H. and R.H.		1939	LBD1485
Home, L.H. and R.H.		1946–47	MBD140
Home, L.H. and R.H.		1949–50	MBD140
Home, L.H. and R.H.		1951	S700
Home, L.H. and R.H.		1952–53	S700
Home, L.H. and R.H.		1953	S700
Home, L.H. and R.H.		1954	P700
Home, L.H. and R.H.	Later	1954	F700
Export		1946–47	MBD140-1
Export		1949–50	MBD140-1
Export, dip left, L.H. and R.H.		1951	S700
Export, dip left, L.H. and R.H.		1952–53	S700
Export, dip left, L.H. and R.H.		1953	S700
Export, dip left, L.H. and R.H.		1954	P700
Export, dip left, L.H. and R.H.	Later	1954	F700
Export, dip right, L.H. and R.H.		1951	S700
Export, dip right, L.H. and R.H.		1952	S700
Export, dip right, L.H. and R.H.		1953	S700
Export, dip right, L.H. and R.H.		1953–54	S700
Export, dip right, L.H. and R.H.	Later	1954	F700
Export U.S.A., L.H. and R.H.		1951–53	S700
Export U.S.A., L.H. and R.H.	Later	1953	S700
Export U.S.A., L.H. and R.H.		1954	F700
Export Europe (not France), L.H. and R.H.		1951–53	S700
Export Europe (not France), L.H. and R.H.		1953–54	S700
Export Europe (not France), L.H. and R.H.	Later	1953–54	F700
Side		1939	1130
Side	Later	1939	1130
Side		1946–52	1130
Side	Later	1952–54	516
Stop tail and number plate		1939	ST38D
Stop tail and number plate		1946–50	ST52

Description	Year	Part No
Stop tail	1951–54	ST52
Panel light bulb holder	1946–54	PL31
Starting motor	1939–47	M35G
Starting motor	1949–50	M35G
Starting motor	1951–54	M418G
Switches		
Dipper	1949–54	FS22-1
Flasher	1954	SD84
Foglamp	1951–54	PS19
Knob		
Lighting and ignition	1939–50	PLC2
Lighting and ignition	1951–54	PRS3
Knob		PRS2
Panel	1939–50	PS7-1
Knob		
Panel	1951–54	PS16
Knob		
Starting motor	1939–50	ST10
Starting motor	1951–54	ST18
Starting motor, solenoid operating	1951–54	SS9
Knob		
Stop lamp	1939–50	54C
Spring		39C
Trafficator	1939–48	TS82
Trafficator	1949–54	SD84
Windshield wiper (Coupé)	1951–54	PS19
Knob		PPG1
Trafficators	1939	SF80
Trafficators	1946–54	SF80
Warning lights		
Ignition	1939	WL3BN
Ignition	1946–50	WL3-1LI
Ignition	1951–54	WL11
Main beam	1951–52	WL12LI
Main beam	1953–54	WL11
Windshield wipers	1939	SW4AD9
(Coupé)	1946–50	SW4AD9
(Coupé)	1951–52	CR4DC113
(Coupé)	1953–54	CR4DC113
(2 and 4-seater)	1946–50	CW1AD16
(4-seater)	1951–54	CW1AD18
(2-seater)	1951–53	CW1AD18
(2-seater)	1954	CRT18AD18
Arm, wiper	1939	735099
Arm, wiper (Coupé)	1946–50	735099
Arm, wiper R.H.D. (Coupé)	1951–54	4737651
Arm, wiper L.H.D. (Coupé)	1951–54	4737678
Arm, wiper L.H. (2-seater)	1946–53	735103
Arm, wiper R.H. (2-seater)	1946–53	735102
Arm, wiper L.H. (4-seater)	1946–50	735111
Arm, wiper R.H. (4-seater)	1946–50	735110
Arm, wiper L.H. (4-seater)	1951–52	735103
Arm, wiper R.H. (4-seater)	1951–52	735102
Arm, wiper L.H. (4-seater)	1953	735107
Arm, wiper R.H. (4-seater)	1953	735106
Arm, wiper L.H.D. (2 and 4-seater)	1954	737814
Arm, wiper R.H.D. (2 and 4-seater)	1954	737815
Blade, wiper	1939	727784
Blade, wiper (Coupé)	1946–50	727784
Blade, wiper (Coupé)	1951–54	*737653
Blade, wiper (2-seater)	1946–53	727809
Blade, wiper (4-seater)	1946–52	727809
Blade, wiper (4-seater)	1953	727784
Blade, wiper (2 and 4-seater)	1954	737811

Plus 4 1939-1954

Description		Year	Part No
Ammeter		1946–50	*36036A
Ammeter		1951–60	36023F
Battery			
(2-seater and Coupé)		1950–51	SLG11E
(2-seater and Coupé)		1952–55	SG9E
(2-seater Tourer)		1956–58	STXW9E
(2-seater Tourer)		1959–60	STGW9E
(4-seater)		1946–55	SLG11A
(4-seater)		1956–58	SLTW11A
(4-seater)		1959	SLG11A
(4-seater)		1960	STGK9E
Bulb holder, panel light		1946–60	39007A
Bulb		—	987
Coil, ignition		1946–57	*45012A/D
Coil, ignition		1958–60	*45054J
Control box		1946–48	*332202
Control box		1949–50	37057E
Control box		1951–53	*37048E
Control box	Later	1953–55	*37138A/D
Control box		1956–59	37182H
Control box		1960	37290D/E
Distributor		1946–47	*406345
Distributor		1948–50	*409639
Distributor		1951–54	*40201A/B
Distributor (TR 2/3 Engine)		1955–57	*40403D
Distributor (TR 2/3 Engine)		1958–59	*40480A
Distributor (TR 2/3 Engine)		1960	40698A
Distributor (TR 2/3 Engine)	Later	1960	40698B
Distributor ('Vanguard' Eng.)		1956–57	40365D
Distributor ('Vanguard' Eng.)		1958	40559A
Distributor ('Vanguard' Eng.)		1959–60	40548D
Flasher unit			
(Export, Coupé & Special Orders)		1955–57	*35003A
Flasher unit			
(Export, Coupé & Special Orders)		1958–59	35010A
Flasher unit			
(Export, Coupé & Special Orders)		1960	35020A
Fuse box		1951–53	37134A
Fuse box	Later	1953–59	033240
Fuse box		1960	033285
Generator		1946–48	*228137
Generator		1949–50	22449A
Generator		1951–53	*22250E/F
Generator	Later	1953–55	22258D
Generator		1956–60	22258E/H
Horn		1946–52	069344
Horn		1953–55	*69012F/J
Horn		1956–60	69047D/F
Horn push			
(2 and 4-seater)		1946–54	*76208A
(Coupé)		1946–54	*76211A
Horn push		1955–60	*76211B/D
Inspection lamp sockets		1946–60	39507A
Junction box		1946–50	78266A
Junction box		1951–55	78264A
LAMPS			
Flasher, front			
(Export, Coupé and Special Orders)		1955–58	*52246A
(U.S.A.)		1960	52345A
Flasher, front and rear		1955–60	52337A
Flasher, rear (U.S.A.)		1960	53600B
Fog (U.S.A. and Special Orders		1946–51	*55063B
Fog (U.S.A. and Special Orders)		1952–55	*55049B
Fog			
(U.S.A. and Special Orders)	Later	1955–60	55128B
Head, Home			
R.H.D., non-dip		1946–47	*50102A
R.H.D., non-dip		1948–50	50753A
R.H.D., non-dip		1951	*50798D
R.H.D., dip left		1952–53	*50995A
R.H.D., dip left	Later	1953	*51353A
R.H.D., dip left		1954	51344A
R.H.D., dip left	Later	1954–55	51336A
R.H.D., dip left		1956–60	51336B/J
Head, Export			
Vertical dip		1946–47	*50291A
Veritcal dip		1949–50	*50458A
R.H.D., dip left		1951	*50821D
R.H.D., dip left		1952–53	*50995A
R.H.D., dip left	Later	1953	*51353A
R.H.D., dip left		1954	51344A
R.H.D., dip left	Later	1954–55	51336A
R.H.D., dip left		1956–60	51336B/J
L.H.D., dip right		1951	*A50821D
L.H.D., dip right		1952	50877B
L.H.D., dip right		1953	*50775E
L.H.D., dip right	Later	1953–54	*51355A
L.H.D., dip right		1954	*51337A
L.H.D., dip right		1955	*51337B
L.H.D., dip right		1956–59	*51337D/E
L.H.D., dip right	Later	1959–60	*51337F/K
Head, Export Europe			
(not France)		1951–53	*50996A
(not France)	Later	1953–54	51354A
(not France)		1954–55	51339A
(not France)		1956–59	51339B/H
(not France)	Later	1959–60	58230A/B
Head, Export U.S.A.		1951–53	*50775E
Head, Export, U.S.A.	Later	1953	*51355A
Head, Export U.S.A.		1954–60	51509A/E
Number plate illumination		1955–60	53093E/J
Side		1946–52	52030A
Side	Later	1952–60	*52182B
Stop tail and number plate		1946–50	*53031D
Stop and number plate		1951–54	53178D
Stop tail		1955–58	53178E/H
Stop tail		1959–60	53330D
Starting motor		1946–48	*250231
Starting motor		1949–50	25017A
Starting motor		1951–54	25526A/J
Starting motor (TR 2/3 Eng.)		1955–60	25541H
Starting motor ('Vanguard' Eng.)		1959–60	25550A
SWITCHES			
Dipper		1949–59	31284A/D
Dipper		1960	31800A
Fan		1955	31248B
Knob		—	315042
Flasher (when fitted)		1955–60	31476A/B
Foglamp (when fitted)		1951–57	31248A/B
Foglamp (when fitted)		1958–60	31434B
Knob		1951–54	*316374
Knob		1955	315042
Knob		1956–57	*316374
Knob		1958–60	313619
Ignition and lighting		1946–50	*34027B
Ignition and lighting		1951–60	31270A/B
Knob		1951–54	*316425
Knob		1955–60	316276
Panel		1946–50	31201B
Panel		1951–57	31248A/B
Panel		1958–60	*31557A
Knob		1946–50	314026
Knob		1951–54	*316373
Knob		1955–57	315042

Description	Year	Part No
Knob	1958–60	312955
Solenoid, starting motor	1951–59	*76411D/E
Solenoid, starting motor	1960	*76465B
Solenoid, operating	1951–58	31253A/B
Solenoid, operating	1959–60	31377A
Knob	1951–54	*317364
Knob	1955–58	317362
Starting motor	1946–50	76428A
Stop lamp	1946–50	31281B
Spring	—	315543
Trafficator	1946–48	31379A
Trafficator (when fitted)	1949–60	*31306A
Windshield wiper	1951–57	31248A/B
Windshield wiper	1958–60	*31557A
Knob	1951–54	*316375
Knob	1955–57	315042
Knob	1958–60	316393
Trafficators (when fitted)	1946–55	*54039A
Trafficators (when fitted)	1956–60	54044E
Warning lights		
Ignition	1946–50	*38013B
Ignition	1951–57	38059A
Ignition	1958–60	38145A
Main Beam	1951–52	38052B
Main Beam	1953–57	38059A
Main Beam	1958–60	38150A
Windshield wipers		
Motor (Coupé)	1946–50	†75066A
Motor (Coupé)	1951–52	*75074J
Motor (Coupé)	1953–55	75144A
Motor (2 and 4-seater)	1946–50	*75059A
Motor (4-seater)	1951–54	*733147
Motor (2-seater)	1951–53	*733147
Motor (2-seater)	1954	75150A
Motor (2 and 4-seater)	1955	75150A
Motor (All Models)	1956–60	75150A/B
Shaft, short (Coupé)	1946–50	737243
Coupling (Coupé)	1946–50	729036/S
Coupling bar		
(2 and 4-seater)	1946–50	727293
(4-seater)	1951–54	732968
(2-seater)	1951–53	732968
Bearing, passenger side		
(2 and 4-seater)	1946–50	732583
(4-seater)	1951–54	731721
(2-seater)	1951–53	731721
Wheelbox		
(Coupé)	1951–54	*72548A/K
(Coupé)	1955	72662A
(2-seater)	1954	72629A/D
(2 and 4-seater)	1955	12629A/D
(All Models)	1956–60	72629A/D
Crosshead and rack assy.		

Description	Year	Part No
(Coupé)	1951-55	736177
(2-seater)	1954	736307
(2 and 4-seater)	1955	736307
(All Models)	1956–60	736307
Outer casing		
Motor to wheelbox		
(Coupé)	1951–52	*739855
(Coupé)	1953–55	739870
(2-seater)	1954	739868
(2 and 4-seater)	1955	739868
(All Models)	1956–60	739868
Wheelbox to wheelbox		
(Coupé)	1951–55	*739956
(2-seater)	1954	*739979
(2 and 4-seater)	1955	*739979
(All Models)	1956–60	*739979
End tube	1951–60	740031
Grommet (Coupé)		
(Housing or wheelbox)	1946–55	729313
(Scuttle)	1946–50	729378
Arm		
L.H.D and R.H.D.		
(Coupé)	1946–50	735099
L.H.D. (Coupé)	1951–55	737678
R.H.D. (Coupé)	1951–55	737651
L.H. (2-seater)	1946–53	735103
R.H. (2-seater)	1946–53	735102
L.H. (4-seater)	1946–50	753111
L.H. (4-seater)	1951–52	735103
L.H. (4-seater)	1953	*735107
R.H. (4-seater)	1946–50	735110
R.H. (4-seater)	1951–52	735102
R.H. (4-seater)	1953	*735107
L.H.D. (2 and 4-seater)	1954–55	*737814
R.H.D. (2 and 4-seater)	1954–55	*737815
L.H.D. (Coupé and 4-seater)	1956	742159
R.H.D. (Coupé and 4-seater)	1956	742158
L.H.D. and R.H.D.		
(Coupé and 4-seater)	1957–60	742291
L.H.D. (2-seater Tourer)	1956	742156
R.H.D. (2-seater Tourer)	1956	742157
L.H.D. and R.H.D.		
(2-seater Tourer)	1957–60	742295
Blade		
(Coupé)	1946–50	727784
(Coupé)	1951–55	*737653
(Coupé and 4-seater)	1956–60	737680
(2-seater)	1946–53	727809
(4-seater)	1946–52	727809
(4-seater)	1953	727784
(2 seater and 4-seater)	1954–55	737811
(2-seater Tourer)	1956–60	737811

4/4 1946-1948
Plus 4 1950-1960

Description	Year	Part No
Ammeter	1949–50	*36036A
Battery		
(4-seater)	1949–50	SLTW11A
(2-seater and Coupé)	1949–50	SLTW11E
Coil, ignition	1949	402032
Coil, ignition	1950	*45012A/D
Control box	1949–50	37057D
Distributor	1949–50	*409639
Generator	1949	*228139
Generator	1950	22449A
Horn	1949–50	069344
Horn push		
(Coupé)	1949–50	76205B
(2 and 4-seater)	1949–50	76218A
Inspection lamp sockets	1950	39507A/B
Junction box	1949	*770017
Junction box	1950	78266A
LAMPS		
Fog or pass	1949	*55049A
Fog or pass	1950	*55063A
Head, Home		
Non-dip	1949	*50490A
Non-dip	1950	50753A
Head, Export		
Vertical dip	1949–50	*50458A
Side	1949–50	52030A
Stop tail	1949–50	*53031B
Starting motor	1949–50	25017A
SWITCHES		
Dipper	1949–50	*360007

Description	Year	Part No
Ignition and lighting	1949–50	34055A
Panel	1949	*314014
Panel	1950	031218
Starting motor	1949	*764251
Starting motor	1950	76428A
Stop lamp	1949–50	*315725
Trafficator	1949	*310656
Trafficator	1950	31190A
Trafficators (Coupé)	1949–50	*54018D
Warning light (ignition)	1949–50	*38013B
Windshield wipers		
Motor (Coupé)	1949–50	†75066A
Motor (2 and 4-seater)	1949	*75059A
Motor (2 and 4-seater)	1950	732480
Clutch housing	1949–50	737412
Coupling bar		
(2 and 4-seater)	1949	*725151
Coupling bar		
(2 and 4-seater)	1950	727293
Bearing	1950	732583
Arm		
(Coupé)	1949–50	735099
L.H. (2-seater)	1949–50	735103
R.H. (2-seater)	1949–50	735102
L.H. (4-seater)	1949–50	735111
R.H. (4-seater)	1949–50	735110
Blade		
(Coupé)	1949–50	727784
(2 and 4-seater)	1949	727804
(2 and 4-seater)	1950	727809

4/4 1949-1950
(Standard engine)

4/4 Series II 1956-1960
(Ford engine)

Description	Year	Part No
Ammeter	1956–60	36226A
Battery	1956–58	GLT7A
Battery	1959–60	BT7A
Coil, ignition	1956–60	*45053D/K
Control box	1956–60	37182H
Distributor	1956–60	*40402H/N & Q/T
Flasher unit (Coupé, Export and Special Orders)	1956–57	*35003A
Flasher Unit (Coupé, Export and Special Orders)	1958–60	35010A
Flasher relay	1956–58	33117D
Fuse box	1956–60	033240
Generator	1956–60	22258E/H
Horn	1956–57	*69012J
Horn	1958–60	69047F
Horn Push	1956–60	*76211B/D
Inspection lamp sockets	1956–60	39507A
Junction box (L.H.D.)	1956–57	78264A
LAMPS		
Flasher, front (Coupé, Export and Special Orders)	1956–58	*52246A
Flasher, front (Coupé, Export and Special Orders)	1959–60	52337A
Flasher, front (U.S.A.)	1960	52345A
Flasher, rear	1959–60	52337A
Flasher, rear (U.S.A.)	1960	53600B
Fog (L.H.D. and optional R.H.D.)	1956–60	55128B
Head, Home		
R.H.D., dip left	1956–60	51336B/J
Head, Export		
R.H.D. dip left	1956–60	51336B/J
L.H.D., dip right	1956–60	*51337D/E & F/K
Head, Export Europe		
(not France)	1956–58	51339E
(not France)	1959–60	58230A
Head, Export France	1956–60	*51590A/H
Number plate illumination	1956–60	53093H
Side	1956–60	*52182D
Stop tail	1956–58	53178F
Stop tail	1959–60	53330D

Description		Year	Part No
Starting motor		1956–58	25022F/L
Starting motor		1959	25022M
Starting motor		1960	25022N/Q
SWITCHES			
Dipper		1956–60	31284A/D
Flasher (when fitted)		1956–60	31476A/B
Foglamp (when fitted)		1956–57	31248B
Knob		—	*316374
Foglamp (when fitted)		1958–60	31434B
Knob		—	313619
Ignition and lighting		1956–60	31270B
Knob		—	316276
Panel		1956–57	31248B
Knob		—	315042
Panel		1958–60	*31557A
Knob		—	312955
Solenoid, starting motor		1956–60	*76411E
Solenoid, starting motor	Later	1960	*76445A
Solenoid, operating		1956–60	31377A
Trafficator (when fitted)		1956–60	*31306A
Windshield wiper		1956–60	*31557A
Knob		—	316393
Trafficators (when fitted)		1956–60	54044E
Warning lights			
Ignition		1956–60	38145A
Main beam		1956–60	38150A
Windshield wipers			
Motor		1956–60	75150A/B
Wheelbox		1956–60	72629A/D
Crosshead and rack assy. cut to 44$\frac{1}{16}$"		1956–60	736307
Outer casing			
Motor to wheelbox		1956–60	739868
Wheelbox to wheelbox		1956–57	*739956
Wheelbox to wheelbox		1958–60	*739979
End tube		1956–60	740031
Arm, left hand drive		1956–57	742156
Arm, right hand drive		1956	742157
Arm, left hand drive and right hand drive		1957–60	742295
Blade		1956–60	737811

Description	Year	Model	Part No
Ammeter	1961–62	CZU20	*36023F
Ammeter	1963–66	CZU20	36307A
Battery			
2-seater (2 off)	1961–66	SG9E	24023107
2-seater (2 off) (Export, dry charged)	1961–66	STGZ9E	24014107
4-seater (1 off)	1961–66	SLG11A	24040657
4-seater (1 off) (Export, dry charged)	1961–66	SLTZ11A	24021620
Bulb holder, panel light	1961–66	—	39007A
Bulb	—	—	987
Coil, ignition	1961	HA12	*45054J/M
Coil, ignition	1962–63	HA12	*45102D/F
Coil, ignition	1964–65	HA12	45132D
Coil, ignition	1966	HA12	45179A
Control box	1961–66	RB106-2	37290D/E
Distributor	1961–63	DM2	*40698A/B
Distributor Later	1963–66	25D4	40795A
Flasher unit	1961–66	FL5	35020A
Fuse box	1961–66	4FJ	54038034
Generator	1961–65	C40-1	22700D/J
Generator (Alternative)	1961–65	C40	*22704A/H
Generator	1966	C40-1	*22715D
Horn	1961	WT618	*69047D/F
Horn Later	1961	9H	*69112A
Horn	1962–66	9H	*54068008
Horn push	1961	HP19	*76211B/D
Horn push	1962–66	HP19	76205D
Inspection lamp socket	1961–66	—	39507A/E
LAMPS			
Flasher, front (U.S.A.)	1961–66	594	52345A/B
Flasher, front and rear	1961–66	594	52337A/B
Flasher, rear (U.S.A.)	1961–66	594	53600B/E
Fog (optional)	1961	SFT576	55128B
Fog (optional)	1962–63	SFT576	55174E
Fog (optional)	1964–66	SFT576	55187D
Head			
R.H.D., dip left	1961–63	F700	†51336B/J
R.H.D., dip left Later	1963–64	F700	†58618A
R.H.D., dip left	1965–66	F700	†58903A
L.H.D., dip right	1961–63	F700	†51337D/K
L.H.D., dip right Later	1963	F700	†58775A
L.H.D., dip right	1964–66	F700	†59193A
Head, Export Europe	1961–64	F700	†58230A/E
Head, Export Europe	1965	F700	†51590D
Head, Export Europe Later	1965	F700	†58231E
Head, Export France	1961–64	F700EF	†51590D
Head, Export France	1966	F700	†59450A
Long range (optional)	1965–66	SLR576	55195B
Number plate illumination	1961–66	467/2	53093B/K
Side	1961–66	516	*52182B/E
Stop tail	1961–66	549	53330D
Screenjet (optional)	1966	5SJ	54071313
Accessories kit	—	—	*54711656
Starting motor	1961–66	M418G	25550A

Description	Year	Model	Part No
SWITCHES			
Dipper	1961–64	21SA	31800A/B
Dipper Later	1964–66	103SA	34536B
Direction indicator	1961–62	57SA	*34385A
Direction indicator	1963	57SA	*34393E
Direction indicator	1964–66	57SA	*34366B
Direction indicator (special orders)	1965–66	TPS1/1	31904A
Knob	—	—	54331875
Foglamp (when fitted)	1961–62	65SA	34338A
Foglamp (when fitted)	1963–66	65SA	*34521A
Ignition and starting	1961–62	47SA	*34384A
Ignition and starting	1963–66	47SA	*31953D/E
Lighting	1961–66	58SA	34340A/D
Knob	1961–62	—	316144
Knob	1963–66	—	316154
Panel light	1961–62	65SA	34338A
Panel light	1963–66	65SA	*34521A
Solenoid, starting motor	1961	ST950	*76465B
Solenoid, starting motor	1962–66	2ST	*76464D/F
Windshield wiper	1961–62	57SA	34339A
Windshield wiper	1963	57SA	*34366A
Windshield wiper Later	1963	57SA	34350E
Windshield wiper	1964–66	57SA	*34393E
Warning lights			
Direction indicator	1961–66	WL15	38191A
Ignition	1961–65	WL15	38189A
Main beam	1961–63	WL15	38190A
Main beam	1964–66	WL15	54360500
Bulb holder	1961–66	—	54945043
Bulb	—	—	281
Windshield wipers			
Motor	1961–63	CRT15	75150A/B
Motor Later	1963–66	DR3A	75414B
Wheelbox	1961–63	—	72629A/D
Wheelbox Later	1963–66	—	*72674B
Rack	1961–63	—	736307
Rack Later	1963–66	—	743235
Outer casing			
Motor to wheelbox	1961–63	—	739868
Motor to wheelbox Later	1963–66	—	745028
Wheelbox to wheelbox	1961–63	—	*739979
Wheelbox to wheelbox Later	1963–66	—	745043
Short, wheelbox	1961–63	—	740031
Short, wheelbox Later	1963–66	—	747046
Ferrule, casing retaining	1961–63	—	738410
Ferrule, casing retaining Later	1963–66	—	740715
Arm (Coupé)	1961	—	742295
Arm (Coupé)	1962	—	742291
Arm (2-seater)	1961–62	—	742295
Arm (all models)	1963	—	742295
Arm (all models) Later	1963–66	—	54712085
Blade (Coupé and 4-seater)	1961–66	—	737680
Blade (2-seater)	1961–66	—	737811

Plus 4 1961-1966

Description	Year	Model	Part No
Ammeter	1961–64	CZU34	*36226A
Ammeter (G.T. and, later, standard)	1964–66	CZU20	*36191B
Battery	1961–66	S7	54028340
Battery (Export, dry charged)	1961–66	SZ7	54028342
Bulb holder, panel light	1963–66	—	39010A
Bulb holder, speedometer light	1963–66	—	554734
Bulb	—	—	987
Coil, ignition	1961–62	LA12	*45111A/D
Coil, ignition	1963–66	LA12	45141B
Control box	1961–66	RB106-2	37290E
Distributor	1961	DM2P4	40652F/H
Distributor	1962	DM2P4	40744A
Distributor	1963–66	25D4	40857F
Distributor (G.T.)	1964–66	25D4	40927D
Flasher unit	1961–66	FL5	35020A
Fuse box	1961–66	4FJ	54038034
Generator	1961	C40-1	22700D
Generator (alternative)	1961	C40	*22704A
Generator	1962–63	C40L	22716B/E
Generator (1500c.c. engine) Later	1963–66	C40-1	22700B/J
Generator (1500c.c engine) alternative) Later	1963–66	C40	*22704A/H
Generator (G.T.)	1964–66	C40L	22716J
Horn	1961	WT618	*69047F
Horn Later	1961	9H	*69112A
Horn	1962–66	9H	*54068008
Horn push	1961	HP19	*76211B/D
Horn push	1962–66	HP19	76205D
Inspection lamp socket	1961–66	—	39507A/E
LAMPS			
Flasher, front (U.S.A.)	1961–66	594	52345A/B
Flasher, front and rear	1961–66	594	52337A/B
Flasher, rear (U.S.A.)	1961–66	594	53600B/E
Fog (Export L.H.D. only)	1961	SFT576	551288
Fog (Export L.H.D. only)	1962–63	SFT576	55174E
Fog (Export L.H.D. only)	1964–66	SFT576	55187D
Head			
R.H.D., dip left	1961–63	F700	†51336B/J
R.H.D., dip left Later	1963–64	F700	†58618A
R.H.D., dip left	1965–66	F700	†58903A
L.H.D., dip right	1961–63	F700	†51337D/K
L.H.D., dip right Later	1963	F700	†58775A
L.H.D., dip right	1964–66	F700	†59193A
Head, Export Europe (except France)	1961–64	F700	†58230A/E
Head, Export Europe (except France)	1965	F700EF	†51590A/H
Head, Export Europe (except France) Later	1965	F700	†58231E
Head, Export Europe (except France)	1966	F700	†59450A
Head, Export France	1961–64	F700EF	†51590A/H
Long range (optional)	1965–66	SLR576	55195B
Number plate illumination	1961–66	467/2	53093H/K
Side	1961–66	516	*52182D/F
Stop tail	1961–66	549	53330D
Screenjet (optional)	1966	5SJ	54091313
Accessories kit	—	—	*54711656
Starting motor	1961	M35H	25071B/D
Starting motor	1962	M35G-1	*25075D
Starting motor	1963–66	M35G-1	25083D/F
SWITCHES			
Dipper	1961–64	21SA	31800A/B
Dipper Later	1964–66	103SA	34536B
Direction indicator	1961–63	57SA	*34385A
Direction indicator Later	1963–66	57SA	*34366A/B
Foglamp (when fitted)	1961–63	65SA	34338A
Foglamp (when fitted) Later	1963–66	65SA	*34521A
Knob	—	—	316102
Ignition and starting	1961–62	47SA	*34384A
Ignition and starting	1963–66	47SA	*31953D/E
Lighting	1961–66	58SA	34340A/D
Knob	1961–64	—	316144
Knob Later	1964–66	—	316154
Panel light	1961–63	65SA	34338A
Panel light Later	1963–66	65SA	*34521A
Solenoid, starting motor	1961	ST950	*76465A
Solenoid, starting motor	1962–66	2ST	*76464D/F
Windshield wiper	1961–63	57SA	34339A/B
Windshield wiper Later	1963–66	57SA	*34393E
Warning lights			
Direction indicator	1961–63	WL15	38207A
Direction indicator Later	1963–66	WL15	38191A
Ignition	1961–66	WL15	38189A
Main beam	1961–64	WL15	38190A
Main beam Later	1964–66	WL15	54360500
Bulb holder	1961–66	—	54945043
Bulb	—	—	281
Windshield wipers			
Motor	1961–63	CRT15	75150A/B
Motor Later	1963–66	DR3A	75414B
Wheelbox	1961–63	—	72629A/D
Wheelbox Later	1963–66	—	*72674B
Rack	1961–63	—	736307
Rack Later	1963–66	—	743235
Outer casing			
Motor to wheelbox	1961–63	—	739868
Motor to wheelbox Later	1963–66	—	745028
Wheelbox to wheelbox	1961–63	—	*739979
Wheelbox to wheelbox Later	1963–66	—	745043
Short, wheelbox	1961–63	—	740031
Short, wheelbox Later	1963–66	—	740746
Ferrule, casing retaining	1961–63	—	738410
Ferrule, casing retaining Later	1963–66	—	740715
Arm	1961–63	—	742295
Arm Later	1963–66	—	54712085
Blade	1961–66	—	737811

Description	Year	Model	Part No
Ammeter	1964–66	CZU20	36307A
Battery	1964–66	SLG11A	24040657
Battery (Export, dry charged)	1964–66	SLTZ11A	24021620
Bulb holder, panel light	1964–66	—	39007A
Bulb	—	—	987
Coil, ignition	1964–65	HA12	45132D
Coil, ignition	1966	HA12	45179A
Control box	1964–66	RB106-2	37290E
Distributor	1964–66	25D4	40795A
Flasher unit	1964–66	FL5	35020A
Fuse box	1964–66	4FJ	54038034
Generator	1964–65	C40-1	22700J
Generator (alternative)	1964–65	C40	*22704H
Generator	1966	C40-1	*22715D
Horn			
Low note	1964–66	H9	*54068009
High note	1964–66	H9	*54068008
Horn push	1964–66	HP19	76205D
Inspection lamp socket	1964–66	—	39507A/E
LAMPS			
Flasher, rear (amber)	1964–66	691	*53956A/B
Fog (optional)	1964–66	SFT576	55187D
Head			
R.H.D., dip left	1964	F700	†58618A
R.H.D., dip left	1965–66	F700	†58903A
L.H.D., dip right	1964–66	F700	†59193A
Head, Export Europe (except France)	1964	F700	†58230E
Head, Export Europe	1965	F700EF	†51590D
Head, Export Europe Later	1965	F700	†58231E
Head, Export Europe	1966	F700	†59450A
Head, Export France	1964	F700EF	†51590D
Long range (optional)	1965–66	SLR576	55195B
Number plate illumination	1964–66	550	54170A
Side and flasher	1964–66	686	52535B
Side and flasher (N. Ireland and Germany)			
R.H.	1964–66	584	52506E
L.H.	1964–66	584	52507E
Stop tail	1964–66	692	*53955B
Reflex reflector	1964–66	RER22	57098A
Screenject (optional)	1966	5SJ	54071313
Accessories kit	—	—	54711656
Starting motor	1964–66	M418G	25550A
SWITCHES			
Dipper	1964	21SA	31800B
Dipper Later	1964–66	103SA	34356B
Direction indicator	1964–66	57SA	*34366B
Direction indicator (special orders)	1965–66	TPS1/1	31904A
Knob	—	—	54331875
Foglamp	1964–66	65SA	*34521A
Ignition and starting	1964–66	47SA	31953E
Lighting	1964–66	58SA	34340B/D
Knob	—	—	316154
Panel	1964–66	65SA	*34521A
Solenoid, starting motor	1964–66	2ST	*76464E/F
Windshield wiper	1964–66	57SA	*34393E
Warning lights			
Direction indicator	1964–66	WL15	38191A
Ignition	1964–66	WL15	38189A
Main beam	1964–66	WL15	54360500
Bulb holder	—	—	54945043
Bulb	—	—	281
Windshield wipers			
Motor	1964–66	DR3A	75450A
Wheelbox	1964–66	—	72774B
Rack	1964_66	—	743213
Outer casing			
Motor to wheelbox	1964–66	—	745028
Wheelbox to wheelbox	1964–66	—	745043
Short, wheelbox	1964–66	—	740746
Arm	1964–66	—	54710008
Blade	1964–66	—	54711283

Plus 4 Plus 1964-1966

	'Plus Four' (TR4 Engine)	'Four Four' (Ford 1500c.c. Engine) (Ford 1500 c.c. G.T. Engine)
Ammeter	36307A	*35191B
		(36099)
Battery	SLG11A	BT7A
	(24040657	(24016517)
Battery (Dry charged – Export)	SLTZ11A	BTZ7A
	(24021620)	(24016527)
Battery, 2-seater and Coupé	SG9E	—
	(24023107)	
Battery, (Dry charged – Export) 2-seater and Coupé	STGZ9E	—
	(24014107)	
Bulbholder, panel	39007	39010
Bulbholder, speedo	—	554734
Bulb	987	987
Cable sundries (see page 3)		
Coil, ignition	45179A	45141B
Control box	37190E/F	37290E/F
Distributor	40795A	40984A
Distributor (G.T. only)	—	41146A
Flasher unit	35020A	35020A
Fuse box	54038034	54038034
Generator	22715E	22700L
	(22700)	
Generator (G.T. only)	—	22716J
Horn	54068086	54068086
Horn push	76205D	76205D
Inspection lamp socket	39507A/E	39507A/E
LAMPS		
Head		
Right hand drive, dip left	58903A/B	58903A/B
Serviced by: Lamp	58811	58811
Serviced by: Rim	554440	554440
Serviced by: Gasket	54521487	54521487
Left ahd drive, dip right,	59193A/B	59193A/B
Serviced by: Lamp	59191	59191
Serviced by: Rim	554440	554440
Serviced by: Gasket	54521487	54521487
Export Europe	59450A	59450A
Serviced by: Lamp	58815	58815
Serviced by: Rim	554440	554440
Serviced by: Gasket	54521487	54521487
Flasher, clear	52345B	52345B
Flasher, amber	52337B	52337B
Fog (optional)	55187D	55187D
Long range (optional)	55195B	55195B
Number plate illumination	53093H/K	53093H/K

	'Plus Four' (TR4 Engine)	'Four Four' (Ford 1500c.c. Engine) (Ford 1500 c.c. G.T. Engine)
Side	*52182D/E	*52182E/F
	(52436)	(52436)
Stop tail and reflex	53330D	53330D
Screenjet (optional)	54071451	54071451
Starting motor	25550A/B	25083F/H
SWITCHES		
Dipper	34536B/E	34536B/E
Flasher	*34366B	*34366B
	(31743)	(31743)
Foglamp (when fitted)	*34521A	34521A
	(34352)	(34352)
Ignition and starter	31953E	31953E
Lighting	34340B/D	34340B/D
Knob	316154	316154
Panel	*34521A	*34521A
	(34352)	(34352)
Solenoid, starting motor	*76464H	*76464H
	(76445)	(76445)
Windshield wiper	*34393E	*34393E
	(31836)	(31836)
Warning lights		
Flasher	38191A	38191A
Bulbholder	†54945043	†54945043
Bulb	†281	†281
Ignition	38189A	38189A
Bulbholder	†54945043	†54945043
Bulb	†281	†281
Main beam	54360500	54360500
Bulbholder	†54945043	†54945043
Bulb	†281	†281
Windshield wipers		
Motor	75414B	75414B
Wheelbox	72674B	72674B
Crosshead and rack (cut to 44 5/32")	743235	743235
Outer casing		
Motor to wheelbox (2-seater only)	54717032	54717032
Motor to wheelbox (4-seater and coupé)	54700378	54700378
Wheelbox to wheelbox	54713402	54713402
End tube	740746	740746
Ferrule	740715	740715
Arm	54712085	54712085
Blade (2-seater only)	737811	737811
Blade (4-seater and Coupé)	737680	737811

BRAKE, CLUTCH AND SHOCK ABSORBER EQUIPMENT (Serviced by Girling Ltd.)

	Morgan 'Four Four'		**Morgan 'Plus Four'**	
	Front	**Rear**	**Front**	**Rear**
Complete brake or caliper, left hand	64032702	64027510	64032702	64027510
Complete brake or caliper, right hand	64032703	64027512	64032703	64027512
Lined shoe or pad assembly				
Leading, left hand	64325422DT	GB45796	64325422DT	GB45796
Trailing, left hand	64325422DT	GB47989	64325422DT	GB47989
Leading, right hand	64325422DT	GB45796	64325422DT	GB45796
Trailing, left hand	64325422DT	GB47989	64325422DT	GB47989

*When ordering, quote part number shown in brackets. †Order separately

Plus 4 1965-1969 (Triumph TR4A engine)
4/4 Series V 1966-1968
Ford engine)

APPENDIX III

SPECIALISTS, MANUFACTURERS AND DISTRIBUTORS

Below are details of Morgan specialist companies trading in the UK, USA and Canada. The vast majority of them only offer services in connection with four-wheeler Morgans but, where they include three-wheelers, or are only operating a service for three-wheelers, this is indicated.

All of the Morgan main distributors worldwide carry some spares and many have extensive stocks. All are willing to give advice. Companies are therefore only listed in this section if they offer any special service other than normal spares. By listing these companies neither the author or the publisher is endorsing or recommending them.

UK

Angel Fabrications
Castle Bank Cottage
Ewloe
Deeside
Clwyd
England
Telephone 0244 532714
Three-wheeler chassis and exhaust systems.

Black Phey Ltd
Raleigh Cottage
The Street
Takeley
Nr. Bishops Stortford
Hertfordshire
England
Telephone 0279-870698
Wooden body frames and restoration.

Brands Hatch Morgans
Brands Hatch Circuit
Dartford
Kent DA3 8NG
England
Telephone 0474-874147
Competition tuning repairs. Spares, sales transporting.

Hamilton Brooks
Ullingswick
Herefordshire
England
Telephone 0432-820320
Morgan pedal cars for ages 2-4.

David Browne
'Woodview'
Jenkins Lane
Tring
Hertfordshire HP23 6NW
England
Telephone 0240-29644
Three-wheeler bearings.

Burlen Services
Spitfire House
Castle Road
Salisbury
Wilts SP1 3SA
Telephone 0722 412500
Carburettor specialists.

Busby Engineering
1, Desford Lane
Ratby
Leicestershire LE6 0LE
England
Telephone 0533-393564
Three-wheeler castings & machining.

F.H. Douglass
1A, South Ealing Road
London W5 4QT
England
Telephone 081-567-0570
Spares, service, repairs, sales for three- and four-wheelers including Series I 4-4s.

Packer Duncan Ltd
Waterloo Garage

Hagley Road
Oldswinford
Stourbridge
West Midlands DY8 2JB
England
Telephone 0384-395186
Spares and used sales.

Ivan Dutton
Unit 12
The Metropolitan Centre
Derby Road
Greenford
Middlesex
England
Telephone 081-578-3810
Reproduction castings for three-wheelers.

Fred Gudgeon
'Cassis'
Avon Castle Drive
Avon Castle
Ringwood
Hampshire
England
Telephone 04254-77368
Repairs & restoration, particularly flat radiator models.

Roy Halford
11, Larksfield Close
Caterton
Oxford
England
Telephone 0993-841993
All electrical components.

Harpers
The Bothy
Essex Lane
Hunton Bridge
Kings Langley
Herfordshire WD4 8PN
England
Telephone 09277-60299/60290
Coachwork & trim specialists. Restoration, spares.

Janet Hill Ventures
Woodfidley
Brockenhurst
Hampshire SO42 7QL
England
Telephone 0590-22091
Scale Morgan models & toys. Morgan jewellery, books, goodies, etc. (list available).

W.H.Hunter (Motors) Ltd
Braid Road
Edinburg EH10 6AP
Scotland
Telephone 0314-471707
Spares, servicing, tuning.

Libra Motive Ltd
2 Rosemont Road
Hampstead
London NW3 6NE
Telephone 071 435 8159/794 7009
Race tuning, repairs, spares.

Light Car & Cyclecar Restoration Co
Unit 226
Artic Trading Estate
Droitwich Road
Hartlebury
Worcestershire
England
Telephone 0564-826891 (evenings)
Three-wheeler restoration badges, spares, some four-wheeler work.

I & J Macdonald Ltd
Maiden Law Garage
Lanchester DH7 0QR
Telephone 0207 520916
Competition, tuning, spares.

Morgan Motor Company Limited
Pickersleigh Road
Malvern
Worcestershire WR14 2LL
England
Telephone 0689-3104/5
All parts including body frame & panels. Three- and four-wheelers.

Morgan Sports Car Club
Spares Registrar Secretary
Bob Cragg
1, Manor Farm
Culham
Abingdon
Oxon
England
Members only

Morgan Three-Wheeler Club
Spares Manager
Nev Lear
1, Hollies Close
Middlezoy
Bridgewater
Somerset
England
Telephone 0823-69591
Members only

Colin Musgrove & Associates
Lanes End Garage
Kenilworth Road
Meer End
Nr Kenilworth
Warwickshire CV8 1PT
England
Telephone 0676-33120/33014
Competition, tuning, restoration. Used sales

Phaeton Engineering
Palmers Road
Emsworth
Hampshire
England
Telephone 0243-372040.
Servicing, repairs.

George Proudfoot
49 King George Road
Porchester
Hampshire
England
Telephone 0705-327184
Repairs and restoration

Rutherford Engineering
Crackling Farm
Agden Lane
High Leigh
Cheshire WA16 6NY
England
Telephone 0565-830224
Restoration, spares, service.

Melvyn Rutter Ltd
The Morgan Garage
Little Hallingbury
Nr Bishops Stortford
Hertfordshire CM22 7RA
England
Telephone 0279-725725
Three- and four-wheeler spares. Used sales (catalogue available).

S G T
Station Road
Taplow
Maidenhead
Berkshire SL6 0NT
England
Telephone 0628-65353/64044
Repairs, spares

John Smith
Inworth
Nr Colchester
Essex
England
Telephone 0376-70438
Gearbox and axle rebuilds.

Colin Wilson
Mount Works
Mill Lane

Horndon-on-the-Hill
Essex
England
Telephone 0375-672313
Three-wheeler spares manufacturer. USA.

USA

Dave Bean Engineering Inc
925 Punta Gorda Street
Santa Barbara
California 93103
USA
Telephone 0101-805-962-8125.
Tuning.

Bonnets Up
5736 Spring Street
Clinton
MD 20735
USA
Telephone 0101-301-297-4759.
Restoration and tuning

British Frame & Engine Co
4831 Ryland Avenue
Temple City
California
USA
Telephone 0101-818-443-0939
Morgan spares

British Racing Green
P O Box 581
Pittston
Pennsylvania 18640
USA
Telephone 0101-717-654-2082
Sales, restoration, parts, service.

Cantab Motors Ltd.
RR1, PO Box 537A
Roundhill
Virginia 22141
USA
Sales & parts

Design With Wood
Brad Davis
P O Box 189A Salisbury Road
Sheffield
Mass 01257
USA
Telephone 0101-413-229-3574.
Wood fabrication.

Engine Room
Greg Solow
318A River Street
Santa Cruz
California 95060
USA
Telephone 0101-408-429-1800,
Mechanical and electrical service.

Ferrari of Los Gatos
Dennis Glavis
66 East Main Street
Los Gatos
California 95030
USA
Telephone 0101-408-354-4000
Sales, spares Plus 4 specialist.

Isis Imports
PO Box 2290
US Custom House
San Francisco
California 94126
USA
Sales and parts.

Lederman Rupp
P O Box 21412
Columbus
Ohio 43221
USA
Accoutrements

Lindskog Balancing Corporation
6 Alfred Circle
Bedford
Mass 01730
USA
Balancing all types of engines, including JAP and Matchless.

Malvern Lane Trading
P O Box 7758
Station G
Colubus
Ohio 43207
USA
Spares

Morgan Fab
P O Box 589
Buford
Georgia 30518
USA
Spares

Morgan Spares Ltd
PO Box 1761
Lakeville
CT 06039
USA
Telephone 0101-518-789-3877
Restoration parts & services

Moss Motors
P O Box M G
Goleta
California 93116
USA
Telephone 0101-805-968-1041.
Wire wheel specialists and TR parts.

Mostly Morgan
23 Baker Street
Brockton
Mass 02402
USA
Restoration parts and services.

Steve Miller
Imported Car Service
1375 Park Avenue
Emeryville
California 94608
USA
Restoration and parts.

Nisonger Corporation,
Smiths Instruments
35 Bartels Place
New Rochelle
New York 10801
USA
Instrument rebuilding.

Olde World Restorations
90 Sunset Lane
Hatborough
Pennsylvania 19040
USA
Telephone 0101-215-441-4151.
Restoration and spares.

Racing International
RD1
Millerton
New York 12546
USA
Restoration tuning information.

Sports & Classics
512 Boston Post Road
Darien
Connecticut 06820
USA
Telephone 0101-203-655-8731
Upholstery and car covers.

TEK Bearing Co Inc
Westwood
MA 02090
USA
Bearing and transmission specialists.

Bill Tracy
3179 Woodland Lane
Alexandria
Virginia 22309
USA
Telephone 0101-703-360-6652.

Electrical spares.

Very British
Locust Avenue Extension
Amsterdam
New York 12010
USA
English spares.

CANADA
CMC Enterprises (1990) Inc
RR3 Bolton
Ontario L7E 5R9
Canada
Parts, sales

Main distributors
UK
Andreas Adamou
Autorapide
Unit 1 Whitehouse Farm
Cocklake
Wedmore
N.Somerset BS28 4HE
England
Telephone 0934 713224

*Brands Hatch Morgans
Brands Hatch Circuit
Fawkham
Kent DA3 8NG
England
Telephone 0474 874147

*Burlen Services
Spitfire House, Castle Road
Salisbury
Wilts SP1 3SA
England
Telephone 0722 412500

Mike Spence
Classic Cars Ltd
Unit 1 Bloomfield Hatch
Mortimer
Reading
Berks RG7 3AD
England
Telephone 0734 333633

Cliffsea Car Sales
Bridge Garage
Ness Road
Shoeburyness SS3 9PG
England
Telephone 0702 295323

Donaldson & Evans
The Wolf Garage
Ashley Road
Hale
Cheshire WA15 9NQ
England
Telephone 061 9291208 (sales)
061 9411916 (service & admin)

*John Dangerfield Garages Ltd
Staplehill Road
Fishponds
Bristol BS16 5AD
England
Telephone 0272 566525/566373

F.H. Douglass
1a South Ealing Road
Ealing
London W5 4QT
England
Telephone 081 567 0570

*Mike Duncan
250 Ikon Estate,
Droitwich Road
Hartlebury
Worcs DY10 4EU
England
Telephone 0299 250025

Robin Kay & Sons
Marine Road
Eastbourne
BN22 7AU
England
Telephone 0323 26462/25563

*Libra Motive Ltd
2, Rosemont Road
Hampstead
London NW3 6NE
England
Telephone 071 435 8159/794 7009

*Lifes Motors Ltd
West Street
Southport PR8 1QN
England
Telephone 0704 531375

*I & J Macdonald Ltd
Maiden Law Garage
Lanchester DH7 0QR
England
Telephone 0207 520916

Otley Motors
Cross Green
Otley
West Yorks LS21 1HE
England
Telephone 0943 465222

Parker of Stepps
Hayston Garage
38 Glasgow Road
Kirkintilloch
Glasgow G66 1BJ
Scotland
Telephone 041 776 1708

*Phoenix Motors
The Green
Woodbury
Exeter
Devon EX5 1LT
England
Telephone 0395 32255

*Station Garage
Station Road
Taplow
Nr Maidenhead SL6 0NT
England
Telephone 06286 605353

Thomson & Potter Ltd
High Street
Burrelton
Blairgowrie
Perthshire PH13 9NX
Scotland
Telephone 08287 247
*Allon White & Son Ltd

The Morgan Garage
High Street
Cranfield
Beds MK43 0BT
England
Telephone 0234 750205

*Wykehams Ltd
6, Kendrick Place
Reece Mews
South Kensington
London SW7 3HF
England
Telephone 071 589 6894/8309
* *May have secondhand Morgans in stock*

AUSTRALIA
Calder Sports Cars
Distributors Australia Pty Ltd
Box 140 PO
Gisborne 3437
Australia

AUSTRIA
Max Bulla
Liebenstr. 40/60/5
A1120 Vienna
Austria

BELGIUM
Garage Albert
Anc Ets Stammet Et Fils
S.P.R.L.
84/86 Rue Osseghemstraat
1080 Brussels
Belgium

CANADA
CMC Enterprises (1990) Inc
RR3 Bolton
Ontario L7E 5R9

CYPRUS
Reliable Sports Cars Ltd
Grivas Digenis Ave
PO Box 5428
Nicosia
Cyprus

DENMARK
Alan Hall
Mollegardsvej 16
GL Hjortkaer
DK-6818 ARRE
Denmark

FRANCE
Jacques Savoye S.A.
237 Boulevard Pereire
75017 Paris
France

GERMANY
Merz & Pabst
Alexanderstr. 46
7000 Stuttgart 1
Germany

K.W. Flaving
4750 Unna
Hochstr. 4
Germany

HOLLAND
BV Nimag
Reedijk 9 3274 Ke
Heinenoord
Holland

IRELAND
Scott Macmillan
Holybrooke Hall
Kilmacanogue
Co. Wicklow
Ireland

ITALY
Ditta
Armando Anselmo
Via Vincenzo Tiberio 64
00191 Rome
Italy

JAPAN
Morgan Auto Takano Ltd
9-25 2 Chome Tsumada-Minami
Atsugi Shi
Kanagawa Ken 243
Japan

LUXEMBOURG
Yesteryear Luxembourg
Import Ltd SA
19 Rue Du Parc
L-8083 Bertrange
Luxembourg

NORWAY
Hallan AS
Malmogt 7
0566 Oslo 5
Norway

PORTUGAL
Manuel F Monteiro & FIlho
Representacoes e Comércio LDA
Rua Dos Correeiros 71
1100 Lisbon
Portugal

SOUTH AFRICA
Angela Heinz (Pty) Ltd
6, The Munro Drive
Houghton 2198
PO Box 2687
Johannesburg 2000
South Africa

SPAIN
Tayre S.A.
Principe de Vergara 253
28016 Madrid
Spain

SWEDEN
A.B. Wendels Bil & Motor
Box 74 Testvagen 10A
S-23200 Arlov
Sweden

SWITZERLAND
Garage De L'Autoroute
Signy S/Nyon
Switzerland

Rolf Wehrlin
Haupstr. 132
Aesch BL
Switzerland

USA
Isis Imports Ltd
PO Box 2290
US Custom House
San Francisco
California 94126
USA

Cantab Motors Ltd
RR1 Box 537A Round Hill
Virginia 22141
USA

** May have secondhand Morgans in stock*

APPENDIX IV

BIBLIOGRAPHY

• THREE-WHEELERS

The Best of The Bulletin
by Barry Davidson, editor of the Morgan Three-wheeler Club's *Bulletin*. Published by The Morgan Three-Wheeler Club, 1973.
Long out-of-print but the Club may be producing a reprint. Useful source for finding out how to adapt parts to fit your Three-wheeler.

Morgan Sweeps The Board
by Dr J D Alderson & D M Rushton. Published by Gentry Books, London, 1978. 250 pages.
The most comprehensive book on Morgan Three-wheeler competition history ever. Contains a lot of detail for those interested in racing, some general history and information regarding various models.

The Vintage Years of the Morgan Three-Wheeler
by Bill Boddy. Published by Grenville Publishing, London, 1970. 28 pages.
Book form of a series of articles from *Motor Sport* magazine, Historical gem covering production of the various models and racing.

The Book of the Morgan Three-Wheeler
by R.M. Clarke. Published by Pitman & Sons Ltd, London.
An absolute necessity for any Three-wheeler owner who intends to carry out their own maintenance and running repairs.

Morgan Three-Wheelers 1930-1952
by R.M. Clarke. Published By Brooklands Books, Surrey. 100 pages.
Copies of road test reports and much more useful information for Three-wheeler owners. A good companion to *The Book of the Three-Wheeler.*

Morgan Three-Wheeler Gold Portfolio. 1910-1952
by R.M. Clarke. Distributed by Brooklands Book Distribution Ltd, 1989.
Compilation of road and comparison tests, technical data, etc., compiled from various motoring magazines.

J.A.P. The Vintage Years
by Jeff Clew. Published by Haynes Publishing Group, England, 1985.
A detailed history of this famous engine manufacturer with many references to Morgans & H.F.S. Morgan..

J.A.P. The End of an Era
by Jeff Clew. Published by Haynes Publishing Group, England, 1988.
The continuing history of the J.A.P. Company with many references to Morgans.

No More Twins
by Clarrie Coombes. Published by the Morgan Three-wheeler Club, circa 1965.
History of the Morgan Three-wheeler Club.

Morgan 75 Years on the Road
by Ken Hill. Published by Blandford Press Ltd, Dorset. 1984. 160 pages.
Traces the history of Morgans -Three- and Four-wheelers - through the Morgan Motor Company's advertising. Reproduction of advertisements with linking text.

Three-Wheelers
by Ken Hill. Published by Shire Publications Ltd. 1986. 32 pages.
Traces the history of Three-wheelers from 1885 to date, with a complete chapter on the Morgan. Compares the marque with others.

Illustrated Morgan Buyers Guide
by Ken Hill. Published by Motorbooks International, Wisconsin, USA, 1989, (second edition 1990).
Detailed study of each model and its investment potential.

The Book of the Morgan
by G.T. Walton. Published by Pitman & Sons Ltd. 1932. 109 pages.
Contemporary book containing a lot of good general information, especially relating to pre-1932 Three-wheelers. Much of the content is reproduced in either of R.M. Clarke's books. (See also under author Harold Jelly).

ABC Of the Morgan
Compiled and published by the Morgan Three-wheeler Club.
Practical tips on owning and running a Three-wheeler - a must for all Three-wheeler owners.

Morgan, History of a Famous Car
Written and published by the Morgan Motor Company Ltd. Three editions, first with hard cover, second soft (two editions).
Also contains information on the history of Four-wheelers.

The Book of the Morgan
by Harold Jelly. Published by Pitman & Sons Ltd, 1933 (first edition); 1934 (second edition); 1935 (third edition. Each edition is a yearly update.
Contemporary book containing a lot of good general information, especially for pre-1936 Three-wheeler owners. Much of the content is reproduced in either of R.M. Clarke's books. (See also under author G.T. Walton).

The Three-Wheeler
By Brian Watts. Published by The Morgan Three-Wheeler Club. 1970. 68 pages.
Another historical gem which details all Three-wheeler models.

• FOUR-WHEELERS

Morgan Sports: From Three Wheels to Four
Published by I P C Transport Press, Ltd. 1977. 80 pages.
Reproduced articles from *Autocar* & *Motor Cycle* magazines from 1910-1977, including road test reports on various models. Interesting.

Morgan 1910-1980
Published by the combined Morgan Clubs (MSCC & MMTWC). 57 pages.
Produced as part of the celebration of the 70th anniversary of Morgan. Contains a general history including competition events and personal reminiscences from prominent Morgan personalities over the years.

The Great Cars/Morgan: 'On Four Wheels'
by Bill Boddy. Part of issue No.71, pages 1409-1419 of a motoring encyclopaedia published in monthly parts. Collectable by the enthusiast but not of too much interest to the prospective buyer.

Morgan - First and Last of the Real Sports Cars
by Gregory Houston Bowden. Published by Gentry Books, London. 1972. 192 pages. Reprinted 1973. New and revised second edition 1986.
General development of the marque, including business and competition history. Contains a lot of useful information.

More Morgan
by Gregory Houston Bowden. Published by Gentry Books, London. 1976. 224 pages. Described as 'A Pictorial History of the Morgan Sports Car'.

Morgan Autobook One
by R. Clarke. Published by Auto Press Ltd, Brighton, England. 1968. 168 pages.
Workshop manual for four-wheeler Morgans up to 1967.

Brooklands Books Series
Morgan Cars 1936-1960.
Morgan Cars 1960-1970.
Morgan Cars 1969-1970.
Edited by R.M. Clarke. Published by Brooklands Books, Surrey, England.
All these books are an absolute must for any Morgan owner who intends to carry out their own maintenance and running repairs. An invaluable source of technical information, equivalent to workshop manuals.

Morgan Cars Gold Portfolio 1968-1989
by R.M. Clarke. Distributed by Brooklands Book Distribution Ltd. 1989.
Compilation of road and comparison tests, technical data, etc., compiled from various motoring magazines.

Morgan Plus 4 & 4/4 Gold Portfolio 1936-1967
by R.M. Clarke. Distributed by Brooklands Book Distribution Ltd. 1993.
Compilation of road and comparison tests, technical data, etc., compiled from various motoring magazines.

Morgan Four Owners Workshop Manual
by R.M. Clarke. Distributed by Brooklands Book Distribution Ltd. 1989.
Updated version of *The Four-wheeler Workshop Manual*. Contains useful information on all models up to and including 1981, except for the Plus 8.

The Morgan Four-Wheeler Workshop Manual
by John Dowdeswell, Distributed by Brooklands Book Distribution Ltd.
Contains very useful information up to and including Series V 4/4s.

The Plus Four Morgan
by Eric Dymock. Published By Profile Publications, Surrey, England. 1967.
A brief but surprisingly detailed history of the model.

Morgan: The Last Survivor
by Chris Harvey. Published by Oxford Illustrated Press. 1987. 273 pages.
An extensive technical history of four-wheelers which is interesting and informative even though many of the 'facts' and photograph captions are either incorrect or misleading.

The Four-Wheeled Morgan
Volume No.1: The Flat Radiator Models
by Ken Hill. Published by Motor Racing Publications. 1977. 136 pages.
Comprehensive history of early Four-wheelers, including competition history, production figures, model specifications and servicing and restoration details.

The Four-Wheeled Morgan
Volume No.2: The Cowled Radiator Models
by Ken Hill. Published by Motor Racing Publications. 1980. 144 pages.
Comprehensive history of later Four-wheelers, including competition history, production figures, model specifications and Morgan clubs worldwide.

Morgan 75 Years on the Road
by Ken Hill. Published by Blandford

Press, Dorset, England. 1984.
(See under Three-wheeler books).

Illustrated Morgan Buyers Guide
by Ken Hill. Published by Motorbooks International, Wisconsin, USA. 1989. (Second edition 1990).
Detailed study of each model and its investment potential. *(See also under Three-wheeler books).*

Morgan (Famous Car Factory Series)
by Bengt Ason Holm. Published by Motorbooks International, Wisconsin, USA 1992.
A brief outline of the history of the Morgan factory from 1909 to date, followed by an in-depth study of the present day manufacturing process, stage-by-stage through the factory.

Morgan. Auto Histoire No. 38.
by F. Wilson McComb. Published by EPA, Paris, France. 1985. 69 pages.
Printed in French, this small format book contains a surprising amount of information.

Moggie: The Purchase, Maintenance and Enjoyment of Morgan Sports Cars
by Colin Musgrove. Published by Quills Publishing, Herts, England. 1980. 112 pages.
A must for any serious Morgan buyer, coverage includes before you buy, what to look for, where to look for it and what to do if you find it.

The Morgan Year Book 1980-81
by Colin Musgrove. Published by The Magpie Publishing Co, Hong Kong. 1981.
The first of what was intended to be a yearly publication. Contains details of Morgan events worldwide for the year in question.

Morgan: Buying & Restoration Book
Published by *Practical Classics & Car Restorer* magazine, 1992.
Reproduction of various articles previously printed in *Practical Classics & Car Restorer* magazine. A useful addition to any practical Morgan owner's bookshelf.

The Morgan Cartoon Book
by Terry Quirk. Published by Quills Publishing, Herts, England. 1980.
Cartoons and poems about Morgans. Pure 'coffee table' book with little to recommend it other than the artistic skill of the author.

Postwar MGs & Morgans
by Henry Rasmussenand John Blakemore. Published by Picturama Publishing, California, USA. 1979.
An interesting comparison of the two marques.

Morgan Plus 8
by Graham Robson. Published by Osprey, London, 1984. 135 pages.
Comprehensive book on the model

Morgan Sport-Und-Tourenwagen 1935-1981
by Halwart Schrader. Published by Schrader Verlag, Germany. 1991. 96 pages.
Attempts to trace the history of the Four-wheeler Morgan through the company's sales brochures. Captions tend to be inaccurate and should not be used as a reliable guide by collectors of Morgan sales literature. German text.

Morgans in the Colonies
by John H. Sheally. Published by Jordan & Co. Virginia Beach, USA. 1978.
A photographic study - with text - of Morgans in America.

Morgans, Pride of the British
by John H. Sheally. Published by TAB Books Inc. Pennsylvania, USA. 1982.
Another photographic study of Morgans in America but with far more useful text and information.

Original Morgan
by John Worrall & Liz Turner. Published by Bay View Books Ltd, Devon, England. 1992.
A detailed examination of each of the Morgan Four-Wheeler models produced by the Morgan Motor Company. Excellent colour photographs of various components, with fairly detailed technical information. By one of the leading Morgan authorities, this is a must for the restorer looking for complete authenticity.

Plus 4 Super Profile
by John Teague. Published by G.T. Foulis, England. 1987.
Detailed study of the model including specifications and production details.

APPENDIX V

MORGAN MINIATURE MODELS

Three-wheelers

For some strange reason three-wheelers have never seemed particularly attractive to model companies as subjects, especially those which mass-produced models. Back in the 1960s a number of Bubble Car models came on the market, but there were no Morgans. Inevitably, a few private model hobbyists who happened to be Morgan owners or enthusiasts have produced scratch models of their favourite cars, one famous example of which is a Grand Prix which was made in silver by a professional silversmith and carried as mascot on his car. Unfortunately, the whereabouts of this model is now unknown.

The following is a list of manufacturers of known Three-wheelers models, complete with comments where applicable.

• THE MECCANO THREE-WHEEL SPORTS CAR

Produced in the 1930s to a scale of approximately 1/16. It is illustrated on page 82 of Gregory Houston Bowden's book *More Morgan*.

Although very rarely found at model swap meets, examples, when they do appear, tend to fetch £40-£60, depending on condition.

• GAIETY TOYS (A CASTLE ART PRODUCTION UK)

This rather crude metal representation of an air-cooled, Beetle-back Super Sports was produced in the mid-to-late '50s in the UK to an approximate scale of 1/24. The models were made in both clockwork and push-along versions, both types being sold in blue, green, red and all chrome plate. Again, a very collectable model and one in mint boxed condition would easily fetch £120 at a model swap meet, other examples being valued pro-rata dependent on condition.

• GAKKEN OR RIKO

Both Japanese companies produced the same model in the late 1960s/early 1970s, a 1/16 scale plastic kit of a 1935 Matchless-engined Super Sports.

• ENTEX

'A lot of fun from Entex'; this Japanese model, carrying registration number MOG 750 and featuring vinyl seat covers, appears to be similar to the previous Matchless-engined model. Marketed as "The MOG".

• MINICRAFT

Produced in the late 1980s in the USA under licence, this appears to be based on the Gakken/Riko model already described. It was produced for a very limited period only and is unlikely to be available new in the UK.

• ACORN MODELS

Around 1969-70 a cast metal model of a 1933-34 Super Sports with top mounted spare wheel was produced in the UK to a scale of 1/57th in a very limited quantity. Two versions were made: one with hood up and low exhaust and the other a racing version with high exhausts, wire mesh aero-screens and complete with a driver and riding mechanic, racing overalls, helmet and goggles. It is believed only two or three examples of each model were produced and that they were originally intended as prototypes for a possible Matchbox model. The same company also produced a very poor vacuum-cast plastic model of a Three-wheeler.

• BRUMM

In the late 1970s and early 1980s this

A selection of various Three-Wheeler models. Back row: Brumm models of Morgan and Sandford Three-Wheelers. Front row, l-r: South Eastern Finecast 1932 Super Sports, scratch-built 1930's Super Sports, Acorn racing with driver and riding mechanic and Acorn Super Sports. The last two were prototypes for possible Matchbox production.

Italian company produced three versions of a 1/43 scale die-cast metal model of a 1923 Morgan Cycle Car, all of which, for some inexplicable reason, were finished with gold-coloured brightwork. One version had a white body with black hood erected and red interior; the second was similar but fitted with aero-screens and single racing number 3 on its tail and the third finished in British Racing Green, with brown interior, had the racing number 3 on both sides of the tail. These models were used as the basis for the 7500 Limited Edition series produced in July 1990.

Brumm also made three models of the 1929 Darmont Morgan, all with chromed brightwork. One was finished in red with the hood erected and brown interior and two more in blue with brown interiors and aero-screens and with a choice of racing numbers 9 or 13 on both sides of the tail.

The white metal die-cast range

• SOUTH EASTERN FINECAST (FORMERLY WILLS FINECAST)

A selection of 1/43 and 1/57 scale Morgan models produced by various companies in the past twenty years. Back row, l-r: Acorn Plus 4, Mikansue 1938 Le Mans Majorette, Grand Prix 1962 Le Mans Plus 4 Majorette and BKL 1949 4/4 Drop Head Coupé Czechoslovakia magnetic drive 4/4. Front row: a selection of 1/57 scale toys produced by various manufacturers worldwide.

Below: BKL 1/24 scale 1979 Kent-engined 4/4, two- and four-seater with removable bonnet revealing full engine detail. Wire wheel option. Fully finished or kit form in diecast white metal.

Two models - a 1/24 scale 1934 Super Sports and a 1/43 scale 1932 Super Sports - are offered in kit form only. Both models are very good on detail in relation to scale.

• SCALELINK
1/76 scale 1928 Aero Super Sports, either as a kit or made up.

• BKL AUTOREPLICA
1/72 scale 1939 'F4'. 1/48 scale 'F2' and 'F4' either as a kit or made up.

• ABS
1/76 scale 1933 Super Sports, offered as a kit or made up.

Of the last three models listed, Scalelink's has the most detail which, inevitably, tends to mean it is also more fragile.

Four-wheeler models

Models of Four-wheeler Morgans have been in much more plentiful supply from the mass-produced model manufacturers. This is especially true of the 1/57 scale models manufactured in Japan and Hong Kong for the toy market, while there has also been a far larger choice available in the die-cast white metal models, in particular from UK manufactures.

White metal die-cast range
1/43 scale

• WESTERN MODELS
Produced an excellent model of a Flat Radiator Plus 4 from 1980 to 1987.

In 1994 this company produced a limited edition of a 1993 Plus 8, 100 each of red or British Racing Green with side screws. The models were made exclusively for the Dutch company Automobile Collection.

• MOTORKITS
Offered a model of a Plus 4 Plus which is no longer produced. (see *B.K.L. Autoreplica*).

• GRAND PRIX
A model of the 1962 Le Mans Plus 4 - again, out of production.
• MIKANSUE
A model of the 1938 Le Mans 4/4, now no longer produced.

• ANDRE-MARIS RUF (AMR)
A model of a 4/4 Four-seater, no longer produced.
• METAL 43
A 4/4 Four-seater, as AMR above and no longer produced.

• BKL AUTOREPLICA
1970s Plus 8 (no longer produced).
1980s PLUS 8.
1960s/80s 4/4 Four-seater, Cowled Radiator.
1963 Plus 4 Plus Fixed Head Coupé.
1950 4/4 Drop Head Coupé, Flat Radiator.
1946 4/4 Two-seater, Standard Special engine.
1938 4/4 Two-seater, Coventry Climax engine.

1938 4/4 Four-seater, Coventry Climax engine.

• AUTO FANTASY
1971 4/4 four-seater and 1971 Plus - made up only.

Resin range 1/43 scale
• PROVENCE MOULAGE
Resin kit of 1962 Plus 4 Le Mans car number 29. Also will be producing a Plus 8 in autumn 1994.

1/24 scale
• BKL AUTOREPLICA
A 1979 4/4 Ford Kent-engined Two-or Four-seater with lift-off bonnet revealing full engine detail (approximately 160mm long).

• METAL 24
A Plus 8 (no longer produced).

• SOUTH EASTERN FINECAST
A 1983 Plus 8 in kit form only.

1/86 scale
• BKL AUTOREPLICA
A 1960s/80s Cowled Radiator Two-seater (approximately 50mm long).

Plastic die-cast range
1/16 scale
• POLISTIL
Die-cast plastic ready-made models, approximately 250mm long, of a Plus 8 Roadster and a Plus 8 Competition were produced in a choice of colours from 1980 to 1992 but are no longer in production.

• L/S MOTORISED
An electric-powered model, offered as a kit only, of a Plus 8 Roadster or Competition, but no longer produced.

1/20 scale (approx)
• MASUDAYA
Representation, with friction drive, of a Two-seater Morgan with a nodding figure of Mickey Mouse, Donald Duck or Goofy as driver.

• AREMONTOIR
Hong Kong-produced copy of the Masudaya models, but with a figure of a helmeted driver.

1/43 scale
• ACORN MODELS
In production in the UK for about two years from early 1969 were very thin, plastic die-cast models of a Plus 4 and a 4/4 with removable hood.

• MAKER UNKNOWN
Believed to date from the late 1960s, a model of a Plus 4 in grey plastic with a large raised figure 8 moulded on top of the bonnet has been found in Norway.

Die-cast metal toys

1/57 scale (approx)
• MAJORETTE
A 75mm long toy model of a cowled-

The contrast of 1/43 and 1/86 scale diecast white metal models ... 1/43 scale, l-r: BKL Plus 8, BKL 4/4 Series I four-seater, Western Models Flat Radiator Plus 4, Grand Prix 1962 Le Mans Plus 4 Super Sports, BKL 4/4 Series I two-seater and BKL Cowled Radiator four-seater. 1/86, l-r: BKL 'F4' and BKL Plus 8. All were or are still available in fully finished or kit form.

radiator 4/4 was produced in a variety of colours and packaging from about 1980 but is no longer in production.

• SIKU
A Plus 8 produced in blue or British Racing Green.

• TOMICA
A Plus 8 with Miss Piggy as driver (not to be confused with the Corgi Miss Piggy, which is not a Morgan.) Tomica also produced a toy Plus 8 in a variety of colours, but none of these described are still produced.

• TOMY
A Plus 8 in various colours no longer produced.

• OLD TIMER MODELS
A Plus 8, no longer produced.

• ZEE
A Plus 8 in various colours, no longer produced.

• DYNAWHEELS
Same model as Zee, above, no longer produced.

• CORGI ROCKETS
A Plus 8, no longer produced.

• WIZZWHEELS
Plus 8, no longer produced.

• L/B MODELS
Plus 8, no longer produced.

• PLASTIC MAGNETIC DRIVE
Very basic 4/4 made in Czechoslovakia in the late 1960s which is no longer produced.

• MANUFACTURER UNKNOWN
A tin-plate 1/24 model of a Plus 8.

• FUNRISE
Vaguely Morgan shaped, very poor representation, about 25mm long, made in China.

Other Morgan model-based memorabilia

Silver or gold jewellery
Cowled Radiator Two-seater, 42mm long.
Cowled Radiator Two-seater, 20mm long
(These models are supplied as keyrings, earrings, necklaces, brooches, etc).

Silver-plated pewter
1/57 scale miniatures of a 1932 Super Sports and a 1980s Plus 8, each mounted on a wooden plinth.

Crystal glass
Replica of a Plus 8, mounted on a wooden plinth. No longer produced.

Wooden models
Representations in wood of a Three-wheeler Tourer and Racer and Flat Radiator Four-wheeler.

Toiletries
Bath soap-on-a-string moulded in the shape of a Plus 8. Originally produced for Marks & Spencer but now only for the Morgan Motor Company.

Pedal car
A Morgan 4/4 suitable for children aged 3-7 years.

Famosa electric car
A Morgan 4/4 suitable for children aged 3-7 years, powered by battery and approximately 1.09m long.

Key to countries

CHI	-	China
CZ	-	Czechoslovakia
F	-	France
GB	-	Great Britain
HK	-	Hong Kong
I	-	Italy
JAP	-	Japan
KOR	-	Korea
N	-	Norway
NL	-	Holland
THAI	-	Thailand

The tables on the following four pages contain a complete listing of all known Morgan models ever produced and is reproduced here courtesy of Mr Cliff Baker.

MODEL MAKER	No.	SCALE	YEAR	MODEL	DESCRIPTION	Made
A.B.S.	R715	72nd	1933	S S	3 Wheeler	GB
Acorn		43rd			3 Wheeler	GB
		43rd		4/4	2 Seater	GB
Automobile		43rd	1994	+ 8	2 Seater hood down red	NL
Collections		43rd	1994	+ 8	2 Seater sidescreen up B.R.green	NL
B.K.L.		72nd	1939	F4	3 Wheeler	GB
Autoreplica		72nd	1968	+ 8	2 Seater	GB
	8	86th	1960	4/4	2 Seater Cowled Rad. can be +8	GB
	AR44	43rd	1938	4/4	2 Seater Coventry Climax	GB
	AR38		1938	4/4	4 Seater Coventry Climax	GB
modified	AR44		1946	4/4	2 Seater Standard Special	GB
	AR25		1950	+ 4	2 Seater D.H.C.	GB
modified	AR 7		1960	4/4	2 Seater (adaption of + 8)	GB
	AR11		1960	+ 4	4 Seater	GB
	AR10		1963	+4+	2 Seater Coupé	GB
	AR 7		1968	+ 8	2 Seater	GB
	AR47	43rd	1940	F 2	3 Wheeler	GB
	AR48	43rd	1940	F 4	3 Wheeler	GB
	AR1007	24th	1979	4/4	2 Seater	GB
	AR1005		1979	4/4	4 Seater	GB
Brumm	1	43rd	1923		Morgan 3 Wheeler green open	I
(can have	2		1923		3 Wheeler white hood up	I
a variety	3		1929		Darmont 3 Wheeler blue open	I
of racing	4		1929		3 Wheeler red hood up	I
numbers)						
SS			1933		Morgan 3 Wheeler red/black open	I
LIM.EDIT.			1933		3 Wheeler yell/black open	I
MODELTIME			1933		3 Wheeler black open	I
Corgi Rocket	921	55th	1968	+ 8	☼ Met.Red white rad chrome wheels	GB
	921		1968	+ 8	☼ Met.Red silver rad black wheels	GB
Corgi Junior	64		1968	+ 8	☼ Red ☼(all with	GB
	64		1968	+ 8	☼ Yellow black interior)	GB
	64			+ 8	☼ Red with label 6 roundal	GB
					seen with number 20 on all of door	GB
Czechoslo.					Blue Plastic magnet powered	CZ
					Green (Made in the 1970's)	CZ
					Red	CZ
					Yellow	CZ
Dynawheels/	D69	57th	1968	+ 8	Metallic Green {also marketed	CHI
DynaClassics	D69		1968	+ 8	Metallic Red by Zee models}	CHI
French		43rd		4/4	4 Seater made by Andre Marie Ruf	F
Gaiety Toy	850429				3 Wheeler Red	GB
(Castle Art					3 Wheeler light Blue	GB
Products)					3 Wheeler Chrome	GB
					3 Wheeler light Green	GB

MODEL MAKER	No.	SCALE	YEAR	MODEL	DESCRIPTION	Made
Gakken					3 Wheeler	
Glass Cryst.		24th	1969	+ 8	2 Seater	GB
Grand Prix		43rd	1962	+ 4	Le Mans 2 litre Class Winner	GB
Hong Kong	201	57th	1968	+ 8	Yellow tan inter. manu.unknown	HK
			1968	+ 8	Yellow black inter.	HK
L/B Models		57th	1968	+ 8	2 Seater	HK
L/S	02002	16th	1979	+ 8	Roadster	JAP
Motorised	2,400		1979	+ 8	Competition	JAP
Majorette						
250 series	261	50th	1969	+ 8	Blue chrome	F
	261		1969	+ 8	Green chrome	F
200 series	261		1968	+ 8	Pink	F
	261		1968	+ 8	Colour changes with temperature	F
	261		1968	+ 8	with case bubble pack (Girls)	F
	261		1969	+ 8	Red /black roof/black interior	F
	261		1969	+ 8	Red /black roof/grey interior	F
	261		1969	+ 8	Red /cream roof/cream interior	F
	261		1969	+ 8	Red /cream roof /cream interior	F
	261		1969	+ 8	Red metallic	F
	261		1969	+ 8	Green / brown roof/brown interior	F
	261		1969	+ 8	Green / cream roof/cream interior	F
	261		1969	+ 8	Green bright " "	F
	261		1969	+ 8	Green mint " "	F
	261		1969	+ 8	Green metallic " "	F
	261		1969	+ 8	Blue metallic	F
	261		1969	+ 8	Mauve metallic (purple)	F
"Crazy						
Roadsters"	455		1969	+ 8	Mauve/black with Big Bebert	THAI
450 series	456		1969	+ 8	Yellow/black with Elliot the Cat	THAI
Masudaya				4/4	Yellow with Mickey Mouse	JAP
Corp.1986				4/4	Pink with Mickey Mouse	JAP
				4/4	Red with Mickey Mouse	JAP
				4/4	with Goofy	JAP
				4/4	Pink with Donald Duck	JAP
				4/4	Red with Donald Duck	JAP
				4/4	Yellow with Donald Duck	JAP
Model World	3301	???		4/4	Light Green multicoloured parts	CHI
wind up	3301			4/4	Yellow multicoloured parts	CHI
(also	3301			4/4	Light Blue multicoloured parts	CHI
marketed	3301			4/4	Pink multicoloured parts	CHI
by	3301			4/4	Blue gold bumper	CHI
Aremmontoir)	3301			4/4	Light Blue gold bumper	H K
{Copies of	3301			4/4	Red gold bumper	H K
Disney	3301			4/4	Yellow gold bumper	H K
originals}	3301			4/4	Green gold bumper	H K
	3301			4/4	White gold bumper	H K
	3301			4/4	Brown gold bumper	H K
	3301			4/4	Purple gold bumper	H K
	3301			4/4	Purple light gold bumper	H K
Meccano					3 Wheeler	GB

MODEL MAKER	No.	SCALE	YEAR	MODEL	DESCRIPTION	Made
Mikansue		43rd	1939		Le Mans	GB
Minicraft	1513	16th	1935	S S	3 Wheeler ACADEMY {American firm}	KOR
Motorkits		43rd		+4+	Coupé	
Norway				+ 4	8 embossed on bonnet	
Old Timer		57th		+ 8	2 Seater	
Piccolino		76th	1920		3 Wheeler	I
Polistil (Tonka)	TG2 TG2 TG2 TG2 TG2	16th	1968 1968 1968 1968 1968	+ 8 + 8 + 8 + 8 + 8	Racer Light Blue Racer Blue + Black Road Yellow + Black Wings Racer Yellow + Silver Wings Road British Racing Green	I I I I I
Provence Moulage	K715	43rd	1962	+ 4	Le Mans 2 Seater	F
Riko					3 Wheeler	
Scalelink	SLC122	72nd	1928	S S	Aero Super Sports	GB
Siku	1062 1062 0836 0836	55th	1968 1968 1968 1968	+ 8 + 8 + 8 + 8	Metallic Blue British Racing Green Metallic Green Metallic Blue (differ.wheels)	D D D D
South East Finecast (Wills fine)	A0 36 A0 20 A2 16	43rd 24th 24th	1934 1934 1984	S S S S + 8	3 Wheeler 3 Wheeler 2 Seater	GB GB GB
Standard Soap Co.				+ 8 + 8	wire wheel brown(Marks & Sparks) wire wheel cream(Morgan Factory)	GB GB
Tomica (Tomy)	26 26 16	57th	1980 1980 1980 1980	+ 8 + 8 + 8 + 8	Red Black interior Met.Blue Tan interior Red Black interior with Miss Piggy	JAP JAP CHI CHI
Western		43rd 24th	1951 1968	+ 4 + 8	2 Seater Flat Rad. 2 Seater (made for Danhausen out of production since 1982)	GB GB

MODEL MAKER	No.	SCALE	YEAR	MODEL	DESCRIPTION	Made
BRUMM	7 8		1922 1922		Sandford 3 Wheeler silver open 3 Wheeler gold hood up	I I
Japanese Tin plate				4/4 4/4	yellow badged as MG red with roof badged as MG	JAP JAP

Other models of interest (not actual Morgans)

Morgan

APPENDIX VI

SAFETY, JACKING & WORKING PROCEDURES & GLOSSARY OF TERMS

Safety, jacking & working procedures

This book contains considerable advice concerned with practical aspects of maintenance, repair, renovation and modification. This information is intended for the use of qualified engineers and mechanics. You must not undertake any such practical work unless you are suitably qualified and confident of your ability to carry out the work safely and competently. We must draw your attention to the following - ***IMPORTANT NOTE.*** *While the author has prepared the practical information in this book with diligence and care, it is possible that it contains errors or omissions. It is not possible to foresee every possible area of personal danger or possibility of mechanical damage and warn readers accordingly. The author, publisher and retailers therefore accept no responsibility for personal injury or mechanical damage which is in any way related to the use of this publication. If you use the practical information contained within this book, you accept full personal responsibility for the consequences.*

• SAFETY

You must work responsibly at all times and always take appropriate precautions to avoid damage or personal injury. Always be aware of potential dangers whenever working on or around your car. Don't think just in terms of the obvious, go a little deeper. Here are just five examples of the level of awareness you should maintain. 1) Remember that most fluids used by the car or for maintenance/cleaning are in some way dangerous - so read the maker's instructions and don't make assumptions. 2) Old engine oil has been proved to cause skin cancer in mice - so protect your skin with gloves or barrier cream and wash contaminated areas quickly. 3) If you're working on the car's fuel system in your garage, remember that if the garage contains a house heating boiler the ignition sequence may occur while there are gas (petrol) fumes in the air - it's also possible the boiler has a constant pilot flame. 4) Slipping on spilled oil, or other fluids, could put you in hospital for a long time. 5) A metal watch strap, bracelet or even a ring could cause a short between a live terminal and earth (ground).

• JACKING AND SUPPORTING THE CAR

There are a number of maintenance, repair and restoration operations which require work under the car, and you will need to make some provision for this. In an ideal world, we would all have vehicle lifts, but most of us must make do with less exotic equipment.

Unless you happen to have an access pit in the floor of your garage, you'll need to raise the car high enough to permit good working access below the body, while preserving a good level of safety. A normal small wheeled hydraulic jack and two large axle (safety) stands are a minimum requirement . You can purchase a suitable jack from tool or auto parts stores, or from mail order sources. We recommend that you avoid the smallest of the jacks supplied for home use. These have a capacity of 1500kg/1.5 tons, which is adequate, but have a restrictive lift range of only around 200mm/8inches. The type with a capacity of 2000kg/2 tons and a lift range of just under 300mm/12 inches is more useful. You'll find that the lift range is significant when you need to raise the car high enough to work underneath. The further the jack can lift, the less you will need to jack in stages, which will save you a lot of time. As rule of thumb, buy the biggest lift capacity you can afford.

Axle stands come in various sizes too, and again, the rule is the bigger the better. These are not only stronger than the smaller types, they allow you to raise the vehicle much higher, and every extra inch helps. There are a few operations on the car which will require around 460mm/18inches clearance underneath, and in all cases, more space means more comfortable working conditions. Remember that if you use a crawler board, you lose a little clearance.

Warning! Don't forget that a car is heavy - if it collapses while you're under it, you will be killed or badly injured, so don't take chances. **Never work beneath a car that is not supported by at least two axle (safety) stands.**

Warning! Never use jacks or axle stands on sloping ground and always ensure that the jack and axle stand pads are bearing on suitably strong structures such as the car's front cross-axle, rear axle or reinforced sections of the chassis.

Warning! If you are only jacking

one end or side of the car, you must chock all those wheels which remain on the ground (back and front of the tyre).

Warning! If the vehicle has to be raised in stages, don't be tempted to jack too much at each stage: the vehicle will become unstable.

• WORKING PROCEDURES

Before you start any practical work on your car, read through the relevant details carefully. Check any drawings or photographs relating to the task, and compare this information with your own car. If necessary re-read the section until you are confident that you know what you will be doing. Not only does confident familiarity minimise mistakes, it will actually save you time when you do the job for real - you won't need to constantly refer back to the book for procedural details.

While working in the open is fine for minor jobs if a major repair or overhaul is envisaged you should make strenuous efforts to find somewhere under cover to work. If absolutely unavoidable, you could remove major assemblies outside, but any dismantling work should take place under some sort of roof.

Most of us will be working inside a normal domestic garage with restricted floor area and ceiling height. Before you start work, make sure that you have sufficient room to work comfortably. If you have the same kind of garage that most of us have to use, clear out as much junk as you can (especially on or around the bench area), and arrange your garden tools, bicycles, ski equipment or whatever neatly, and well away from your work area. Try to keep at least one side of the garage clear of domestic items.

You will need plenty of light as well. Make sure you have adequate general lighting, with extra lighting over the workbench, plus one or two inspection lights for use around and under the car and a flashlight for use in confined spaces.

Don't forget that you will need somewhere to store removed parts safely, and this requirement will vary according to what you intend to do. In most cases, you can lay out the parts on shelves or in boxes, though some of the bigger assemblies, especially things like the dash panel, seats or the hood will obviously require much more space.

If you are going to undertake a major overhaul which entails removal of the engine and transmission, you will need around twice the area required to park the car, plus adequate access to allow the engine/transmission assembly to be removed. You will need to assess the likely space requirement and make suitable arrangements before you start work.

Set aside an area where you can leave major parts or assemblies safely, with no risk of accidental damage. You will know best how you can arrange this according to your own circumstances. You may be planning to carry out major work over an extended period, maybe over the winter months. If so, plan some kind of long-term storage for removed parts, with dust sheets to cover removed major assemblies or body parts until they are needed again.

The seasoned mechanic will instinctively collect handy containers - plastic tubs and resealable plastic bags are great for tiny, easily lost parts, and cardboard cartons are good for bigger parts. If money is no object, get some stackable plastic bins. Avoid using glass jars for storage - they break too easily.

Use PVC electrical tape or masking tape to identify parts, hose and wiring connections, etc as they are removed - you can write on the tape or on the part itself with a fine-point indelible marker pen. You can also use card or paper labels for this. Have a notepad handy for making simple sketches of how specific parts fit together or how wiring or hoses are routed. Avoid pens that use water-based inks which will rub off easily. If you have the time and the equipment, a photographic record of your work can be very useful.

The value of marking things cannot be overstressed - you will be surprised how easily it is to get confused about which wire or bolt goes where, just a few hours after you disconnected them. In many cases you can make life easier for yourself by refitting bolts and screws as you remove them, either screwing them loosely back into position after removing the component they retain, or in the case of covers, pushing the screw back through the cover and through a piece of corrugated card. When you get round to installing the part, the fasteners will be in the correct relative positions - this is especially important where you have similar screws or bolts of varying length.

American/English glossary of automotive terms

American	English
A-arm	Wishbone (suspension)
Antenna	Aerial
Axleshaft	Halfshaft
Back-up	Reverse
Barrel	Choke/venturi
Block	Chock/wedge
Box end wrench	Ring spanner
Bushing	Bush
Clutch hub	Synchro hub
Coast	Freewheel
Convertible	Drop head
Cotter pin	Split pin
Counterclockwise	Anti-clockwise
Countershaft	Layshaft (of gearbox)
Crescent wrench	Open-ended spanner
Curve	Corner
Dashboard	Facia
Denatured alcohol	Methylated spirit
Dome lamp	Interior light
Driveaxle	Driveshaft
Driveshaft	Propeller shaft
Fender	Wing/mudguard
Firewall	Bulkhead
Flashlight	Torch
Float bowl	Float chamber
Freeway, turnpike, etc.	Motorway
Frozen	Seized
Gas tank	Petrol tank
Gas pedal	Accelerator pedal

American	English
Gasoline (gas)	Petrol
Gearshift	Gearchange
Generator (DC)	Dynamo
Ground	Earth (electrical)
Header/manifold	Manifold (exhaust)
Heat riser	Hot spot
High	Top gear
Hood	Bonnet (engine cover)
Idle	Tickover
Intake	Inlet
Jackstands /Safety stands	Axle stands
Jumper cable	Jump lead
Keeper	Collet
Kerosene	Paraffin
Knock pin	Roll pin
Lash	Freeplay/Clearance
Latch	Catch
Latches	Locks
License plate /tag plate	Number plate
Light	Lamp
Lock (for valve spring retainer)	Split cotter (for valve cap)
Lopes	Hunts
Lug nut	Wheel nut
Metal chips or debris	Swarf
Misses	Misfires
Muffler	Silencer
Oil pan	Sump
Open flame	Naked flame
Panel wagon/van	Van
Parking light	Sidelight
Parking brake	Handbrake
Piston pin or wrist pin bearing/bush	Small (little) end bearing
Piston pin or wrist pin	Gudgeon pin
Pitman arm	Drop arm
Power brake booster	Servo unit
Primary shoe	Leading shoe (of brake)
Prussian blue	Engineer's blue
Pry	Prise (force apart)
Prybar	Crowbar
Prying	Levering
Quarter window	Quarterlight
Recap	Retread
Release cylinder	Slave cylinder
Repair shop	Garage
Replacement	Renewal
Ring gear (of differential)	Crownwheel
Rocker panel	Sill panel
Rod bearing	Big-end bearing
Rotor/disk	Disc (brake)
Secondary shoe	Trailing shoe (of brake)
Sedan	Saloon
Setscrew, Allen screw	Grub screw
Shift fork	Selector fork
Shift lever	Gearlever/gearstick
Shift rod	Selector rod
Shock absorber, shock	Damper/shocker
Snap-ring	Circlip
Soft top	Hood
Spacer	Distance piece
Spare tire	Spare wheel
Spark plug wires	HT leads
Spindle arm	Steering arm
Stablizer or sway bar	Anti-roll bar
Station wagon	Estate car
Stumbles	Hesitates
Tang or lock	Tab
Taper pin	Cotter pin
Teardown	Strip(down)/dismantle
Throw-out bearing	Thrust bearing
Tie-rod (or connecting rod)	Trackrod (of steering)
Transmission	Gearbox
Troubleshooting	Fault finding/diagnosis
Trunk	Boot
Tube wrench	Box spanner
Turn signal	Indicator
Valve lifter	Tappet
Valve lifter or tappet	Cam follower or tappet
Valve cover	Rocker cover
VOM (volt ohmmeter)	Multimeter
Wheel cover	Roadwheel trim
Wheel well	Wheelarch
Whole drive line	Transmission
Windshield	Windscreen
Wrench	Spanner

Morgan

INDEX

Dear Reader,
We hope you enjoyed this Veloce Publishing production. If you have ideas for books on Morgan or other marques, please write and tell us.
Meantime, Happy Motoring!

THE END